Geography of British Columbia

Geography of British Columbia
People and Landscapes in Transition

4th Edition

Brett McGillivray

UBCPress · Vancouver · Toronto

29 28 27 26 25 24 23 22 21 20   5 4 3 2 1

Printed in Canada.

Library and Archives Canada Cataloguing in Publication

Title: Geography of British Columbia : people and landscapes in transition / Brett McGillivray.

Names: McGillivray, Brett, 1944- author.

Description: 4th edition. | Includes bibliographical references and index.

Identifiers: Canadiana (print) 20200169939 | Canadiana (ebook) 20200169947 | ISBN 9780774864329 (softcover) | ISBN 9780774864336 (PDF)

Subjects: LCSH: British Columbia – Geography. | LCSH: Human geography – British Columbia. | LCGFT: Textbooks.

Classification: LCC FC3811 .M33 2020 | DDC 917.11–dc23

Canadä

UBC Press gratefully acknowledges the financial support for our publishing program of the Government of Canada (through the Canada Book Fund), the Canada Council for the Arts, and the British Columbia Arts Council.

Printed and bound in Canada by Friesens
Set in Frutiger Condensed and Utopia by Artegraphica Design Co. Ltd.
Substantive editor and copy editor: Lesley Erickson
Proofreader: Kristy Lynn Hankewitz
Indexer: Celia Braves
Cartographer: Eric Leinberger
Cover designer: Martyn Schmoll

UBC Press
The University of British Columbia
2029 West Mall
Vancouver, BC V6T 1Z2
www.ubcpress.ca

This book is dedicated to the new generation of geographers:
Keita, Heidi, May, and Sophia

Contents

Preface

Readers will find this fourth edition of *Geography of British Columbia* to be a major departure from the third edition. In my research of British Columbia – which has traced overall population, rural versus urban population, employment in the various sectors (goods and services), as well as provincial revenues over time, all of which are in this new edition – it is apparent that the resource sector of the economy is not nearly as important as it once was. It is also apparent that a great deal of the population is located in the larger municipalities. These changes are reflected in this revised edition.

The text begins with an Introduction. Following, there are now two parts to the text: Part 1 contains revisions to the previous edition's Chapters 1, 2, 3, and 7 (now Chapter 4). All other chapters are a deviation from the previous edition, but the content has not been lost. The major focus of Part 2 is on resources and resource development; however, there are no individual chapters on resources. Rather, the discussion is organized in chapters chronologically. In this manner, I believe, it assesses how resources have been used, exploited, managed, and mismanaged to our present day. "The Tragedy of the Commons" story (see the Introduction) is used throughout a number of the chapters to illustrate attitudes and consequences of resource exploitation. The final "tragedy" may be climate change.

Throughout Part 2, the treatment and mistreatment of First Nations has been assessed, as well as that of Asians; their stories are woven into this historical perspective. Indigenous Peoples and other racialized peoples, as well as the province's resources, were subject to government policies that guided economic, social, and political development. And while provincial and federal policies were influential and often contentious, many other intervening events and conditions, such as technological innovations, wars, recessions, geophysical hazards, and terrorism, were also significant in the transition of the provincial landscape. After Confederation, more and more people moved to the province, more infrastructure was constructed, and new urban patterns were added to the map of British Columbia. These issues are all addressed in this book. This fourth edition concludes with policy directions for a more sustainable future.

Geography of British Columbia

Introduction

The geography of British Columbia is in constant flux. Between 2014 and 2017 alone, the following events occurred, transforming the landscape and the way people engage with it:

- Heat waves shattered temperature records, and wildfires devasted parts of the province, causing thousands to flee their homes.
- Fracking triggered large quakes in the oil and gas patch.
- Tla'amin First Nation implemented a treaty with the provincial government.
- The high-tech sector became a bigger employer than mining, oil and gas, and forestry combined.
- A tragic avalanche took the lives of five snowmobile riders.
- The provincial government formally opposed the Kinder Morgan Pipeline expansion.
- A tent city popped up to protest the lack of affordable housing in Vancouver.
- The wine industry boomed in BC's interior.
- The number of sockeye salmon returning to the Skeena River reached record lows.

As this list suggests, human and physical processes are altering the province's landscape.

The discipline of geography seeks to understand these processes, in the present and in the past. Geography has been defined as the study of "where things are and why they are where they are" (McCune 1970, 454). "Things" can be physical features, people, places, ideas (or human innovations), or anything in the landscape. "Where" questions concentrate on location as well as recognizing physical and human patterns and the distribution of various activities, people, and features of the landscape. Many of these questions can be answered simply by looking at a map. Look at a road map or online map of British Columbia. Where is wine country? Where is the territory covered by the Tla'amin treaty? Where do earthquakes occur, and what towns were affected by the wildfires? Where are the sockeye spawning grounds?

Knowledge of where things are is basic, essential geographical information. To test your knowledge of British Columbia, draw a map of the province from memory and place on it the features you consider important. This cognitive mapping exercise reveals individual landscape experiences (which can be shared with others) and demonstrates the importance of location. Using maps to answer "where" questions is the easiest aspect of geographical study.

Answering the question "Why are things where they are?" is more complicated. "Why" questions are far more difficult than "where" questions and may ultimately verge on the metaphysical. Even so, as you study geography, you'll be encouraged to conduct research about and to analyze the various physical, economic, political, cultural, and historical factors that have shaped a specific location or locational patterns, whether it be the location of a type of vegetation, a community, a group of people, or a resource. Why do grape wines grow so well in the Okanagan Valley? Why is Vancouver where it is, and why has it grown so rapidly? Why did Barkerville become a ghost town? Why were the Japanese removed from the coast of British Columbia? Why did the Nechako River get dammed for hydroelectric power? Why is the Peace River region not part of Alberta? These questions are not easy and often require historical, physical, cultural, political, and economic assessments.

So, too, do "what" questions. Some definitions of geography include the question, "What is the significance of these locational patterns?" (Renwick and Rubenstein 1995, 5). What influence do people have on the environment, and what influence does the environment have on people? Humans are constantly shaping and modifying the landscape to meet the demand for resources – clear-cutting forests, damming rivers, and building power plants that pour emissions into the air and water – and these acts produce an environmental backlash to ecosystems and human health.

All these questions – Where? Why? What? – mean geography is a practical and pragmatic discipline, one that encourages an understanding of the surface of the earth on all geographic scales. Geography is a discipline that lends itself to being out of the classroom and in the environment, where one can read both the physical and human landscapes. Physical geographers are interested in the physical processes that influence the landscape. Human geographers, by contrast, look at where people live, what their activities are, and how they have modified the landscape. Of course, a combination of physical and human processes often modify landscapes, and both sides of the discipline incorporate a spatial perspective. As Figure I.1 shows, geography can be divided into a number of subfields and is associated with many other disciplines, but the spatial element keeps it distinct.

Geography allows us to recognize the range of physical characteristics responsible for mountain building and erosion and for weather and climate patterns. From the viewpoint of physical geography, changes to the landscape are often measured in several hundreds of millions of years, and the BC landscape is no exception. A combination of physical processes produced a spectacular variety of mountains, rivers, lakes, islands, fjords, forests, and minerals in British Columbia. Studying the province's geography allows us to understand why some communities and regions are at considerable risk from floods, forest fires, or avalanches and how these risks can be reduced or eliminated.

Geography of British Columbia will help you develop the critical thinking skills necessary to unravel the complexity of spatial patterns, processes, and relationships, and these skills can up open up many career opportunities. This book will help you understand not only physical processes that led to changes to the landscape but also the processes responsible for settlement and development of the land and why people live and work where they do. You'll come away with an understanding of how past decisions and actions have shaped the landscape of the present. Throughout the book, complex processes are described in simple language, and more complex terms, **highlighted in bold** (on first use), are explained in a glossary, located at the end of the text.

From a European, colonial perspective, British Columbia has a short history of settlement and development compared to eastern Canada or to many other nations in the world. Indigenous Peoples, however, have well over ten thousand years of history with the lands that eventually became British Columbia, and anthropologists and archaeologists are still adding new evidence of their settlement patterns and use of resources. Indigenous Peoples and the explorers, fishers, sojourners, and settlers who began to arrive in the eighteenth century exploited and altered the landscape, sometimes irreversibly.

To understand these developments and their significance over time, we need to consider movement over space. **"Time-space convergence"** (sometimes referred to as time-space compression or collapse) refers to changing technologies of movement that shrink time and space. Today, for example, a flight from London, England, to Vancouver, British Columbia, takes approximately nine hours. In the 1790s, when the maritime fur trade for sea otters began to draw Europeans to the Pacific Northwest – the region that comprises present-day British Columbia, Washington, Oregon, and northern California – a trip to the region from London took nearly seven months and meant sailing around South America. By 1886, when Vancouver was incorporated

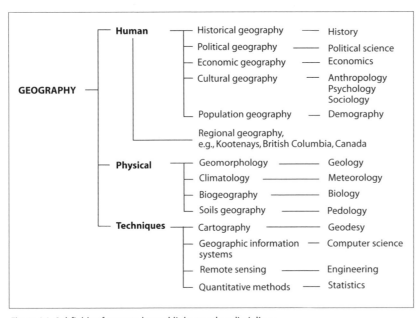

Figure I.1 Subfields of geography and links to other disciplines

as a city, the voyage from London to the new city had been reduced to three weeks with the introduction of steam-driven vessels and the completion of the Canadian Pacific Railway across Canada. Changing transportation routes affected the movement of goods and people, settlement patterns, and resource development. Railways (along with ship transport) promoted the movement of high-bulk, low-value goods such as wheat, lumber, and coal over great distances with relatively low freight rates. By comparison, current satellite systems provide instant global communications that have led to a worldwide reorganization in the production of goods and services and spatial relationships more generally.

The concept of time-space convergence helps explain a shrinking world, but it's important to keep in mind that the shrinking of time and space does not occur evenly. Geographic locations that are connected differ, sometimes greatly, from locations that are not connected. Freeways such as the Coquihalla allow people to travel quickly to communities in southern British Columbia, while northern and coastal communities that have only secondary roads or, in some cases, no roads are significantly more isolated. The construction of airports, railways, port facilities, pipelines, and communications infrastructure have given birth to similar disparities. These transportation developments have played a significant role in settlement, development, and economic advantage throughout the province.

Another important and related concept to keep in mind is **spatial diffusion**, which is employed to trace the movement (or flow) from one location to another of goods, people, services, innovations, and ideas. For example, the spatial diffusion process is used to trace where new innovations in computer software occur and where they are adopted. The spatial diffusion process is also used to describe events, such as the waves of smallpox epidemics that decimated Indigenous Peoples, the evolution and pattern of salmon cannery locations, or the spread of high-speed internet service. All of these movements were influenced by carriers and barriers (Gould 1969). Carriers are instrumental in the spread and adoption of innovations, goods acquisition, or the contraction of diseases; barriers prevent, or block, this movement. "Relocation diffusion" refers specifically to the movement of people (e.g., refugees) from one place to another. The terms "barriers" and "carriers" apply to relocation diffusion, but the terms "push" and "pull" factors are also used. Push factors include the many political, economic, and social forces that cause people to move, such as overpopulation, warfare, or religious persecution. Pull factors are the various conditions that attract people to a new location. Both push and pull factors have been responsible for moving people to British Columbia.

Statistics are also useful in assessing trends and patterns over time. Table I.1 indicates the evolution of British Columbia's population, including its gradual transformation from a rural to an urban province. Isolation was a major factor in prohibiting early non-Indigenous settlement and development. However, time-space convergence overcame the friction of distance in this frequently vertically challenged landscape, and this change is reflected in rapid population growth, particularly after the completion of the transcontinental railway. The conquering of distance also facilitated the global transition, especially in trade and investment, from the Atlantic to

Table I.1

Rural and urban population, 1871–2018

Year	Population	Ten-year change	Percentage	
			Rural	Urban
1871	36,247	–	91.1	8.9
1881	50,387	14,140	88.1	11.9
1891	98,173	47,786	57.5	42.5
1901	178,657	80,484	49.5	50.5
1911	392,480	213,823	48.1	51.9
1921	524,582	132,102	52.8	47.2
1931	694,263	69,681	56.9	43.1
1941	817,861	123,598	45.8	54.2
1951	1,165,210	347,349	47.2	52.8
1961	1,629,082	463,872	27.4	72.6
1971	2,184,621	555,539	24.3	75.7
1981	2,744,465	559,844	22.0	78.0
1991	3,282,061	537,596	19.6	80.4
2001	4,078,447	796,386	18.0	82.0
2011	4,400,057	321,610	14.0	86.0
2018	4,991,687	591,630*	10.7	89.3

* Seven-year change only.

Sources: Dominion Bureau of Statistics (1944), Statistics Canada (2009), Moazzami (2015), BC Stats (2019).

the Pacific, tying British Columbia closely to the Asia-Pacific region.

Time-space convergence and spatial diffusion are basic geographic concepts essential to understanding movement over space and through time, while statistics, graphs, and maps are the key tools of the geographic trade (the "techniques" shown in Figure I.1). They allow geographers to begin to understand the dynamics of the where, why, and what questions. *Geography of British Columbia* employs these concepts, tools, and techniques in a comprehensive exploration of British Columbia's peoples and landscapes and their transition over time. The book is divided into two parts. Part I offers a foundational understanding of the geography of British Columbia, especially for those with little or no knowledge of the discipline. Chapter 1 underscores the importance and relevance of adopting a spatial approach to understanding the development of the province, including settlement patterns in the eight provincial regions. Chapter 2 introduces some of the basic processes of physical geography with a particular focus on the profound impact that weather, climate, and the physical features of the surface of the earth have had over the span of tens of millions of years.

Chapter 3, in turn, explores how these basic processes of physical geography gave birth to geophysical hazards – extreme events such as floods, wildfires, avalanches, and earthquakes. In British Columbia, the combination of a rugged, vertical landscape with mainly prevailing westerly winds has led to considerable property damage and loss of life, for which corrective and preventative measures had to be (and must be) taken. Finally, Chapter 4 sets the stage for Part 2 by defining what geographers mean by **"resources"** and introducing several models for examining the implications of resource harvesting and the importance of resource management, including how resources have been used to grow the economy over time.

Part 2 unfolds chronologically rather than thematically to show the role that resources have played in shaping and reshaping the province while at the same time avoiding giving the impression that British Columbia remains a resource-dependent province. Since the 1970s, there has been a fundamental transformation of the province's economy, to the point where resources now contribute few jobs or provincial revenues. Today, British Columbia is a **service economy**. Adopting a historical approach allows for a better understanding of how this major transition took place and, tragically, why resources have been mismanaged in the province. And it makes it easier to trace how and why the shift away from resources has had a negative impact on some regions more than others.

Resources – especially renewable resources such as furs, fish, and forests – played a significant role historically, but, unfortunately, both settlers and governments treated them as never-ending in a boundless "frontier," an approach that can best be understood as part of the tragedy of the commons, an economic theory that explains why people use natural resources to their advantage without considering the good of society as a whole. In a classic essay published in *Science* in 1968, Garret Hardin explained the theory by telling the story of two tribes who measure their wealth in cattle. In the beginning, the cattle feed on the commons – a finite, shared territory or **common property resource**. Over time, war and cattle poaching cause grief, but the commons remain sustainable and support the cattle economy. The tragedy occurs when "peace" is achieved, when individuals from each tribe have the motivation to gain wealth by increasing their herd. In this time of unrestricted capitalism, the commons soon becomes overgrazed, bringing tragedy to all.

A similar process has occurred in British Columbia again and again. Since the arrival of European colonizers to the northwest coast of America, renewable resources have suffered from species extinction, or near extinction, through colonial and then provincial exploitation. Once one resource was depleted, colonizers, governments, or companies have moved on to exploit another. These physical-human interactions prompt us to look at the landscape another way, namely, from an environmental perspective.

Chapter 5, which opens Part 2, reveals how the accidental discovery of a lucrative market for furs in China became the lure for Russian, Spanish, British, and American interest in the Pacific Northwest. At first, the maritime fur trade, which focused on the sea otter, led to conflict between the British and the Spanish and between colonizers and Indigenous Peoples. But it was two nonrenewable resources – coal and gold – that ultimately led to the British acquiring colonial control over the **territory**, which in turn led to the arrival of more settlers, declining Indigenous populations, and, eventually, Confederation, when British Columbia became part of Canada in 1871.

Chapter 6 begins with the promises made by the federal government during Confederation and ends in the early twentieth century. One of the key promises made was a transcontinental railway. Although the railway was delayed by a scandal, it was eventually constructed but not without tremendous amounts of racism being directed towards Chinese contract labourers and anguish being felt among residents in Victoria who wanted the terminus to be in their own city rather than in Vancouver. The completion of the Canadian Pacific Railway led to the construction of regional rails, which were connected to steam paddlewheelers (steamers) on lakes and the ocean, allowing for the movement of resources and people throughout the province and the emergence of a new "boom and bust" economy.

Chapter 7 examines BC's increasing resource dependency in the context of unpredictable global events such as the two world wars. Dependent on resource development, immigration, and British investment in railways, the province was rocked by swings in commodity prices that led to more boom-and-bust cycles, and tough economic times brought on extreme racism towards Asians. But the Second World War catapulted Canada into an industrial nation, fuelled by major inputs from BC's natural resources. As Chapter 8 explains, the greatest economic boom in North American history followed the war, and it was accompanied by a baby boom, high immigration rates, increased disposable income, a demand for single-family dwellings, and the birth of car culture. In British Columbia, as elsewhere, values changed in a rapidly changing world marked by the **Cold War** and fear of nuclear annihilation. British Columbians witnessed the beginnings of the **megaprojects** era as clear-cut logging, open-pit mining, major hydroelectric dams, transmission lines and pipelines, and roads and rails scarred the landscape, giving rise to the environmental movement and renewed protests by First Nations.

But Chapter 9 explores a major shift that occurred in the last three decades of the twentieth century as the province shifted from being profoundly rural and resource-dependent to highly urban and service-oriented.

The transition, fuelled by globalization, was complicated and difficult and required resource industries to restructure themselves and downsize as they were hit by major and unforeseen economic predicaments, including the end of the Cold War, energy crises, recessions, and free trade agreements. On another level, they had to contend with environmental organizations and First Nations demanding, and gaining, recognition of **Indigenous Title** (called "Aboriginal Title" by the Supreme Court of Canada).

Although the Nisga'a Treaty, which recognizes Indigenous Title, was signed in 2000 by the federal government and BC's NDP government, Chapter 10 explores what happened when the Liberal Party came to power from 2001 to 2017, some of the strangest, most contradictory decades in the province's history. One of the first things the new government did was hold a referendum to reverse recognition of Indigenous Title. But by 2005 the Liberals had forged a New Relationship with First Nations that recognized Indigenous Title. However, both Gordon Campbell's and Christie Clark's governments continued to conduct land-use decisions on unceded territories as if treaties and landmark Supreme Court decisions had not occurred. The period was also characterized by some of the most stringent regulations to reduce greenhouse gas emissions, including a **carbon tax** to address **climate change**. As these regulations were put in place, the government simultaneously insisted that resources were "the backbone of the economy," promoted liquefied natural gas as if it were not a fossil fuel, and supported oil pipelines from Alberta. It became embroiled in land-use conflicts, many of which were over resource exploitation.

To conclude, I reflect on the history of resource mismanagement in the province in the context of a present and a future in which climate change is a reality. Geography is about people and the environment, and this edition of *Geography of British Columbia* stresses the political decisions that have affected and moulded the province's landscape. But it also addresses the impact of unforeseen events and conditions, including geophysical hazards, war, recessions and depressions, and, more recently, terrorism. Collectively, these global and local forces helped shape a province that was once reliant on resources but no longer depends on them. In the era of climate change, the regional economies that have developed in British Columbia and the uneven distribution of its population and employment opportunities means that some regions are or will be more vulnerable than others, and both governments and citizens need to be prepared.

REFERENCES

BC Stats. 2019. "Population Estimates: Total Population." https://www2.gov.bc.ca/gov/content/data/statistics/people-population-community/population/population-estimates.

Dominion Bureau of Statistics. 1944. *Canada Year Book 1943–44.* https://www66.statcan.gc.ca/eng/1943-44-eng.htm.

Gould, P.R. 1969. *Spatial Diffusion.* Resource paper no. 4. Washington, DC: Association of American Geographers.

Hardin, Garrett. 1968. "The Tragedy of the Commons." *Science* 162 (3859): 1243–48.

McCune, S. 1970. "Geography: Where? Why? So What?" *Journal of Geography* 7 (69): 454–57.

Moazzami, Bakhtiar. 2015. *Strengthening Rural Canada: Fewer and Older: The Population and Demographic Dilemma in Rural British Columbia.* http://strengtheningruralcanada.ca/file/Fewer-Older-The-Population-and-Demographic-Dilemma-in-Rural-British-Columbia1.pdf.

Renwick, W.H., and J.M. Rubenstein. 1995. *An Introduction to Geography: People, Places, and Environment.* Englewood Cliffs, NJ: Prentice Hall.

Statistics Canada. 2009. "Selected Population Characteristics, Canada, Provinces and Territories." Table 17-10-0118-01. https://www150.statcan.gc.ca/t1/tbl1/en/tv.action?pid=1710011801&pickMembers%5B0%5D=1.11.

Part 1
Geographical Foundations

British Columbia, a Region of Regions

1

British Columbia is a large province that encompasses nearly 950,000 square kilometres. To put this in context, many countries or nation-states are significantly smaller. Although the province is larger than many countries, it has a small population (5.0 million people in 2018) relative to its size. It has a population density of only 5.3 persons per square kilometre. Great Britain, by contrast, is four times smaller in size but has 307.5 persons per square kilometre (World Population Review 2017). To take these comparisons one step further, it's fair to say that few nation-states have such a variety of landscapes as British Columbia. The province truly is a **region** of regions that can be divided and subdivided on the basis of both its physical and cultural characteristics. In this chapter, you'll learn how to adopt a spatial approach to understanding the development of the province, including settlement patterns in the eight provincial regions.

A REGIONAL GEOGRAPHY APPROACH

Adopting a **regional geography** approach means dividing the province or region of British Columbia into geographic areas that have common physical or human/cultural characteristics (Gregory 2000, 687). On the physical side, river drainage systems, plateaus, mineral deposits, forests and vegetation, frost-free days, latitude, elevation, and precipitation are the criteria by which the province can be divided into distinct regions. And from a regional perspective, British Columbia is a unique province within Canada for a variety of reasons. Its physical characteristics in particular set the province apart from all others. The province's highly indented coastline, "punctured by fjords," spans some 41,000 kilometres (Dearden 1987, 259), and BC has the youngest and highest mountains in the country. It is often described as a vertical landscape. It also has the greatest amount of fresh water in Canada, an essential resource for Pacific salmon and a potential source of hydroelectric power.

The province's relatively mild, wet west coast has the warmest winter temperatures in Canada; in contrast, the considerably colder and drier Interior has desert conditions in the southern river valleys. If you travel from west to east in the province, you'll discover that vertical change due to high mountain ranges dropping off to valleys of rivers and lakes results in a combination of climate, soils, and vegetation that produces distinctive geographic patterns. The same is true if you travel from south to north: the province stretches across an eleven-degree span of latitude, from the forty-ninth parallel in the south to sixtieth parallel in the north, and there are significant and distinctive weather and climate patterns across the span that also influence soils and vegetation.

Distinctive physical characteristics and a unique global location have also influenced the province's human characteristics. On the human or cultural side, features such as a common language or religion can demarcate regions (making them **cultural regions**), as can other political, economic, and social factors. For example, for many people who live in a particular area and have shared historical experiences, such as Indigenous Peoples, there is a sense of place, or "nationalism," that comes with a connection to the land. To give more recent examples, district and health board boundaries represent the organization of space based on political decisions, whereas fishing zones, tourist areas, forestry regions, newspaper-circulation areas, policing jurisdictions, and school districts are regions derived more from economic and social functions.

British Columbia's population and settlement patterns were also unique. The precontact population of Indigenous Peoples, particularly in coastal locations and along salmon-bearing rivers, was greater than anywhere else in Canada (Muckle 1998). And unlike in the rest of Canada, non-Indigenous "discovery" and settlement came from the west rather than from the east, with the province's abundant supply of resources serving as the main attraction for migrants and the reason for its rapid growth. Yet the province's physical characteristics initially made resource extraction and export to distant markets difficult, resulting in regionally differentiated patterns of settlement and development (Robinson 1972). British Columbia went through distinctive territorial struggles to become a British colony and further political struggles to establish its present boundaries. Again differentiating it from the rest of Canada, its connection to the Pacific, particularly to Asia, increased as transportation systems developed. No other province has such a long history of immigration by Asians – first Chinese, and later Japanese and Sikhs. Nor has any other province gained such a reputation for being so adamantly racist.

Although regional geography can be employed to divide British Columbia into parcels, or regions, to examine its characteristics critically and make sense of its diversity, this approach has some limitations. Critics point out that regions are not islands unto themselves but are linked in ways that can't be captured by traditional regional geography. Moreover, the characterization of a region may be appropriate only at one point in time, meaning that regions must be reconfigured as conditions change.

To address these concerns, some geographers have distanced themselves from the traditional approach by engaging in what has been referred to as "reconstructed regional geography" (Pudup 1988). Using a host of analytical tools and borrowing from other disciplines, they take into account a number of factors, including the many complex relationships within any landscape, the interactions that link adjacent regions, and even global conditions. They recognize that regions can change over time and that regions may overlap. For example, from a cultural and historical perspective, territorial boundaries or seminomadic regions divided Indigenous Peoples in the Pacific Northwest. These regional boundaries shifted because of warfare, scarcity of resources, and changes in technology. The greatest change of all, however, came with the arrival of Europeans, who reorganized the landscape into different regions and placed First Nations on small parcels of land called reserves. The political boundaries of British Columbia have been drawn a number of times, but it is only since the 1990s that provincial politicians have recognized the historical territories of Indigenous Peoples. **Place-name geography** largely erased oral, traditional Indigenous names throughout the province and replaced them with mainly British ones. In recognition of the countless injustices done to First Nations over time, many regions are now recognizing and using the Indigenous names for places, such as Haida Gwaii, Salish Sea, and qathet Regional District.

When dividing the province into meaningful regions, external regionalization must also be acknowledged. For example, British Columbia was initially claimed by Spain before coming under the colonial control of Britain. When British Columbia joined Confederation in 1871, it became spatially one region (and the largest province) in an independent Canada, breaking the bonds of colonization and relieving anxiety that the territory would be annexed by the United States. In the late 1980s and early 1990s, the signing of the Free Trade Agreement in 1989 and the North American Free Trade Agreement in 1994 (which has been in dispute with the United States since 2017) and increased trade and investment in the Asia-Pacific region have placed British Columbia, and Canada, in a new, global regional economic alignment, but not without conflict. A regional geography of British Columbia must take all these changing conditions into consideration.

Figure 1.1 divides the province into eight regions, devised by considering historical developments in combination with census subdivisions. It details features such as mountains, rivers, incorporated communities, and transportation systems. (Note that Indigenous Title boundaries are not shown on the map; the colonial system within which the census subdivisions occurred did not recognize these boundaries, which unfortunately do not correspond to the eight regions currently dividing the province.) The map will help you become familiar with the vastness of the landscape, the physical and human features that distinguish each region, and the factors that integrate these separate regions with other parts of Canada and the world.

Table 1.1 provides the population change for these regions, spanning over 135 years. By monitoring regional population change, we can compare and contrast regions with the greatest growth with regions with the least growth, and we can also trace fluctuations within any of the eight regions. Graphing the absolute growth of each region also illustrates the rate of change and gives a sense of historical development. For example, a comparison of just two regions – the Lower Mainland and Vancouver Island–Central Coast – provokes interesting questions about the rate of growth. Why did the Lower Mainland, a geographically smaller area, outstrip the Vancouver Island–Central Coast region so rapidly between 1901 and 1911? Bear in mind that the Lower Mainland includes the City of Vancouver, incorporated in 1886. Now observe the slope of the graph lines between the years 1921 and 2016 in Figure 1.2, which shows the population increases of the two regions. What accounts for the different rates of population growth during this period, and why has the Lower Mainland population continued to grow at a more rapid rate?

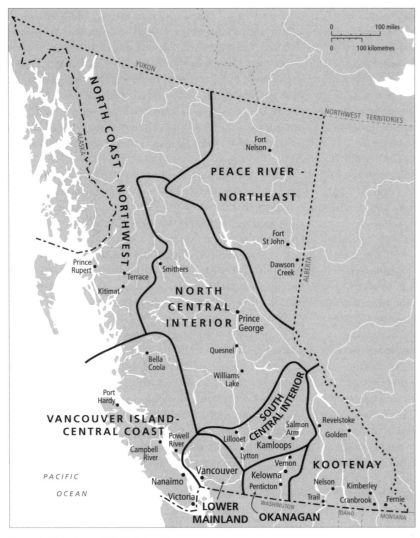

Figure 1.1 Regions of British Columbia

the Second World War were also significant global events that affected each region. Following the war, technologies of time-space convergence, in combination with more global means of producing goods and services, altered the way people made a living and reorganized the value of resources. The Lower Mainland region, with its greater connectivity – roads, railways, a port, an international airport, and conference centres – increasingly gained the greater proportion of the population.

A similar comparison could be made for any two other regions and would require analysis of the many factors that influence regional growth. Census figures can also reveal the ethnic composition of each region, knowledge of which can form the basis for a whole host of social, ethnic, political, and economic questions. These questions, in turn, could lead to geographic analyses of various groups.

These types of statistical analyses can promote further geographic knowledge about settlement and the development of communities. Table 1.2, for instance, ranks the ten largest municipalities in 2016 by population. It is interesting to compare these populations to selected census years in the past (in which the ranking is also given). Many of our present-day communities did not exist before the twentieth century; others have changed boundaries; some were much more significant in the past; and others have lost population. What factors are responsible for the growth or decline of communities, and how are communities connected to their regions and to other communities? The conclusions arrived at following the regional comparison in Figure 1.2 are reinforced if we compare the older centre of

The answers lie partly in the political decision to locate the Canadian Pacific Railway terminus and an international port at Vancouver, decisions that stimulated economic and population growth. Employment opportunities related to fishing, forestry, mining, and agriculture played a role in each region, and the port facility of Vancouver greatly widened the catchment area for exporting resources. The First World War, the opening of the Panama Canal, the Depression of the 1930s, and

Table 1.1

Population by region, 1881–2016

Year	Vancouver Island– Central Coast	Lower Mainland	Okanagan	Kootenay	South Central Interior	North Central Interior	North Coast– Northwest	Peace River– Northeast
1881*	18,777	7,949	1,316	863	4,725	7,550	7,376	923
1891*	39,767	23,543	3,360	3,405	6,390	4,889	16,839	n/a
1901**	54,629	53,641	7,704	32,733	14,563	5,123	9,270	948
1911**	84,786	183,108	21,240	51,993	24,103	9,011	16,595	1,644
1921	119,024	256,579	23,728	53,274	32,232	18,615	18,986	2,144
1931	133,591	379,858	30,919	63,327	37,621	23,236	18,689	7,013
1941	164,751	449,376	40,687	72,949	37,394	26,272	18,051	8,481
1951	233,250	649,238	62,530	93,256	50,363	41,324	20,854	14,395
1961	312,160	907,531	86,230	107,466	57,346	89,085	38,203	31,061
1971	415,254	1,256,425	130,498	125,643	106,993	133,906	63,080	45,155
1981	517,536	1,434,739	196,774	145,412	139,175	186,992	68,376	55,463
1991	611,654	1,829,537	240,291	141,480	142,628	190,141	67,975	58,355
2001	717,997	2,378,240	310,602	163,502	167,364	210,621	66,673	63,448
2011	772,912	2,725,573	347,551	162,927	169,995	197,106	57,201	67,882
2016	805,082	2,928,729	362,861	164,662	173,854	194,409	56,118	72,196

*   Some approximations for regions, as the province was divided into only five electoral areas.
** Some approximations for regions, as the province was divided into only seven electoral areas.
Sources: Census of Canada (1951, 1971), BC Stats (1986, 1991, 2012, 2017).

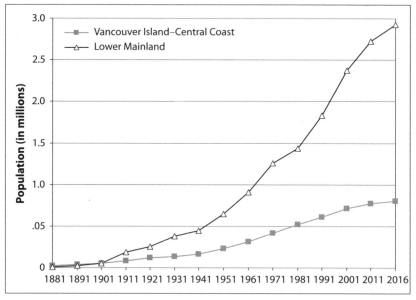

Figure 1.2 Vancouver Island–Central Coast and Lower Mainland populations, 1881–2016
Sources: Census of Canada (1951, 1971), BC Stats (1986, 1991, 2012, 2017).

Victoria with the newer city of Vancouver. The extremely rapid rise of Vancouver's population since its incorporation in 1886 illustrates the powerful influence of the developments discussed above. In contrast, as the provincial capital, Victoria experienced the growth of government and services along with infrastructure developments that linked it to Vancouver Island's resources and to the mainland, but these economic links were not nearly as extensive as those of Vancouver.

Other communities throughout British Columbia have changed their population ranking over time through the expansion of transportation infrastructures, resource development, and processing, as well as a shift to service industries. Beginning in the late twentieth century, the new urban-growth dynamics of the tourism and retirement industries, along with technologies that shrink time and space, began to affect some communities more than others. These communities, and their growth (or decline), are intimately tied to the regions in which they are located. To lay the foundations for discussions and analyses of these trends, the remainder of this chapter outlines the eight regions of British Columbia in terms of their distinctive physical characteristics and historical development.

VANCOUVER ISLAND–CENTRAL COAST

This region combines Vancouver Island with the Central Coast, which extends from Powell River north to Bella Coola. Vancouver Island has a rugged spine of mountains, referred to as the Insular Mountains. The Central Coast is part of the Coast Mountains range, which has peaks reaching higher elevations than in the Canadian Rockies

Table 1.2

Population of municipalities for selected years, 1881–2016

Municipality	2016	2011	2001	1921	1901	1891	1881
Vancouver (CMA)	(1) 2,558,029	(1) 2,394,270	(1) 2,073,681	(1) 163,220	(1) 27,010	(2) 13,685	–
Victoria (CMA)	(2) 365,631	(2) 358,672	(2) 325,569	(2) 38,727	(2) 20,919	(1) 16,841	(1) 5,925
Kelowna (CMA)	(3) 197,017	(3) 191,114	(4) 100,496	2,520	–	–	–
Abbotsford-Mission (CMA)*	(4) 181,169	(4) 180,945	(3) 153,801	–	261	–	–
Nanaimo (CA)	(5) 107,462	(6) 92,361	(6) 76,185	(5) 6,304	(4) 6,130	(4) 4,595	(2) 1,645
Chilliwack (CA)	(6) 104,662	(8) 80,892	(8) 65,672	(9) 3,161	277	–	–
Kamloops (CA)	(7) 104,449	(5) 92,882	(5) 80,655	(7) 4,501	(8) 1,594	–	–
Prince George (CA)	(8) 83,494	(7) 83,225	(7) 75,567	2,053	–	–	–
Vernon (CA)	(9) 62,169	(9) 55,418	(9) 34,957	(8) 3,685	802	–	–
Courtenay (CA)	(10) 56,359	24,099	18,304	–	–	–	–
Penticton (CA)	33,761	(10) 33,553	(10) 32,339	– **	–	–	–
New Westminster	– ***	– ***	– ***	– ***	– ***	(3) 6,641	(3) 1,500
North Vancouver	– ***	– ***	– ***	(3) 7,652	–	–	–
Prince Rupert (CA)	11,261	12,802	15,282	(4) 6,393	–	–	–
Nelson (C)	11,249	10,371	9,703	(6) 5,230	(5) 5,273	–	–
Rossland (C)	3,639	3,614	3,804	2,097	(3) 6,156	–	–
Fernie (C)	4,333	4,532	4,812	2,802	(6) 1,640	–	–
Revelstoke(C)	7,316	7,287	7,826	2,782	(7) 1,600	–	–
Trail (C)	7,376	7,801	7,905	(10) 3,020	(9) 1,360	–	–
Greenwood (C)	688	710	695	371	(10) 1,359	–	–

Note: CMA = census metropolitan area; CA = census area; C = city. The numbers in parentheses indicate the municipality's size rank.
* Boundary change and incorporated as a city in 1995.
** Incorporated in 1909, but population less than 1,000.
*** Included in Vancouver CMA.
Sources: Census of Canada (1951, 1971), BC Stats (1986, 1991, 2012, 2017).

and includes one of the highest mountains in British Columbia, Mount Waddington (4,016 metres). These mountains run mainly north-south and influence weather and climate conditions significantly. The prevailing westerlies often result in torrential rains on the west side of Vancouver Island and on the mainland along the coast where it is exposed to the open Pacific. The Olympic Mountains of Washington State and the Insular Mountains of Vancouver Island create a **rain shadow effect** on the east side of Vancouver Island and on the southern end of the Central Coast. The Pacific Ocean at these latitudes (approximately 48°30' to 52° north) is considerably warmer than the Atlantic Ocean on the east coast of Canada. Precipitation variations caused by the rain shadow also influence vegetation and, in particular, the growth of Douglas fir in the drier areas. There are no large rivers on either Vancouver Island or the mainland, but the many small rivers and streams are important for fish habitat and some hydroelectric power.

Historically, the Vancouver Island–Central Coast region has been home to many Indigenous Peoples. The peoples in this region experienced the longest exposure to non-Indigenous peoples in the province, and diseases took a huge toll. It was here that the Spanish and British squared off in the 1780s over territorial claims for colonization and the valuable sea otter trade. By the early 1800s, British Columbia was embroiled in an overland fur trade struggle between the aggressive North West Company and the Hudson's Bay Company. In 1821, the merger of these two companies resolved the dispute, leaving the Hudson's Bay Company with a monopoly over the territory.

Non-Indigenous settlements were sparse and temporary until a series of political and economic events occurred. The first was the discovery of coal at the north end of Vancouver Island in the mid-1830s. Later, the **Oregon Treaty** of 1846 annexed the British territory and forts south of the forty-ninth parallel, and as a consequence, Victoria was established at the south end of Vancouver Island (see Figure 1.3). By 1849, discoveries of even richer coal deposits had occurred in the Nanaimo region, which attracted more European settlers, but in 1858, gold was discovered on the lower reaches of the Fraser River, triggering an avalanche of miners seeking their fortune. Victoria became the main port of entry for much of this activity. Further discoveries of gold in the

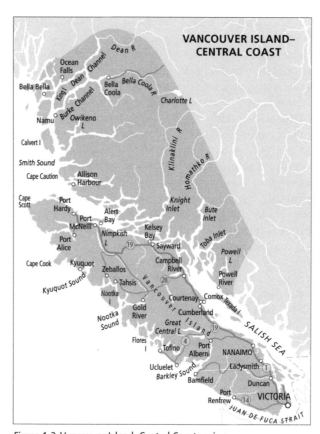

Figure 1.3 Vancouver Island–Central Coast region

Cariboo region enhanced Victoria's position as the area's main administrative and service centre. The two separate colonies of Vancouver Island and the mainland were amalgamated in 1866. Victoria became the capital, a position it has maintained ever since.

The location of the capital attracted settlers to the southern end of Vancouver Island. The rest of the island was opened up in response to the discovery of resources, transportation developments, and technological change. For example, coal discoveries enhanced Vancouver Island's status as a major energy producer, and the discovery of copper led to the construction of copper smelters. Later, iron ore was discovered on Texada Island. Alongside the whaling industry, salmon fishing and canneries sprang up on the coasts of both the island and mainland, and farming settlements emerged mainly in the southeast portion of the island. By far the most important industry throughout the region was forestry. Large lumber mills such as the one in Chemainus were in operation by the 1880s, and they were followed by pulp mills in the communities of Ocean Falls, Port Alice, and Powell River in the early 1900s.

Historically, railways were the most important means of land transportation on Vancouver Island. The Esquimalt and Nanaimo Railway (E&N) opened in 1886 and was the most important line, as it came with a provincial land grant to over one-quarter of the island, including some of the best forests in the province. Roads appeared first on the southeast side of Vancouver Island, and they eventually linked the south end of the island to the north, even as tentacles stretched across to the few communities on the island's west side (Wood 1979). Ocean-going transport provided the main link between Vancouver Island and mainland coastal communities such as Bella Bella, Bella Coola, Ocean Falls, and Powell River. The highly indented and rugged mainland coastline hindered road development in the early days of settlement. Only Bella Coola was linked by road to the rest of the province.

Today, more than 800,000 people live in this region. The forest industry, which was the economic backbone for much of Vancouver Island's history, has faced economic uncertainty since the 1980s, resulting in many sawmill and pulp-and-paper mill closures and increased unemployment in those industries. Farming, including farmgate wineries, a wide range of recreation and tourism options, commercial fishing, and fish farming (although controversial) are other areas of employment. The Canadian Forces Base in Comox, which employs approximately sixteen hundred people, is significant to the Comox-Courtenay region. The southeastern portion of the island has experienced some unique economic dynamics. It is linked intimately to the Lower Mainland region in the provision of administrative and service functions for western Canada and the Asia-Pacific region, and it has also become home to a huge number of retirees who are drawn to its mild winters, which mean relatively little snow to shovel and year-round recreational activities. This part of the island also attracts the greatest number of tourists because of direct ferry links with the Lower Mainland.

LOWER MAINLAND

The climate of the Lower Mainland is similar to that of the southeast coast of Vancouver Island, but it has higher precipitation, including more snow in winter because of its proximity to the North Shore Mountains and Burnaby Mountain. The Fraser River, the largest river system in the province and most significant to the salmon-fishing industry, is an important physical feature of this region (see Figure 1.4). Historically, it was the major salmon runs on the Fraser River that attracted many Indigenous Peoples such as the Hul'q'umi'num', Squamish, Tsleil-Waututh and Stó:lō, to the region; they shared this treasured resource with other First Nations from Vancouver Island. For non-Indigenous people, gold was the main attraction following its discovery on the Fraser in 1858. Agricultural settlements soon followed, but securing these rich agricultural lands from the threat of floods has not been easy. Before the establishment of Vancouver, major sawmills on Burrard Inlet exported lumber, and canneries operated at the mouth of the Fraser River.

Transportation has been a major factor in the growth and development of this region. The completion of the CPR at Port Moody and its extension to Vancouver in 1886 was the catalyst for the rapid population growth observed in Table 1.2. Vancouver, with its national railway and international port, was the main centre for this relatively small geographic region and was largely responsible for the growth of the adjacent Fraser Valley to the east, the Squamish-Whistler-Pemberton corridor to the north, and the Sunshine Coast to the northwest. The mountains and valleys framed the transportation links and settlement patterns for the region.

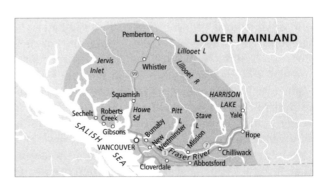

Figure 1.4 Lower Mainland region

The Sunshine Coast has a linear settlement pattern following the Strait of Georgia and is connected to Vancouver via ferry at Horseshoe Bay in West Vancouver. The Pacific Great Eastern Railway (renamed the British Columbia Railway, or BCR, in 1972 and taken over by CN in 2004) began running between Squamish and Quesnel in 1921 and was then extended south to North Vancouver in 1956 and north beyond Quesnel. This railway line has been an important transportation link to the ports at Squamish and North Vancouver. But today the Sea-to-Sky Highway is the main transportation system as it winds its way beside Howe Sound to Squamish and then follows valleys leading past Whistler to Pemberton, Lillooet, and the interior of the province. Vancouver, Whistler, and all the communities between them, as well as the route that links them, have gained a lot of attention, economic investment, and population increases since hosting the 2010 Winter Olympics.

In the Fraser Valley, the river served as the original transportation system. The construction of the CPR and later the Canadian National Railway and British Columbia Electric Railway made the region even more accessible. Road systems were built in the early 1900s, and eventually the construction of the Trans-Canada Highway and other highways linked Vancouver and the Fraser Valley to the rest of the province and south to the United States.

Favourable climate, superb natural features, highways, railways, international port facilities, an international airport, educational facilities, and commercial links to the rest of Canada, Asia, and the world make Vancouver a world city. This region has become the focus of the high-tech and film industries along with tourism, international banking, finance, insurance, real estate, the head offices of resource-based industries, and most international immigration to the province. Forestry, fishing, and agriculture, the region's main industries historically, are no longer the main sources of employment. The Lower Mainland has over 60 percent of the province's population, and this margin will increase in the future.

OKANAGAN

The Okanagan Valley lies between the Cascade Mountains to the west and the Monashee Mountains to the east. There are several lakes in the valley, Okanagan Lake being

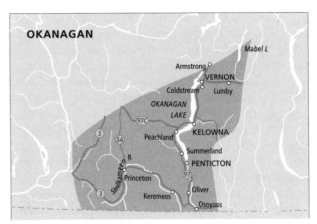

Figure 1.5 Okanagan region

the largest (see Figure 1.5). The region's southern location between two large mountain chains means it enjoys a continental climate – hot summers and relatively cold winters. The vegetation of this arid valley consists mainly of grasses, sagebrush, and few trees, but forests grow on the moister mountain slopes.

The region is the home of the Syilx (Okanagan) First Nations. Some non-Indigenous settlement occurred during the fur trade, but much more took place as the region became recognized for its farming potential as a fruit-growing area. The Okanagan is one of the few places in Canada where apples and "soft fruit" such as peaches and cherries grow, but irrigation is necessary in this dry belt. A number of communities evolved to serve the growing agricultural settlement, and Vernon, Kelowna, and Penticton, all on Okanagan Lake, became the most prominent. Prior to the arrival of railway lines, paddlewheelers (also known as steamers) on Okanagan Lake served an important transportation function. The Kettle Valley line, built in 1915, provided the link between the Kootenays and the Lower Mainland, giving access to the southern Okanagan, while a branch line of the CPR built in 1925 linked Kamloops to Kelowna and served the northern portion of the region.

Several changes occurred after the 1960s. Tourism, which had mainly been a summer activity, expanded into a year-round endeavour centred on golf, skiing, and mountain biking. The dry climate and four distinct seasons, combined with relatively low land and housing prices and easy access to the Lower Mainland, also made the valley a favourable location for retirement. These characteristics led to increased population, urban sprawl, and land-use conflicts, especially over agricultural land, until **Agricultural Land Reserves** were implemented in 1973. These reserves were designed to stop relatively rare agricultural land for from being converted to urban, industrial, commercial, and recreational uses. Fishing and pleasure-boat motors carried Eurasian watermilfoil (an invasive aquatic plant) into the region, where it spread through the water systems, converting once sandy beaches to a mass of weeds and jeopardizing the growing tourism industry. Eradication programs in the 1970s made use of herbicides that provided some control but also raised concern about the potential carcinogenic effects.

Other industries expanded in the region, including mining and forestry, both of which increased employment and the population base. With the signing of the Free Trade Agreement between Canada and the United States in 1989, agriculture changed rapidly. Fruit crops continued to dominate, but to compete with US producers who had access to inexpensive labour, BC producers developed new varieties of apple trees that required considerably less labour. The grape and wine industry, forced to compete globally, met the challenge by growing new grape varieties. New rules permitting the sale of wine from farms accounted for a large part of the success of the industry, which now attracts thousands of tourists each year.

The Okanagan has many physical assets and a fairly diversified economy, making it one of the province's rapid-growth regions (Table 1.1). Kelowna has become the most important service, administrative, and manufacturing centre, and its regional airport and a highway link it to the coast via the Coquihalla Highway. Forest fires, however, have been particularly hazardous for the region. The firestorm of 2003 incinerated 238 homes, and the firestorms of 2009, 2017, and 2018 forced the evacuation of thousands of residents (CBC 2003; Price 2009; EmergencyInfoBC 2017; CBC News 2018).

KOOTENAY

Mountains, rivers, and valleys are the main physical features that define the Kootenay region. The rugged

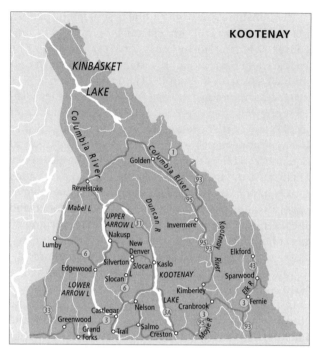

Figure 1.6 Kootenay region

mountain chains run north-south, from the Monashees in the north, through the Selkirks and the Purcells to the east, to the Rockies in the south. All rivers and lakes in these valleys form part of the Columbia River system, which exits British Columbia and flows into the United States at Trail (see Figure 1.6). Climatically, this region is similar to the Okanagan, but it is slightly colder in winter, not as hot in summer, and has slightly more precipitation.

Several Indigenous Peoples reside in this region: the Sinixt and Ktunaxa (Kootenay) to the east, the Syilx (Okanagan) to the west, and the Shuswap First Nation to the north. Census Canada has traditionally divided the region into east and west, but both sides share a valuable mineral resource base. Gold, silver, coal, copper, lead, and zinc were all discovered and became the lure to settlement and development following the fur trade era. When the Crow's Nest Pass Agreement was signed in 1897, the CPR became the principal landowner, provider of rail transportation, and developer of resources in the region. The agreement was a deal struck between the federal and provincial governments and the CPR to run a branch line from Lethbridge, Alberta, through the Crowsnest Pass to the mineral-rich Kootenays, ending initially at Nelson. Through the agreement and its subsequent purchase of railway grants, the CPR acquired millions of acres of land, coal deposits, metal mines, a major smelter at Trail (Cominco), and West Kootenay Light and Power. The CPR exerted enormous control over this region. Other railway companies built lines and acquired land grants, but the CPR purchased most of them over time, consolidating its hold on the economy. By 1901, five of the largest communities in the province were located in this region, but population statistics (see Table 1.2) don't reflect the boom-and-bust cycles experienced by individual mines and smelters. The list of **ghost towns** in the Kootenays is sufficient to warrant several books and articles (Barlee 1970, 1978a, 1978b, 1984).

Historically, settlement patterns in the area have been influenced mainly by mineral exploitation, transportation developments, and agricultural opportunities. But other factors led to settlement. For example, the Doukhobors migrated to the Kootenays between 1908 and 1913 in an attempt to escape religious persecution in Europe and political persecution in Saskatchewan. The area had good agricultural land for their communal lifestyle and appeared to be relatively isolated from government interference. In the early 1940s, many Japanese families, evacuated from the coast, were sent to communities and work camps throughout the Kootenays such as Greenwood, Sandon, New Denver, and Slocan City.

Forestry has been another resource activity. The Kootenays experienced forestry expansion after 1961 when pulp mills were built in West Kootenay at Castlegar and in East Kootenay at Skookumchuck. In the 1960s, the provincial government also became involved in hydroelectric megaprojects. Through the Two Rivers Policy, the Peace River, in the northeast, and the Columbia River, in the Kootenays, were developed for hydroelectricity simultaneously. The dams also provided flood protection for cities in Washington State and Oregon State. Other dams, such as the Revelstoke Dam, were later constructed to fulfill a perceived demand for electricity by British Columbians.

Although resources such as mining, forestry, and hydroelectricity are significant employers, both mining and forestry have faced serious setbacks. But sawmill and mine closures have, to some degree, been offset by a vibrant tourist industry, which offers some diversification. The region has many hot springs and lakes, and opportunities abound for skiing, hiking, sports fishing, hunting, mountain biking, and sightseeing. Investments in tourist infrastructure such as ski facilities (e.g., Panorama, Fernie, and Kimberley), golf courses, and casinos have increased tourism. Nevertheless, because the region is a considerable distance from major urban populations in Alberta, British Columbia, and the United States, it will likely show slow growth in the future.

SOUTH CENTRAL INTERIOR

The South Central Interior is largely identified by the Southern Interior Plateau. The region extends west of Lillooet to Revelstoke on the Columbia River, but it is the Thompson River system that mainly defines the region (see Figure 1.7). The Thompson River valley is hot in summer and cold in winter, and precipitation occurs mainly on the surrounding mountains. The Secwépemc of the Interior Salish have been the traditional users of the land.

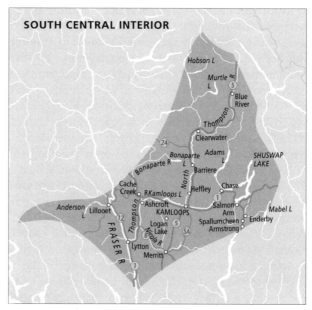

Figure 1.7 South Central Interior region

Early European interest in this region stemmed from furs and gold. The Thompson River was an important "highway" for the fur trade, and Kamloops was established as a fur trade post in 1812. Small amounts of gold were discovered in the region in the mid-1850s but not in amounts significant enough to create a gold rush. The gold rush that followed gold discoveries along the Fraser River up to the Cariboo in the north in the 1860s, however, attracted cattle drives and cattle ranching to these interior grasslands.

When the CPR main line was built, Kamloops became the main supply and service centre for the region. Its location, where the North and South Thompson Rivers meet, was enhanced by later transportation developments such as the extension of the Canadian National Railway (CNR) down the North Thompson and through Kamloops on its way to Vancouver. With the development of road systems, it gained a spot along the Trans-Canada Highway, and it is currently at one end of the Coquihalla Highway, which runs to the coast.

Farming and ranching were the region's main industries until the 1960s, when the forestry, mining, tourism, and retirement industries began to develop. Of them, forestry was the most important. A pulp mill was established in Kamloops in the mid-1960s along with sawmills and other forest-product manufacturing endeavours throughout the region. Mining also played a significant role when a copper mine and smelter was built in Kamloops. Although it closed in 1997, the mine reopened in 2007. A considerably larger copper/molybdenum mine in the Highland Valley from Ashcroft to Merritt is also in production. Tourism is also increasing in importance, largely because of transportation routes such as the Coquihalla Highway (which no longer has tolls) that lead to Kamloops and then branch north to Jasper and Edmonton and east to Banff and Calgary. The region's accessibility and other favourable features have also attracted retirees. Although the region has experienced relatively rapid population increases (see Table 1.1), it is dominated by one urban centre – Kamloops. This centre – which includes Thompson Rivers University and other institutions that perform medical, transportation, and other administration functions – has become an important service centre for the South Central Interior region.

NORTH CENTRAL INTERIOR

The North Central Interior is one of the largest regions in the province, and it is defined by the Northern Interior Plateau. The northern half of the Fraser River, with its many tributaries, is a large part of this region, although the area extends westward to include Smithers and the Bulkley Valley (see Figure 1.8). Temperatures in winter are colder than in southern interior locations and not as hot in summer. The mixing of Pacific and Arctic air masses results in increased precipitation, but this varies throughout the region, in relation to the mountain chains.

Historically, the area has been home to many First Nations, such as Tsek'ehne, Nat'oot'en (Lake Babine First Nation), Wet'suwet'en, and Dakelh. Non-Indigenous settlement began with the overland fur trade, which relied on the Fraser River for transportation. Fur trade companies erected forts throughout the region in the early 1800s, but it was not until the Cariboo Gold Rush

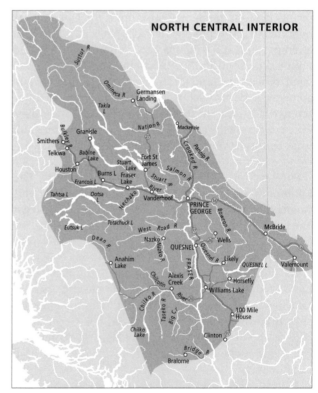

Figure 1.8 North Central Interior region

of the early 1860s that more permanent non-Indigenous settlement began. Miners used a number of routes to gain access to the Barkerville area from the Lower Mainland until the Cariboo Road was constructed in the mid-1860s. The development of Barkerville and other mining communities attracted ranching to the southern part of the region, including the Chilcotin Plateau. (Gold was also discovered in the northern part of the region – in the Omineca Mountains, for example – but these finds were not sustainable.) By the end of the 1860s, however, the era of major gold discoveries had come to an end, and miners began to disappear, along with the mining towns.

This large region remained relatively uninhabited until the CNR connected it to the port of Prince Rupert to the west in 1914. From the south, the Pacific Great Eastern Railway (PGE) was built in a series of stages: Squamish to Quesnel in 1921, Quesnel to Prince George in 1952, Prince George to the Peace River in 1958, and Prince George to Fort St. James in 1968. Prince George, like Kamloops, became a transportation hub of rail lines and highway systems that served central and northern British Columbia.

From 1961 to 2001, as Table 1.1 indicates, there was considerable interest in the region. The mining of copper, molybdenum, and gold brought many workers into the area, but the main industry was forestry. The region's forests remained largely untouched until the 1960s, when an increase in the world demand for forest products saw a massive expansion in the industry throughout the North Central Interior. The manufacturing of plywood, lumber, and other forest products was integrated with pulp-and-paper mills built at Quesnel, at Prince George, and in the new town of Mackenzie.

Unfortunately, many resource industries have shut down since the 2000s. The forest industry in particular has been hard hit by the mountain pine beetle epidemic, which forced the closure of many sawmills. Although, the region's population has declined since 2001, Prince George is growing in population and has emerged as the region's largest centre (see Table 1.2). The city – which is accessible and home to the University of Northern British Columbia – has become an important service centre to central and northern British Columbia. It performs a role similar to that of Edmonton in the province of Alberta.

NORTH COAST–NORTHWEST

The North Coast–Northwest is a large region that has remained relatively isolated because of its rugged, mountainous landscape. The Coast Mountains, which extend the length of the region, are among the highest in British Columbia. The northwest corner consists of the St. Elias range, which includes Fairweather Mountain (4,663 metres), the highest peak in British Columbia. The coast is highly irregular, and its northern tip makes up the Alaska Panhandle (the strip of Alaska that extends south along the coast). The heavily forested Haida Gwaii (Traditional Territories of the Haida and composed of over 150 islands) also forms part of this region. Climatically, much of the area is exposed to the Pacific and to the westerly flows of wind. Its northern location also exposes the region to the Aleutian low-pressure system, which brings a lot of rain in the summer and snow in the winter. The Skeena, Nass, and Stikine are some of the largest river systems in the province that flow to the Pacific (see Figure 1.9).

Within the region, the north coast has been home to high-density populations of Indigenous Peoples, whereas the northwest interior has had much lower populations. A number of factors led to early European development of the North Coast–Northwest region. In the late 1700s, the sea otter trade aroused considerable interest. The Russians, who erected fur trade posts across Alaska and the Panhandle were the first to exploit these valuable furs. They laid territorial claim to Alaska, thus cutting off the northwest from the sea. Subsequently, Russia sold the territory to the United States in 1867. Without access to the Pacific, the northwest portion of the region was left largely to Indigenous Peoples and the overland fur trade.

Other resources encouraged settlement, but it was only temporary. Salmon canneries began to dot the landscape in the late 1800s and early 1900s, especially at the mouth of the Skeena and Nass Rivers and along the coast. However, with the exception of those at the mouth of the Skeena, most canneries disappeared in the 1950s because of improvements in fishing technology. Although small amounts of gold were discovered on Haida Gwaii in 1850 and on the Skeena River in 1863, no gold rush occurred in either case. Coal was mined on Haida Gwaii, but it was only used for coastal steamers in the late 1800s. The Klondike Gold Rush in the Yukon opened up the

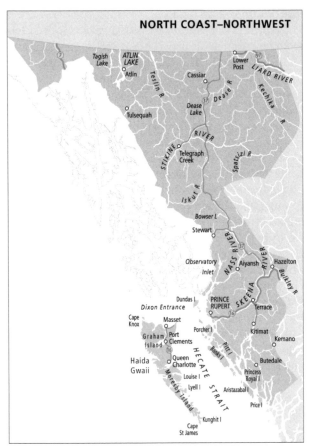

Figure 1.9 North Coast–Northwest region

eventually lead to the mining of the Telkwa coal deposits, which lasted from 1918 to 1984. Prince Rupert also became the site of an early pulp mill, which encouraged growth. A copper smelter was built at Anyox, north of Prince Rupert, but it lasted only until the Depression. Following the Second World War, the construction of the Alaska Highway made the remote northwest portion of the region (e.g., Lower Post and Atlin) more accessible. Another major development for this region, and an icon of industrial development in the frontier, was the construction of the Kenney Dam to supply the energy required for the new community of Kitimat. This planned town was built to house the workers for a new aluminum plant in the early 1950s. The Stewart-Cassiar Highway linked communities along Highway 16 (those from Prince George to Prince Rupert) to the Alaska Highway through the northwest.

As in other regions, the mining and forest industries underwent major expansion beginning in the 1960s. A major copper mine near Stewart, a large open-pit mine for asbestos at Cassiar, and a number of small gold mines brought employment to these rather isolated locations. The development of the Quintette and Bullmoose mines and the new town of Tumbler Ridge in the Peace River region – a coal-mining region known as Northeast Coal – in the early 1980s resulted in the CNR rail line being double-tracked to Prince Rupert and a large coal port being built at nearby Ridley Island. New investments in Ridley Island included the development of new bulk-handling facilities (e.g., for petroleum products and wood pellets) and the ability to handle container shipping. Although a planned expansion of the aluminum smelter at Kitimat, which depends on water from the Nechako River, was cancelled in 1995 for environmental reasons, recent modernizations have increased production and reduced greenhouse gases, and aluminum production continues to be important to the region.

The forest industry has also expanded in a number of directions. Much of the valuable timber from Haida Gwaii was harvested and sent on barges to the mills of the Lower Mainland. A pulp mill was opened at Kitimat in the 1960s, but it closed in 2010. Prince Rupert has emerged as the largest centre for the region, but the downturn in the forest industry resulted in the closure of a number of sawmills in the area and its pulp mill in 2001, putting into

northernmost portion of the region in 1898, leading to the building of the famous Chilkoot Pass, the White Pass, and by 1900 the White Pass and Yukon Railway. Gold seekers passed through the northern tip of British Columbia before entering the Yukon, and although there was plenty of interest in the region, there was little permanent settlement.

It wasn't until the CNR arrived at Prince Rupert in 1914 that permanent settlement increased. Charles Hays, a prominent entrepreneur in the early history of Prince Rupert, was a major shareholder of the railway, an industrial waterfront landowner, and an avid promoter of the town. Unfortunately, he never saw his vision materialize – he was a passenger on the Titanic in 1912 and was not counted among the survivors. But the railway did perspective the importance of the forest industry to this city and the region as a whole (see Table 1.2).

Northeast of Prince Rupert, in 2000, the Nisga'a of the Nass River Valley became the First Nation to enter into a modern treaty in British Columbia, setting a precedent for future treaties throughout the province. The treaty resulted in a cash settlement, land ownership (i.e., their land base is no longer a reserve), an Indigenous-owned forestry company, and a substantial share of the Nass River commercial salmon fishery.

More recently, the Kitimat–Prince Rupert region has been in the spotlight for two controversial resource proposals. The first, the Enridge Northern Gateway Pipelines project, was intended to connect Alberta's Tar Sands oil to Kitimat, where it would be transported to Asia. The second was for the production and export of **liquefied natural gas (LNG)** from Kitimat, Lelu Island (off Prince Rupert), and a number of other locations. The federal government cancelled the oil pipeline to Kitimat in favour of another proposed pipeline to Burnaby, by Kinder Morgan, which is also extremely controversial. The LNG proposals got cancelled but more for economic than political reasons – the price of LNG collapsed in 2015.

Even today, there are few roads or rail lines through this area, and growth continues to be tied closely to resource development. The North Coast–Northwest region is slow growing, and its population has declined in recent years.

PEACE RIVER–NORTHEAST

Most of the Peace River–Northeast region does not fit the broad physical description of British Columbia as a mountainous, vertical landscape. This flat, sedimentary region east of the Rockies is physiographically similar to the Prairies. The two major rivers, the Peace River in the south and the Liard River in the north, are part of the Mackenzie River system (see Figure 1.10), which drains into the Arctic Ocean. The region contains areas of permafrost, bog, and boreal spruce forests. Temperatures are cold in winter and surprisingly warm in summer, when the days are long, inducing **convection precipitation**.

The region covers the Traditional Territories (or parts of the territories) of a number of First Nations, including the Tsek'ehne, Dane-zaa, Saulteaux, Nehiyawak, Kaska

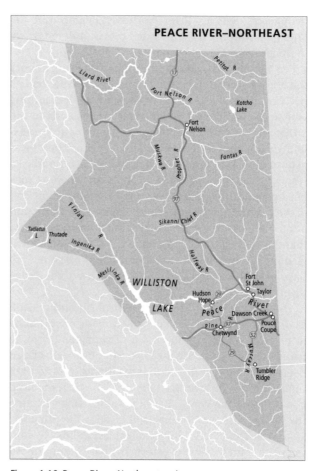

Figure 1.10 Peace River–Northeast region

Dena, and Dene Tha'. The North West Company fur traders were the first non-Indigenous people to enter the region, and it is here that the earliest fur trade forts in British Columbia were erected. Discoveries of gold on the Peace River in the 1860s warranted the inclusion of the territory into British Columbia, but the finds were insufficient to sustain permanent settlement. The Yukon Gold Rush at the end of the nineteenth century led to the signing of Treaty 8, which covers the region north of Edmonton, the northwest corner of Saskatchewan, and the Peace River–Northeast Region of British Columbia. The Canadian government believed that the region would be a route to the goldfields and that the treaty would be a means of avoiding conflict. The Indigenous

Peoples of the Peace River–Northeast Region were the only First Nations to be included in a numbered treaty in the province.

Few agricultural settlers ventured this far north until the homesteads of the south and central Prairies had all been taken up. The development of hardy, early-maturing wheat also facilitated agricultural homesteads in the 1920s and '30s, and problems associated with accessibility were improved when a rail line from Alberta was extended to Dawson Creek in 1930. When the Alaska Highway was constructed during the war, Dawson Creek became Mile 0, helping to open up the region. At the end of the war, the federal government made more farmland available for returning servicemen. But it wasn't until 1971 that the PGE was finally extended to Fort Nelson. Wheat farming on the region's excellent soil and cattle rearing have been the main agricultural activities of the Peace River area.

The discovery and development of oil and natural gas in the 1950s encouraged investment, the building of pipelines, and the movement of considerably more people to the region. By the 1960s, the Peace River had become one target of the massive hydroelectric plan referred to as the Two Rivers Policy. The plan involved constructing the W.A.C. Bennett Dam (which created the largest reservoir, or artificial lake, in the province, Lake Williston) and building transmission lines to connect the dam to southwestern British Columbia. All this activity attracted even more people to the Peace River, and the energy crisis, which began in the early 1970s, sparked another round of oil and gas exploration and development.

The sedimentary basin also contained Northeast Coal, which was developed beginning in the early 1980s. The new town of Tumbler Ridge housed the miners, an electric rail line of the British Columbia Railway (formerly the PGE) was constructed, and millions of dollars were spent upgrading the CNR line from Prince George to Prince Rupert and building a coal port at Ridley Island (as discussed above). The costs borne by both provincial and federal governments to export the coal were massive. Unfortunately, as the world market demand and price for coal declined, so too did contracts with Japanese buyers. The Quintette mine closed in 1999 and the Bullmoose in 2003, leaving Tumbler Ridge struggling to convert to a retirement, tourist, and recreation community. Fortunately, the price of coal rebounded, and a new coal mine, Wolverine, opened in 2006. The discovery of dinosaur fossils has also resulted in tourism related to paleontology. Still, Tumbler Ridge's economy is dependent on the volatile price of coal, which results in the opening and closing of mines. All mines closed in 2015.

Today, Fort St. John and Dawson Creek are the largest centres in the Peace River–Northeast region. The forest industry employs many people in its pulp mills, sawmills, oriented-strand-board plants, and a plywood mill. And although the agricultural sector continues to be important, it, like the coal industry, is unstable. The recession of 2008, for example, resulted in plummeting grain prices; although they rebounded to near historic levels by 2012, they have seen a steady decline since 2017. Tourism has increased in response to a variety of recreational opportunities, the diversity of the landscape, and the discovery of dinosaur footprints at Tumbler Ridge and near the W.A.C. Bennett Dam, and the Alaska Highway continues to be a major tourist attraction.

With all of these developments comes the potential for conflict between resource industries and those interested in preserving the wilderness. For example, the provincial government, through BC Hydro, has recently resurrected and approved a controversial proposal for the Site C Dam on the Peace River near Fort St. John. The project has many farmers, First Nations, environmentalists, and citizens concerned about the consequences of flooding the Peace River Valley. Similarly, although oil and natural gas continue to be important to the region, fracking for natural gas has many in the environmental movement concerned over the amount of water required in the process, the contamination of both surface and ground water, and evidence that it causes earthquakes.

In 1998, when the Muskwa-Kechika Management Area Act was signed, an attempt was made to accommodate all interests in this very large wilderness region, which stretches from Lake Williston north to the Yukon border. The economy and growth of the Peace River–Northeast region is tied to its diverse resource endowment. It is a large region, the population is relatively small, and growth, or decline, is tied to resource demand.

SUMMARY

In a regional geography approach, British Columbia can be divided into eight regions for the purpose of an overview. The regional perspective is important, not only for recognizing British Columbia's unique physical and human attributes but also in assessing the global and external forces that have shaped the province. The divisions are based on distinct physical characteristic in combination with economic activities, settlement patterns, and population data. Each region has a unique history, physical characteristics, Indigenous populations, and economic development.

A common thread in this regional approach is how population increased in each region as a result of resource development, which often relied on political policies, technological development, especially with respect to transportation, and innovations related to industrial productivity. In many of these regions, the resource industries are no longer important employers, and this has resulted in population stagnation and decline. As a consequence, population growth is extremely uneven, favouring the regions with large urban centres. The Lower Mainland is home to over 60 percent of the provincial population, and this number is expected to continue to increase.

REFERENCES

Barlee, N.L. 1970. *Gold Creeks and Ghost Towns of Southern British Columbia.* Summerland, BC: Self-published.

–. 1978a. *The Best of Canada West.* Langley, BC: Stagecoach.

–. 1978b. *Similkameen: The Pictograph Country.* Summerland, BC: Self-published.

–. 1984. *West Kootenays, the Ghost Town Country.* Surrey, BC: Canada West Publishers.

BC Stats. 1986. "1986 Census of Canada: Profiles of British Columbia." bcstats.gov.bc.ca/StatisticsBySubject/Census/1986Census.aspx.

–. 1991. "1991 Census of Canada: Profiles – British Columbia." bcstats.gov.bc.ca/StatisticsBySubject/Census/1991Census/Profiles.aspx.

–. 2012. "British Columbia Municipal Census Populations (1921–2011)." bcstats.gov.bc.ca/StatisticsBySubject/Census/MunicipalPopulations.

–. 2017. "Population Estimates: Municipalities, Regional Districts, and Development Regions, 2011 to 2018." gov.bc.ca/gov/content/data/statistics/people-population-community/population/population-estimates.

CBC, Radio One. 2003. "BC Fire Talk Tape." *The Current,* 14 November.

CBC News. 2018. "Lightning-Caused Wildfires in B.C.'s Okanagan Prompt Evacuations, Traffic Chaos." 19 July. https://www.cbc.ca/news/canada/british-columbia/wildfire-bc-mount-eneas-1.4752971.

Census of Canada. 1951. *British Columbia Population by Census Subdivisions, 1871 to 1951.* Table 6/6–84 to 6–88. Ottawa.

–. 1971. *British Columbia Population by Census Subdivisions, 1961 and 1971.* Table 8/8–92 to 8–96. Ottawa.

Dearden, P. 1987. "Marine-Based Recreation." In *British Columbia: Its Resources and People,* ed. C.N. Forward, 259–80. Western Geographical Series, vol. 22. Victoria: University of Victoria.

EmergencyInfoBC. 2017. "RESCINDED: Evacuation Alert: Okanagan Centre Fire." 27 July. emergencyinfobc.gov.bc.ca/evacuation-order-alert-okanagan-centre-area-of-lake-country-issued-by-central-okanagan-regional-district/.

Gregory, D. 2000. "Regions and Regional Geography." In *The Dictionary of Human Geography,* 4th ed., ed. R.J. Johnston, D. Gregory, G. Pratt, and M. Watts, 687–90. Oxford: Blackwell.

Muckle, R.J. 1998. *The First Nations of British Columbia.* Vancouver: UBC Press.

Price, Mike. 2009. "Revisiting the Glenrosa Fire: Evacuation Modeling with ArcGIS Network Analyst 10." *ArcUser,* Winter. esri.com/news/arcuser/0111/glenrosa.html.

Pudup, M.B. 1988. "Arguments within Regional Geography." *Progress in Human Geography* 12: 369–90.

Robinson, J.L. 1972. "Areal Patterns and Regional Character." In S*tudies in Canadian Geography: British Columbia,* ed. J. Lewis Robinson, 1–8. Toronto: University of Toronto Press.

Wood, C.J.B. 1979. "Settlement and Population." In *Vancouver Island: Land of Contrasts,* ed. C.N. Forward, 3–32. Western Geographical Series, vol. 17. Victoria: University of Victoria.

World Population Review. 2017. "United Kingdom Population, 2017." worldpopulationreview.com/countries/united-kingdom-population/.

Physical Processes and Human Implications

2

Much of the attraction, and beauty, of British Columbia comes from the variety of its physical features. Scenic landscapes of rugged mountains, large and turbulent rivers, diverse flora and fauna, and often isolated settings – all changing from season to season – are both great assets and great challenges. The province has a rich resource base of minerals, energy, forests, fish, agriculture, and tourism. The physical setting has exerted a powerful influence on settlement, the urban system, and transportation corridors, but it has also posed a major risk to the human population.

In this chapter, you'll learn about the physical processes that have shaped the landscape, beginning with the geologic time scale and the distinctive rock structures that make up the crust of the earth. There are three categories of rocks: (1) igneous, (2) sedimentary, and (3) metamorphic. Being able to assess the age of and distinguish among different rock types – even those billions of years old – will give you a sense of the generalized geological makeup of British Columbia. The study of the processes that create these landforms and transform the surface of the earth is known as **geomorphology**.

You'll also learn about the physical processes related to weather and climate. The spatial and seasonal patterns of temperature and precipitation vary greatly throughout the province, with broad distinctions between the coast and the interior and between southern and northern locations.

Finally, you'll learn how soils and vegetation influence each other and are a product of many other physical processes, including variations in climate and geomorphology. The vegetation patterns of British Columbia are distinct and obvious, ranging from huge coastal coniferous forests to southern interior valleys of cactus and sagebrush. These physical landscapes have major implications for human habitation of the province.

From a human perspective, significant changes in terms of dress, popular music, political policies, and employment opportunities tend to occur every ten to twenty years. From a physical perspective, when considering changes to the landscape, you'll need to think in terms of tens, or even hundreds, of millions of years (Table 2.1). The geological makeup of any region is complex, and one means of unravelling this complexity is to assess the age of rocks. Life forms on the earth have been a major influence in dating rocks because their fossils remain as a benchmark of time. Phanerozoic rocks, containing fossils of complex life forms, date from 545 million years ago. As Table 2.1 indicates, there are further divisions within this time period, and all are established through carbon dating. Of course, there are rocks much older than these fossiliferous ones; they are known as Precambrian rocks and were created between 4.6 billion and 545 million years ago, a huge expanse of time.

This province has undergone profound transformations, especially over the last 200 million years. Climatic changes have caused ocean levels to rise and fall and glaciers to scrape and modify the landscape; tectonic processes have been responsible for mountain building, **volcanic activity**, and the addition of islands to British Columbia. The landscape is by no means static.

Table 2.1

Geologic time scale

Eon	Era	Period	Years before present
Phanerozoic	Cenozoic	Quaternary	
		Holocene	10,000
		Pleistocene	1,800,000
		Tertiary	
		Pliocene	5,800,000
		Miocene	23,800,000
		Oligocene	33,700,000
		Eocene	54,800,000
		Paleocene	65,000,000
	Mesozoic	Cretaceous	146,000,000
		Jurassic	208,000,000
		Triassic	248,000,000
	Paleozoic	Permian	280,000,000
		Carboniferous	360,000,000
		Devonian	408,000,000
		Silurian	438,000,000
		Ordovician	505,000,000
		Cambrian	545,000,000
Precambrian			
Proterozoic			2,500,000,000
Archean			3,800,000,000
Hadean			4,600,000,000

THE THREE CATEGORIES OF ROCK AND THE ROCK CYCLE

The first category of rock, **igneous rock,** is created when molten **magma** cools and hardens into a solid state. As the temperature drops, crystals form in the magma, like ice crystals forming in water. These crystals are called minerals and have different compositions. Typical minerals found in igneous rocks include feldspar, quartz, hornblende, and olivine. Because hornblende and olivine contain iron, they are darker than feldspar and quartz. Igneous rocks that contain mostly feldspar and quartz minerals are usually light in colour, an example being granite. Igneous rocks that have mostly iron-rich minerals are often dark grey or black; basalt is an example. Basalt is denser than granite since it contains larger proportions of iron and other heavy minerals. Granite is a typical continental rock, while the majority of oceanic rocks are basalt.

The size of the minerals in an igneous rock reflects the rate of cooling of the magma. If the magma cools relatively fast (over days, years, or even decades), the crystals are so small that a microscope is needed to see them. These quickly cooled rocks are found near or at the earth's surface, primarily as a result of volcanic activity and are known as extrusive or volcanic igneous rocks. Basalt is an extrusive igneous rock and is commonly found in lava flows, dikes, and sills. Dikes and sills form as magma pushes into older rocks; sills form parallel to surrounding rock layers, while dikes cut across the rock structure (Figure 2.1). Basalt's cooling process frequently produces a pattern of cracks resulting in columns of four-, five-, or six-sided rock, creating what is known as columnar jointing. These distinctive columns can be seen along many of British Columbia's highways, an indication of the province's extensive volcanic history.

Magma that cools tens of kilometres below the earth's surface may require millions of years to solidify. During this time, the crystals grow to a size where they can be seen by the human eye. These intrusive or plutonic rocks, of which granite is an example, are often found in huge plutonic masses known as batholiths (Figure 2.1). Much of the Coast Mountains range of western British Columbia is batholithic. Although formed deep below the surface, granitic batholiths may be pushed upward during large-scale crustal movements. As surface **weathering** and

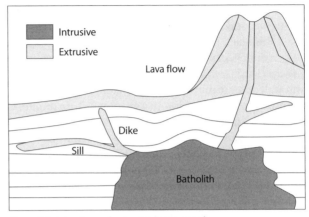

Figure 2.1 Intrusive and extrusive igneous rock

erosion strip off the overlying rock formations, the batholiths are exposed.

Rocks at the surface are exposed to weathering processes caused by water, air, and vegetation growth and decay, and these processes eventually create our second category of rocks, **sedimentary rocks,** or rocks formed from unconsolidated materials bonded together. Weathering may cause a rock to fragment into small, sedimentary pieces called clastics. Weathering by surface water and groundwater may also cause the minerals within rock to dissolve, making water "hard" and oceans salty. After weathering, sediments and dissolved minerals may be moved by streams, waves, wind, glaciers, or gravity to a new location, a process known as erosion. These sediments are eventually deposited, often in layers. The layering effect happens when different sizes and shapes of sediments are deposited, such as small, rounded particles on top of large, angular ones. This usually occurs when the agent of erosion changes, such as from wind to water. As sedimentary layers accumulate, the bottom ones are pushed deeper into the crust, causing compaction and heating. As well, groundwater can infiltrate the sediments and precipitate (or solidify) some of the dissolved minerals.

The precipitated minerals act as a cement to bond the sediments together. Compaction, heating, and cementing change the sediments into rock. Some common examples of clastic sedimentary rocks are sandstone (made from sand-sized sediments) and shale (made from clay and silt-sized sediments). Precipitated minerals may also form a second type of sedimentary rock called evaporatives. Rock salt, for example, forms in the shallow waters of warm oceans; as the water evaporates, the salts are precipitated. A third sedimentary rock type is formed through organic processes. As plant and animal products accumulate, their remains may eventually turn into rock. Coal is an example. Tiny organisms my also utilize minerals dissolved in lakes and oceans to form their shells and skeletons (similar to how humans extract calcium from food to make bones). When the organisms die, their hard parts sink to the bottom of the water body and accumulate to form limestone.

Sedimentary rocks are located throughout British Columbia. Some are found in their original horizontal positions, but many are folded, tilted, and fractured. These **deformations** occur during times of mountain building, a prime example being the Rocky Mountains on the east side of the province.

The third category of rocks, after igneous and sedimentary, is called metamorphic. When rocks are exposed to extreme pressures and temperatures, or to chemical infusions from nearby magma bodies, the minerals within the rocks change in shape and chemistry. The original rock does not melt; it metamorphoses into a new rock. Limestone, for example, will metamorphose into marble. **Metamorphic rocks** are mostly associated with mountain building and igneous intrusions. In British Columbia, metamorphic rocks can be found in the Coast Mountains in conjunction with their huge batholithic cores.

The **rock cycle** reminds us that rocks are not static in geologic time, and that all can be related (Figure 2.2). Beginning with the molten state, igneous rocks are "born." Like all other rocks, however, they are subject to the processes of weathering and erosion. As particles are deposited and/or precipitated, igneous rocks can become sedimentary rocks. Both sedimentary and igneous rocks can be subject to intense heating, pressure, and chemical action,

becoming metamorphic rock in the process. All three rock groups may go back to the molten form.

This basic information about rocks provides the foundations for a generalized geology of British Columbia, one that recognizes that rocks and landforms were created during different eras, as Figure 2.3 demonstrates. Consider the first category of rock, igneous. Much of the Coast Mountains range and the South Central Interior region of British Columbia are made up of intrusive igneous and are approximately 100 million years old. Consequently, the region should be characterized by granitic types of rock. Of course, the actual geology of any region is much more complex, and those familiar with the Coast Mountains just north of Vancouver around Squamish and Whistler will recognize Mount Garibaldi, the Black Tusk, and obvious basalt columns as extrusive igneous formations. Mount Fissile, behind Blackcomb Mountain, however, has sedimentary rocks containing fossils of seashells at over 2,000 metres in elevation. Many mountain-building processes have been at work in the various regions of British Columbia. Figure 2.3 gives an overview only of the dominant rock type.

The interior of the province and the north end of Haida Gwaii are a mix of flat-lying lava and some sedimentary rock, the second category. Much of the interior was under water for considerable periods, and sedimentary rock should be found in these locations. About 40 to 50 million years ago, however, significant volcanic activity

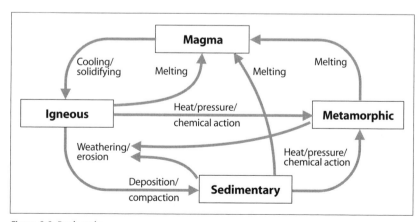

Figure 2.2 Rock cycle

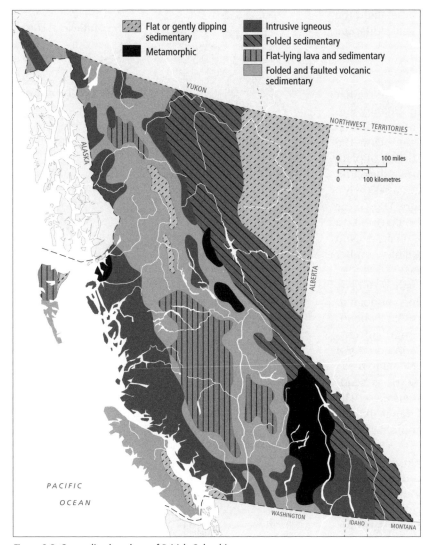

Figure 2.3 Generalized geology of British Columbia
Source: Modified from Ryder (1978).

plains-like landscape of the Peace River–Northeast region, which is structurally part of the Great Plains of North America. The mineral resources found in this region include oil, natural gas, and coal.

The Rockies are also made up of folded sedimentary rock. Some 150 million years ago, there was an enormous amount of pressure on flat-lying sedimentary rock that folded it up to elevations in excess of 4,000 metres. Subsequent mountain-building episodes occurred, the last one approximately 65 million years ago. A drive through the Rockies will not only give you the impression of a vertical landscape but also show you the folding of the sedimentary layers.

Most of Vancouver Island, the south end of Haida Gwaii, and much of the interior are formed of folded and faulted volcanic and sedimentary rock. These areas were covered by water in the last 500 million years and are characterized by sedimentary rock. Approximately 150 million years ago, volcanic activity and pressures were exerted to push these sedimentary and volcanic structures into very rugged landscapes.

The final category, metaphoric rock, is found mainly in southeast British Columbia in the Kootenays. These very hard rocks are the oldest rock structures in the province, dating back 1 billion or more years.

produced huge lava flows that covered the sedimentary levels. In areas such as the spectacular Helmcken Falls on the Murtle River in Wells Grey Park, the much harder basalt rock, twenty to thirty metres thick, can be seen quite clearly over the top of the much softer sedimentary layers. However, flat-lying or gently dipping sedimentary rocks formed some 65 million years ago formed the

Recognizing the processes that formed the three basic rock structures, having some idea of the age of landforms, and being able to generalize about where these rocks and landforms can be found is the beginning of understanding the geomorphology of British Columbia. In the context of millions of years, mountains have been built up and worn down, and it is important to understand the factors that shaped and changed the physical landscape in more detail.

DEFORMATION OF THE EARTH'S SURFACE

"Deformation" is a broad term that describes the rather complex geologic processes of mountain building and changes to the surface of the earth. The forces that fold, tilt, and fracture rocks are produced deep below the earth's surface and are explained by the theory of **plate tectonics**. Figure 2.4 shows a cross-section of the earth, which is made of several components. At the centre is a solid inner core ringed by a liquid outer core. This core is ringed by the mantle, which is solid but behaves like a plastic substance because of the extreme pressures and temperature. The mantle is also composed of a number of layers, each with its own characteristics.

Close to the top of the mantle is the asthenosphere, which has temperatures and pressures high enough to partially melt the mantle rocks. Because the asthenosphere is in a semimolten state, convection currents develop. The moving plumes of molten material put pressure on the solid rocks above, forcing them to break and shift position. The brittle rocks above the asthenosphere belong to a layer known as the lithosphere. The bottom part consists of rocks from the upper mantle, while the top part of the lithosphere contains ocean crust and continental crust.

Each of these regions – core, mantle, and crust – has a different mineral composition. The core is composed mainly of iron; the mantle has large amounts of a rock called peridotite; the ocean crust is primarily basalt; and the continental crust contains mainly granite. Sedimentary and metamorphic rocks are also found throughout the crust. The lithosphere, which contains both mantle and crustal materials, is thinnest beneath the oceanic crust, where rising convection plumes from the asthenosphere can force the lithosphere to fracture and move. The broken parts of the lithosphere are known as plates, and most volcanic and **earthquake activity** occurs at their edges.

The theory of plate tectonics suggests that the earth's lithosphere consists of seven large plates, as well as many small ones. Several of the large plates are capped primarily by oceanic crust. Others contain both ocean and

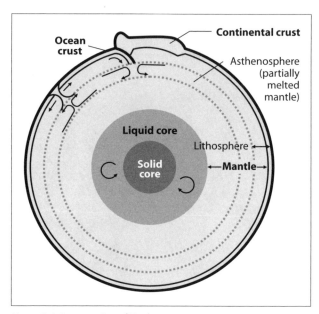

Figure 2.4 Cross section of Earth

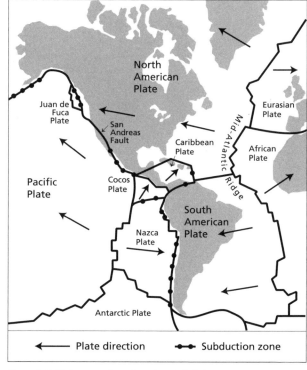

Figure 2.5 Major oceanic and continental plates

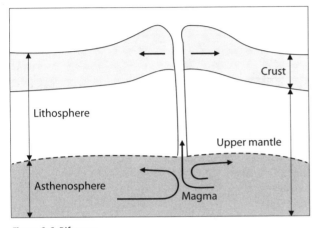

Figure 2.6 Rift zone

continental crust. The main plates making up North and South America are shown in Figure 2.5.

Where magma finds its way to the surface of the earth and splits the lithosphere apart, it creates a divergent boundary or **rift zone**. The cooling magma forms new igneous rock in the rift zone. This rock, in turn, may be broken apart by new magma upwelling from below. Rocks formed and fractured in the rift zone make up what are known as the trailing edges of plates. Roughly half of the fractured rock is attached to the trailing edge of each plate while the other half is attached to the second plate. The new rocks, although solid, are still very hot. They may have temperatures of several hundred degrees Celsius. The high temperatures cause the rocks to expand. As the rocks are forced away from the rift zone, they begin to cool, contracting and shrinking in volume. A look at the cross-section of a rift zone (see Figure 2.6) shows that the ocean becomes deeper on either side of the zone because the cooling crust contracts and thus compresses downward. The oceanic crust also becomes smoother the farther it is away from a rift zone because sediments from the continents settle in the ocean basins, level the irregularities of the igneous crust, and form a continental shelf.

Rift zones separate the plates, but in doing so, they create collisions in other parts of the globe. When two plates are forced together, the edges may fracture and buckle, forming mountain ranges. This process is known as **orogeny**. Prime examples of orogeny occur when the

edge of a plate contains continental crust. The Andes on the west coast of South America were formed in this manner. The Pacific Northwest is another region where orogenic processes are at work. Besides colliding and sliding past each other, sometimes one of the plates – usually a denser, oceanic one – is forced downward and underneath the other. (Recall that oceanic rocks such as basalt are much denser, or heavier, than continental rocks such as granite.) This process is called **subduction**, and the major subduction zones for North and South America are shown in Figure 2.5. Continental rocks rarely subduct, which is why they are some of the oldest in the world – up to 3.96 billion years old. The subducting plate forms a trench in the ocean floor as much as five kilometres deep. As the plate subducts deeper, it warms. Eventually, the temperature is high enough to allow some of the plate to melt. If enough magma is produced, it will flow upward to create plutonic batholiths within the crust and volcanic eruptions at the surface.

The new rocks created at a rift zone (the trailing edge of a plate) are gradually transported to a subduction zone (the leading edge of a plate), where they are eventually destroyed (returned to the magma state). On large plates, it may take an average of 250 million years to form, transport, and finally destroy the rocks. In this way, the plate acts as a giant conveyor belt. This analogy applies only to some parts of the world, however, as in other regions the plates collide and slip past each other to form **transform faults**.

The leading edge of the Juan de Fuca Plate (see Figure 2.7) is in close proximity to Vancouver Island, where it is then subducted under the North American Plate. For the coast of British Columbia, this means that there is a relatively narrow continental shelf and little buildup of sediments. By comparison, the trailing edge of the North American Plate, shown in Figure 2.5, is in the middle of the Atlantic Ocean. Consequently, there is a very wide continental shelf, with a buildup of sediments off the Maritime provinces. Sedimentary buildup also provides potential for oil and natural gas deposits (and the wider the continental shelf, the greater the potential).

Rift zones, subduction zones, and transform faults all occur along the west coast of North America. California is marked by the famous San Andreas Fault, where the

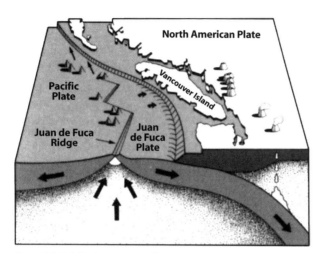

Figure 2.7 Subduction process off the southern BC coast
Source: Energy, Mines and Resources Canada (n.d.), courtesy of Pacific Geoscience Centre, Geological Survey of Canada.

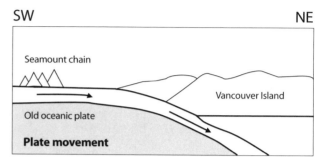

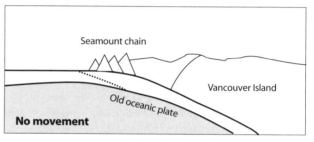

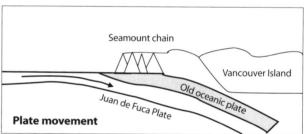

Figure 2.8 Accreted terrane process for southern Vancouver Island
Source: Modified from Monger (1990).

Pacific Plate moves in a northerly direction past the North American Plate. Each movement produces an earthquake, and one of the characteristics of the San Andreas Fault is how it appears to "lock" in specific locations, such as San Francisco. Here, the pressure builds up until a major shift takes place. In 1906, one of these shifts had catastrophic consequences, destroying 80 percent of the city and killing approximately three thousand people.

The Juan de Fuca Plate, the plate that stretches from northern California to the end of Vancouver Island, is much smaller. Its jagged rift zone is only 200 to 300 kilometres west of Vancouver Island (Figure 2.7). The subduction process also produces earthquakes, sometimes large ones, as the oceanic plate slides under the continental plate. Deep oceanic troughs, orogeny, batholiths, and evidence of volcanic activity (some of it fairly recent, such as Mount St. Helens) are characteristic of the northwest Pacific region.

North of Vancouver Island, the Pacific Plate continues its northerly migration past the North American Plate, with a transform fault in immediate proximity to Haida Gwaii. This movement, with its history of major earthquakes (e.g., magnitude 6.1 in 2015), continues all the way to Alaska, where another very active subduction zone occurs. The highest mountains in Canada – the Mount St. Elias range in southwestern Yukon and the northwest corner of British Columbia – attest to the mountain-building pressures of this collision. This region is also one of the most seismically active areas of Canada and the United States.

Subducting plates sometimes carry fragments of oceanic and continental crusts from other regions. These fragments, of differing compositions and ages, are known as **terranes**. Instead of being subducted along with the plate, the terranes are pushed up, or accreted, against the edge of the second plate. Figure 2.8 shows an example of a small terrane, the Seamount chain of mountains, being accreted on to the south end of Vancouver Island. The diagram shows the dynamics of mountain building and easterly migration of the Juan de Fuca Plate. It also shows that the southern end of Vancouver

Island is composed of geologic structures formed in other regions at other geologic times.

Terranes are a big part of British Columbia. Geologic evidence suggests that the whole Cordilleran region, from the Rockies to the Insular Mountains, is a series of accreted terranes. Figure 2.9 organizes these terrane additions into five distinct belts: Foreland, Omineca, Intermontane, Coast, and Insular. The Foreland Belt is represented by the Rocky Mountains, which form the eastern edge of the Cordilleran. The Omineca Belt includes the Omineca Mountains as well as the Purcell, Selkirk, Columbia, Monashee, and Cariboo ranges. The Interior Plateau region of the province makes up the Intermontane Belt, while the Coast Belt includes the rugged Coast and Cascade ranges. The Insular Belt is made up of the mountains on Vancouver Island, Haida Gwaii, and the Alaskan Panhandle. These dynamic tectonic forces have resulted in the highly complex geology of the province, with its rugged landscape rich in minerals.

Piecing together the geologic history of British Columbia is a complex task, but it can be approximated with a basic understanding of the tectonic processes responsible for earthquakes and volcanic activity and for the terranes that accrete onto the continental crust. By meshing geologic time and recognizing rock formations, we know that "two hundred million years ago, there was no British Columbia – at least there was no British Columbia west of the present Rocky Mountains. The low shore of the continent sloped off into the sea perhaps where Calgary and Dawson Creek are today, and the continental shelf extended westward to the Okanagan-Quesnel-Cassiar areas" (Cannings and Cannings 1999, 11–12). Approximately 170 million years ago, the first set of terranes were added onto the craton – the ancient stable geologic formation of the Canadian Shield, including its perimeter of continental shelves. This is the Omineca Belt in Figure 2.9, and the collision of this terrane initiated the process of lifting, folding, and thrusting these sediments "at least 150 km eastward onto the edge of the old continent" to form the Rocky Mountains (Christopherson and Byrne 2009, 377).

By about 85 million years ago, more terranes were added, and the force of these additions transformed the low-lying sediments into the lofty Rockies. The famous Burgess Shale in Yoho National Park, which has plant

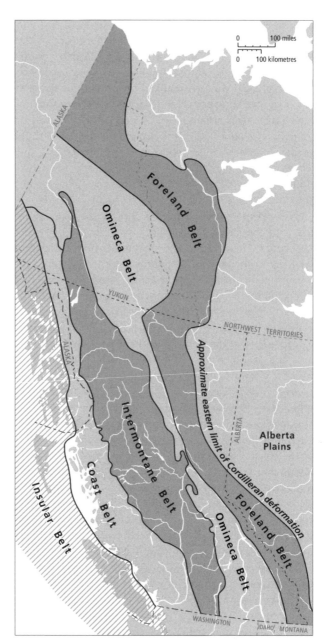

Figure 2.9 Accreted terranes of western Canada
Source: Modified from Natural Resources Canada (2007).

and animal fossils dated at 520 million years, resulted from the collision that uplifted the ocean floor of the craton. Volcanic activity in the interior of British Columbia approximately 60 to 40 million years ago added to the complexity and was caused by a much larger subduction zone than exists at present. Massive amounts of lava flowed over existing sediments to form the Interior Plateau. The subduction zone migrated westward to its present location off the coast of Vancouver Island, but as this was happening, more volcanic activity occurred, more terranes were added, and considerable compression pushed up the Coast Mountains. In the far northwest corner of the province, "the Yakutat Terrane is crunching into the Chugach Terrane. There, some of North America's highest and most spectacular mountains rise virtually from the seacoast, and the force of the impact continues to push them up at the remarkable rate of 4 centimetres a year" (Cannings and Cannings 2004, 25).

Another piece of the puzzle that requires explanation is how metamorphic rock that is over a billion years old ended up in the Kootenays. Of course, over a billion years ago, this landform was not part of British Columbia. The terrane was formed elsewhere in the world and through tectonic forces was moved and joined onto the continental plate. The tectonic forces responsible for moving terranes also moved continental and oceanic plates. Neither British Columbia nor the continental plate were at the present latitude (or longitude) throughout geologic history. A fossil record of great tropical forests and dinosaurs, in a warmer climate, indicates a location much closer to the equator. In the future, further terranes could be added to British Columbia, and it has been speculated that "Vancouver Island will eventually become part of the mainland" (Foster Learning 1997–2004).

Because of this complex geological journey, western North America, British Columbia in particular, remains an extremely volcanic landscape that has experienced eruptions in fairly recent times. Figure 2.10 outlines a number of volcanic belts. To the north, the Pacific Plate is actively moving northward along a transform fault, resulting in the Stikine Volcanic Belt (shown in Figure 2.10), the most active volcanic region in the province and in Canada, with over one hundred volcanoes. The Anahim Volcanic Belt is a series of "hotspot" volcanoes

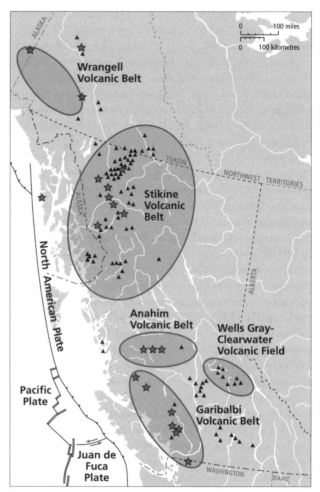

Figure 2.10 Young volcanoes of western Canada
Source: Modified from Natural Resources Canada (2019).

similar to those creating the Hawaiian Islands. The youngest is the Nazko Cone in the Chilcotin region. A series of seismic tremors occurred in this location in 2007, leading to speculation of another volcanic eruption. The Wells Grey–Clearwater Volcanic Field is made up of layers of thick basalt. The volcanoes in the Anahim Volcanic Belt range in age from 12 million years old at the western end to a few thousand years old at the eastern end (Hickson and Ulmi 2006). During the time of their formation, thick glacial ice covered the region at least twice before the Fraser (or Wisconsin) glaciation, which occurred approximately 25,000 and 10,000 years ago. Volcanoes erupted below and through the ice, producing unique glacial volcanoes and deposits (Natural Resources Canada 2019). The most southerly belt, Garibaldi Volcanic, is a product of the subduction zone activity. These statovolcanoes are "the most explosive young volcanoes in Canada" (Hickson and Ulmi 2006; Natural Resources Canada 2010). Although Mount Meager erupted more than two thousand years ago, its eruption was of similar magnitude to that of Mount St. Helens in 1980. Volcanic activity is clearly not only a mountain-building process in this province but also a potential natural hazard.

Deformation of the earth's surface may occur as a result not only of tectonic factors but also of **isostasy**, a process of loading and unloading the surface of the earth with sediments and ice. As sediments erode from higher elevations such as the Rockies (which, keep in mind, are made up of sedimentary rock), these majestic mountains are gradually losing their lofty elevations. Erosion over the past 150 million years might lead one to expect that the "soft" sedimentary rocks would have been reduced to plains. But the Rockies should be thought of instead as a floating raft piled high with cargo: as the cargo, or weight, is removed, the raft rises up. For every three metres of erosion, there is approximately two metres of uplift, or isostasy.

The last glacial age is the most recent and obvious example of isostasy. As the glaciers built up to a depth of 1,500 to 2,000 metres, their enormous weight pushed the surface of the earth down perhaps several hundred metres. Ocean levels fell to produce this volume of ice but receded only tens of metres. As the ice melted – which happened in British Columbia only in the last 10,000 years – the ocean levels rose and, eventually, so did the land. This uplifting of the land, or isostatic rebound, takes thousands of years and still continues today. The lag in time between a relatively rapid rise in sea level and relatively slow rise in the level of the land has produced beach formations at different elevations around the world.

WEATHERING AND EROSION

The term "weathering" describes a number of processes responsible for breaking down rock over time, whereas "erosion" refers to the movement of rock materials via a number of agents – gravity, water, wind, and ice – working singly or in combination. Erosion continues the weathering process and helps to form sediments.

Mechanical weathering involves the physical destruction of rocks into smaller and smaller components. Freezing and thawing represent one common and powerful type of mechanical weathering in British Columbia. Rain or melting snow runs into cracks and fissures of rocks, where it is subject to freezing. The expansion of the resulting ice often fragments the rock particles.

Chemical weathering, as the name describes, refers to a chemical action that breaks rock down. The common substances of water, oxygen, and carbon dioxide, or some combination of these agents, can dissolve and chemically react with the minerals that make up rock. Sands, silts, and clays are often the product of chemical weathering, as is the red "oxidized" colour of soil.

The agents of erosion often work in combination. Gravity, for example, pulls broken rock and other weathered material downward, but when these materials are lubricated with water, they are often carried away much more readily. Gravity is a powerful force and responsible for a host of erosion activities, referred to as mass wasting. These activities can include rockfalls, debris torrents, slumping, soil creep, and major landslides, such as the one near Hope in 1965. In all cases, the landscape is altered by particles moving downslope, sometimes rapidly and sometimes extremely slowly.

Flowing water, in the fluvial process, is an effective agent of erosion in British Columbia because of the many turbulent streams and rivers. Water carries particles ranging from boulders to clay particles, carving up the landscape and depositing the materials along stream beds where the stream slows, into the deltas of lakes, or into the ocean. In the southern half of the province, the Fraser and Columbia River systems dominate, and in the north the Skeena, Nass, and Stikine are the major rivers flowing to the Pacific. The northeastern portion of British Columbia is drained towards the Arctic by the Peace and Liard River systems (Figure 2.11). Where these rivers and streams flow through mountainous terrain, they cut the land into V-shaped valleys. Where there is less relief, the river systems tend to meander, forming broad river valleys.

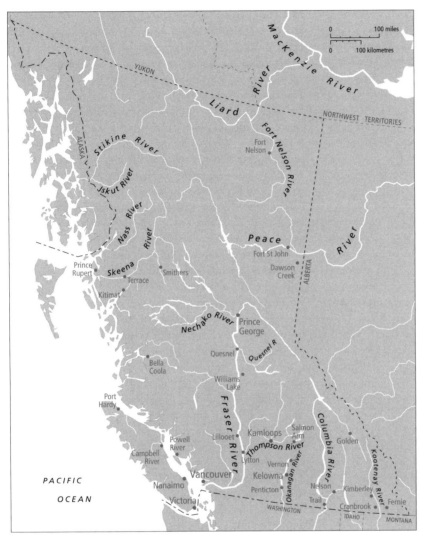

Figure 2.11 River systems of British Columbia

are common. Wind-blown material may also act as an abrasive agent as it "sand blasts" landform structures.

The final factor, and a highly noticeable one because of its recent activity, is glaciation. There have been approximately twenty ice ages in the past 1.8 million years. The most recent began some 75,000 years ago, reaching its peak 18,000 years ago and covering most of British Columbia and Canada (Bone 2002, 48). When the climate began to cool at the beginning of the glacial age, heavy snowfall in the high alpine areas of the mainland gradually compacted and transformed into icy glaciers. These glaciers moved down through stream and river valleys, all the way out to the Strait of Georgia, where they coalesced with others flowing from the mountains of Vancouver Island.

These huge accumulations of ice, which reached 1,500 to 2,000 metres in height, carried a great deal of material with them and carved the landscape with their movement. Glacial erosion acted like a bulldozer. Movement through existing river valleys scoured out the V-shaped valleys and turned them into U-shaped valleys. The process did not end at the coastline but continued along the ocean floor. Subsequent global warming melted the glaciers, which raised sea levels and, despite isostatic rebound, left a coastal landscape of "drowned" glacial valleys called fjords – a significant asset for deep harbour ports and coastal navigation. In the Coast Mountains and other ranges, peaks under 2,000 metres have been rounded off through the movement of glaciers.

Moraines are another common landform in British Columbia. They are the large, linear piles of boulders, gravel, and sand that accumulated around the edges of

British Columbia's indented coastline, along with its many islands, is subject to continual erosion from the actions of the ocean. Waves pound and modify the coastline relentlessly. Currents and tides also assist in the erosion process, wearing down and carrying away materials. In other regions of the coast, where these materials are deposited, islands and coastlines build up.

Wind plays a somewhat minor role in the erosion process. It picks up and disperses fine silts, particularly in the dry interior of British Columbia, where dust storms glaciers. Erratics, large boulders that have been transported by glaciers, appear most obviously on flat landscapes such as the Fraser Valley.

Today, glaciers at high elevation in alpine areas are the remnants of what once covered the province.

WEATHER AND CLIMATE

"Weather" refers to the day-to-day atmospheric conditions that influence plans for any number of outdoor activities or travel. "Climate" refers to the longer-term effect of weather. Climate is revealed by the collection of weather statistics over months and years, which enable climatologists to assess the averages, and extremes, of atmospheric conditions for any location or region. For example, the climate for the southwestern corner of British Columbia, in comparison to the rest of the province or to Canada as a whole, makes this region one of the most attractive in the country. Climatic conditions have been observed to have cycles, some of which are longer than others. El Niño conditions, for example, occur approximately every eight to ten years, resulting in much greater precipitation for coastal British Columbia. An ice age represents a much longer climatic cycle.

Figure 2.12 illustrates the global movement of air currents. The sun is the engine, or driving force, of weather and climate variations throughout the world. At the equator, a huge volume of air rises because of intense surface heating. As the air rises thousands of metres, a low-pressure zone is produced at the surface. This is typically characterized by precipitation, because the warm moist air rises, expands, and cools, causing condensation. High above the surface, the rising air diverges to the north and south poles. At approximately thirty degrees north and south of the equator, these air masses descend, producing high-pressure zones as they fall and compress. These subtropical high-pressure belts are the driest regions in the world. In the northern hemisphere, some of the descending air moves back along the surface in a southerly

direction, attracted to the low-pressure zone. Other portions of the air masses continue in a polar direction. The surface air mass collides with an air mass driven towards the equator by polar high-pressure zones. Where these collisions take place, a low-pressure zone is produced, referred to as the polar front. Thousands of metres above the polar front are high-velocity, easterly moving winds known as the jet stream.

The prevailing wind patterns shown in Figure 2.12 are influenced by flows of air between high- and low-pressure zones, by the rotation of the earth, and by the seasons of the year. During summer in the northern hemisphere, the days are longer, the sun is higher in the sky, and the wind patterns shift northward. During the winter, there is less sun, and the wind patterns shift southward. British Columbia is largely influenced by winds known as the westerlies and by two pressure zones: the Aleutian Low (producing cloudy and wet weather) to the north and

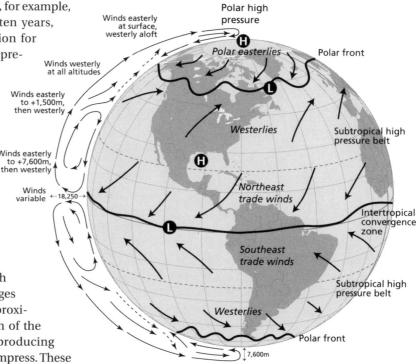

Figure 2.12 Global climate
Source: Modified from map by W. Heibert in Welsted, Everitt, and Stadel (1996), with permission.

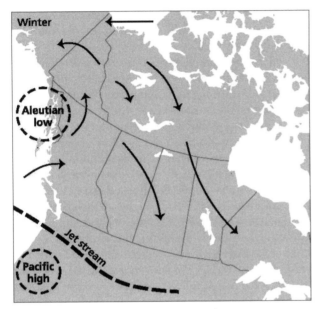
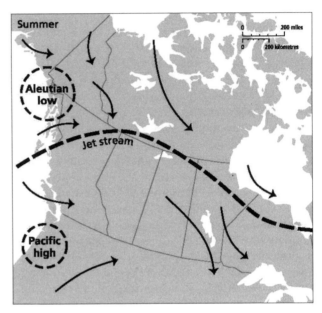

Figure 2.13 Jet stream influences in winter and summer

Figure 2.14
Climate regimes
of British Columbia
Source: Modified from
Hare and Thomas (1974).

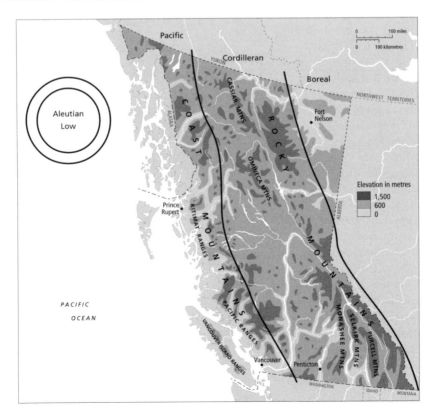

the Pacific High (producing clear and dry weather) to the south. The jet stream demarcates these two pressure zones and, as Figure 2.13 shows, moves in seasonal patterns.

There are other influences on the weather and climate of British Columbia. The relatively warm Pacific Ocean directly affects the coastal regions. Farther inland, the land mass and mountains have the greatest influence. Land heats up more rapidly than water and to higher temperatures, and it cools down more rapidly and to lower temperatures. As Figure 2.14 illustrates, the province has three climate regimes, or climate regions: Pacific, Cordilleran, and Boreal (Hare and Thomas 1974). Pacific Canada is the coastal region. Unique in Canada for its mild winters, it is often described as having a modified Mediterranean climate. The Cordilleran is defined by the many mountain chains between the Coast Mountains and the Rockies. The Boreal regime occurs in the plain-like northeast region of the province. The term "boreal" also defines a vegetation of scrub forest typical of northern climates. Both the Cordilleran and Boreal climate regimes are less subject to maritime influences than to continental air masses, which give them far greater extremes.

In Figure 2.14, Prince Rupert, Vancouver, Fort Nelson, and Penticton represent north and south coastal locations and north and south interior locations. Table 2.2 gives their mean monthly temperature and precipitation. A graph of the data would instantly show the similarities and differences between these communities and the regions they represent.

Temperature differences between the coastal and interior locations are greatest in the winter months, when Vancouver and Prince Rupert have considerably milder temperatures than Penticton and Fort Nelson. Winter temperatures in the coastal region are mainly influenced by a relatively warm, large body of water – the Pacific Ocean – and the prevailing westerlies. In the summer, the ocean is slow to heat up and therefore provides a modifying influence. One of the most noticeable and dramatic influences on coastal temperatures occurs in the winter months. Frigid polar air, being dense and heavy,

Table 2.2

Climate data for selected communities

	Vancouver 49°11' N 123°10' W Elevation: 4.3 m		Prince Rupert 54°17' N 130°26' W Elevation: 35 m		Penticton 49°28' N 119°30' W Elevation: 342 m		Fort Nelson 59°50' N 122° W Elevation: 382 m	
	Daily average (°C)	Precipitation (mm)	Daily average (°C)	Precipitation (mm)	Daily average (°C)	Precipitation (mm)	Daily average (°C)	Precipitation (mm)
January	4.1	168.4	2.4	276.3	−0.6	26.9	−20.3	21.5
February	4.9	104.6	2.7	185.6	1.0	19.8	−15.2	14.9
March	6.9	113.9	4.2	199.6	5.0	23.6	−7.6	18.8
April	9.4	88.5	6.4	172.4	9.1	26.0	3.0	18.9
May	12.8	65.0	9.0	137.6	13.9	39.3	9.7	49.0
June	15.7	53.8	11.6	108.8	17.7	46.3	15.1	63.0
July	18.0	35.6	13.4	118.7	21.0	28.7	17.1	78.4
August	18.0	36.7	13.8	169.1	20.4	28.3	15.1	71.3
September	14.9	50.9	11.5	266.3	15.1	24.6	9.2	40.2
October	10.3	120.8	8.0	373.6	8.8	26.0	0.5	32.6
November	6.3	188.9	4.3	317.0	3.2	28.1	−12.8	25.6
December	3.6	161.9	2.7	294.2	−1.1	28.6	−18.7	18.0

Note: This table presents climate normals for the years 1981–2010.
Source: Environment Canada (2019).

moves as a high-pressure air mass from the Arctic down through the interior of British Columbia and funnels through the mountain passes to the coast, bringing freezing temperatures to Vancouver and Victoria. As one travels inland, the moderating effect of the Pacific Ocean becomes less influential and the speed and intensity with which land can heat and cool becomes more influential. Consequently, temperatures in Fort Nelson and Penticton have greater variation than those in Prince Rupert and Vancouver.

Winter temperatures differ considerably between the two interior communities of Penticton and Fort Nelson, mainly because of latitude. Fort Nelson, just south of the sixtieth parallel, has only a few daylight hours in winter, and, even then, the sun is at an extremely low angle. Penticton, just north of the forty-ninth parallel, gets a good deal more incoming solar radiation because of the increased number of sunlight hours in the winter. Fort Nelson is also much closer to the freezing Arctic high-pressure air masses, which are not hindered by any physical barriers, while Penticton is protected by a number of mountain ranges that act as barriers to these Arctic air masses. Extremely cold winter temperatures (below minus 25 degrees Celsius) are relatively rare in Penticton.

In a vertical landscape such as British Columbia, elevation also influences temperature. If you participate in alpine activities, you can experience cool temperatures almost year-round. At very high elevations (over 3,000 metres), extremely cold temperatures can be experienced even in summer.

Considerable differences in precipitation can be seen throughout the year between Vancouver and Prince Rupert even though both have coastal locations. A number of factors are at play. For the Vancouver area, the rain shadow effect has a significant role (Figure 2.15). The relatively warm prevailing westerlies move over the Pacific, absorbing moisture as it evaporates from the ocean. This air mass is forced to rise (a phenomenon known as the **orographic effect**) over either the Olympic Mountains of Washington State or the Insular Mountains of Vancouver Island, where the mass cools and contracts. Cooling and contraction cause condensation, which results in a great deal of precipitation on the western slopes – some of the highest in Canada. As the air mass descends the eastern slopes, it expands and warms and has the ability to absorb more moisture as it crosses the Strait of Georgia and Vancouver. The North Shore Mountains and other ranges of the Coast Mountains then force the air mass to repeat the orographic effect. The rain shadow region receives considerably less precipitation. It is worth noting that within Greater Vancouver, significant differences in precipitation exist between south Delta and North Vancouver, just thirty-five kilometres apart (Figure 2.16).

Even though Prince Rupert has Haida Gwaii to intercept the westerlies, there is little rain shadow effect. Haida Gwaii does not have high mountains like Vancouver Island, and the distance from Haida Gwaii over the Hecate Strait to the mainland is considerably farther than across the Strait of Georgia. Consequently, the air mass is still laden with moisture as it is forced to rise up the rugged Coast Mountains, which form the backdrop to Prince Rupert. The Aleutian Low brings rainy and turbulent weather, but its position is changeable, influenced by seasonal changes in the jet stream (see Figure 2.13).

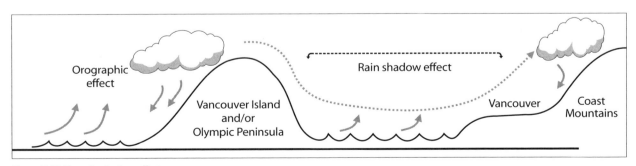

Figure 2.15 The rain shadow effect

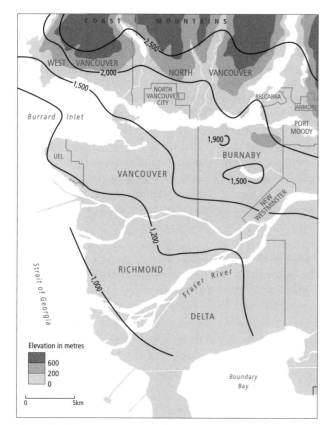

Figure 2.16 Vancouver area annual precipitation in millimetres
Sources: Modified from Wynn and Oke (1992).

In the summer, Prince Rupert is still to the north of the jet stream wave and very much under the influence of the Aleutian Low, whereas Vancouver is influenced by the Pacific High.

Precipitation patterns in the interior are very different from those of the coast, both in quantity and by season. The greatest influence in the summer is from incoming solar radiation, which heats up the land rapidly, leading moisture to evaporate, rise, cool, and then condense into dark thundershowers. This convection process is particularly common on the prairie-like landscape of Fort Nelson. Penticton is nearly a desert. Its location in the southern Okanagan Valley is well inland, and the many mountains between the Pacific and Penticton wring out most of the moisture. This rugged topography ensures that most of the precipitation falls on the mountain peaks

and that relatively little reaches the dry valley bottom. What moisture does fall in the summer months is mainly due to convection.

During the winter, precipitation in the interior is predominantly in the form of snow. In northern, flat locations such as Fort Nelson, the surface has high reflectivity. Incoming solar radiation is therefore reflected back rather than absorbed, and the atmosphere has little chance to build up moisture. It is difficult for Pacific flows of moisture to penetrate this far inland. Penticton, on the other hand, does have a marginal increase in precipitation during the winter months. The more southerly swing of the Aleutian Low at this time of the year can, and does, bring more moisture. The Monashee Mountains to the east of Penticton accumulate the precipitation in the form of snow, which serves the skiing industry.

THE INFLUENCE OF CLIMATE CHANGE

The Kyoto Protocol, an agreement to reduce greenhouse gases below 1990s levels by 2012, was ratified by the Canadian Parliament in 2002, thus acknowledging the negative global impact of increasing our carbon footprint. The protocol was followed by Al Gore's film and accompanying book, *An Inconvenient Truth,* in 2006, which raised worldwide awareness of climate warming. The consequences of climate change in British Columbia will be significant: rising temperatures will cause increased evapotranspiration and less precipitation, and glaciers will lose mass or, in many cases, disappear over the next century (Walker and Sydneysmith 2008). These developments will directly influence hydroelectric production, especially in the winter months, when there is increased demand. Natural hazards (e.g., firestorms, coastal flooding, high winds, avalanches, and hail) will increase in frequency, and agriculture will be hit by serious water shortages (drought) in regions such as the Okanagan. Other regions, however, will increase the range of crops produced.

Warmer waters will place both ocean and freshwater fish in jeopardy, and warmer weather and hotter, drier summers will subject forests to pest infestations and other adverse effects (Wilson and Hebda 2008). Today, warmer-than-average winter temperatures have set the stage for the pine beetle to have killed a billion or more pine trees in the interior of the province. Because of unusually dry conditions, firestorms have also become worse. The firestorms of the summers of 2017 and 2018 were the largest in the province's history and were accompanied by the greatest number of evacuations. Tragically, the fires in 2017 "emitted an estimated 190 million tonnes of greenhouse gases into the atmosphere – a total that nearly triple[d] B.C.'s annual carbon footprint" (Hernandez and Lovgreen 2017).

Seventy-five percent of Canada's mammal and bird species, 70 percent of its freshwater fish, 60 percent of its evergreen trees, and thousands of other animals and plants make their home in British Columbia (Pojar 2010, 5). The spatial pattern of these plants and animals is influenced primarily by climate change, and Jim Pojar reminds us that "ecosystems do not migrate – species do" and "most species cannot disperse (move) quickly enough to keep pace with the projected changes" (Pojar 2010, 5). Vegetation change is already occurring and, with it, animal habitats. To ensure that species will survive, we will need to study and understand new migration paths.

Understanding the consequences of climate change is an important step towards establishing solutions. And solutions do exist. The science behind climate change and its consequences is sound, and international environmental organizations such as 350.org have been lobbying governments for many years to reduce their levels of carbon dioxide to below 350 parts per million to save the planet. In a 2019 report, the Intergovernmental Panel on Climate Change looked at what it would take to limit warming this century to 1.5 degrees Celsius above preindustrial levels. It found that it would be possible, but worldwide greenhouse gas emissions would have to be halved by 2030, reach net-zero by 2050, and be carbon-negative thereafter (Intergovernmental Panel on Climate Change 2019).

Al Gore's most recent film, *An Inconvenient Sequel: Truth to Power,* produced in 2017, also emphasizes the urgent need to reduce our dependency on fossil fuels. There are a number of options, including a major shift in energy use from nonrenewable resources and big hydro to green, renewable energy sources such as wind, solar, geothermal, and small hydro. The auto industry must produce more efficient, less emission-producing vehicles – for instance, hybrid vehicles or vehicles that use biodiesel and hydrogen. More funding for public transit will encourage people to use their cars less, and, on an individual level, biking, walking, carpooling, and conserving energy are measures that can reduce each person's environmental footprint.

SOILS AND VEGETATION

Soils are the most fundamental element for growing plants, but fertile soils are in short supply in British Columbia, approximately two-thirds of which consists of mountains, rocky land, and water. The parent materials of the soils of the other third have been formed primarily by glacial drifts (Dalichow 1972, 9). Soils consist of mineral and rock fragments that have undergone varying

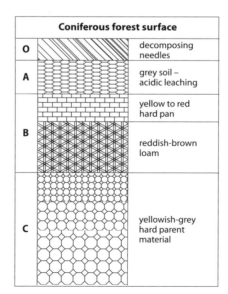

Figure 2.17 Typical coastal soil horizon
Sources: Modified from Valentine and Lavkulich (1978) and Witherick, Ross, and Small (2001).

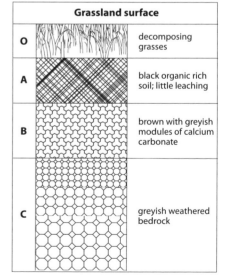

Figure 2.18 Typical interior soil horizon
Sources: Modified from Valentine and Lavkulich (1978) and Witherick, Ross, and Small (2001).

degrees of physical and chemical weathering along with decaying organic matter, resulting in many combinations of minerals, water, and air. The organic matter, or vegetation, is greatly influenced by climate. Soils are also influenced by the bulldozer effect of glaciation where material (till) has accumulated. As a consequence of these processes, soils have a variety of textures and are continually changing over time. Two main soils are found throughout British Columbia: forest-related soils on the coast and semi-arid grassland soils in the central interior.

A cross-section, or profile, of a typical coastal soil reveals its characteristics in various layers or horizons (Figure 2.17). The top layer of the coniferous forest soil is made up of undecayed organic matter (O layer) and gives way to the A-horizon, where the organic matter decays and is added to weathered rock. These uppermost layers are subject to weathering and erosion more than the others. Below the A-horizon is the B-horizon, which accumulates minerals that have been filtered, or leached, out of the A-horizon. The C-horizon is the most stable layer and is often called the parent material (Lavkulich and Valentine 1978). Somewhat different characteristics make up the soils of the dry interior grasslands, where considerably less leaching through the horizon occurs. Figure 2.18 is a profile of soil typical of areas such as the Okanagan. Owing to glaciation, the parent material in much of British Columbia has been transported and mixed so that it rarely bears much resemblance to the underlying bedrock.

Vegetation is tied closely to soils and climate, and given that both vary greatly throughout the province, it

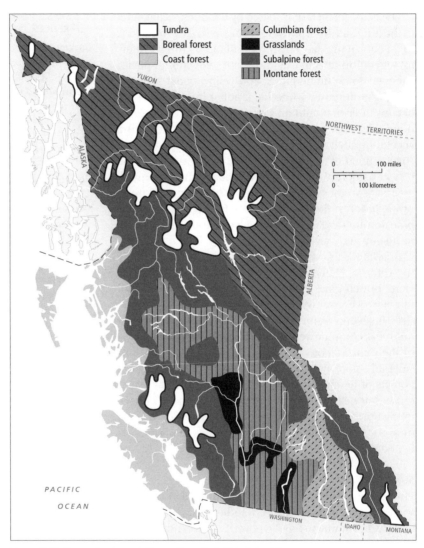

Figure 2.19 Generalized vegetation patterns in British Columbia
Sources: Modified from Barker (1977), Dalichow (1972), and Jones and Annas (1978).

will come as no surprise that there are also many vegetation zones. British Columbia is known for its forests, particularly the coniferous forests that dominate many regions. There are considerable differences between the types of forests on the coast and in the interior. Forests also differ from north to south with latitude change, often mirroring the changes to forest species wrought by altitude (Figure 2.19).

The coast forest is dominated by hemlock, cedar, and, in the drier southern portions, Douglas fir. The trees grow to an enormous size. The term "Subalpine forest" refers to vegetation affected by mountain elevation. Species such as alpine fir, hemlock, and yellow cedar grow on coastal mountains, while lodgepole pine and spruces occupy interior mountain elevations. Even higher elevations, with extreme climate and scarce soil, have tundra vegetation such as moss and lichens. The montane forest covers much of the south and central interior. Because it is considerably drier than the coast and has much greater extremes in temperatures, this region is more prone to forest fires. To the south, forests of ponderosa pine dominate, while varieties of spruce are found in the more northerly regions.

Grasslands are also part of this landscape and occupy the Interior Plateau along with southern arid valleys. Here, bunch grass was the dominant vegetation until massive overgrazing occurred from the 1860s, Cariboo Gold Rush on. Sagebrush, which cattle do not eat, is now the most common vegetation. Columbian forest grows in the mountainous Kootenay region, which has considerably more moisture than the montane forest region. Forests of hemlock, cedar, and Douglas fir grow there, but not at nearly the size of those on the coast. Finally, to the north is the boreal forest, characterized primarily by spruce and aspen, whose size and distribution is affected by latitude and permafrost.

SUMMARY

Physical processes such as rock formation, deformation, weathering and erosion, and weather and climate have had, and continue to have, a powerful influence on British Columbia's landscape. These processes must also be taken into account when we examine the human development of the landscape. The historical use of the land by Indigenous Peoples, and later by the first non-Indigenous people to the region, was greatly influenced by the physical landscape. Over time, with the creation of new technologies, the physical environment has played less of a role. Yet our modern economy is still very much tied to the land, just as our activities are tied to daily weather conditions. And whenever there is a flood, earthquake, avalanche, or windstorm, the power of nature is apparent.

REFERENCES

Barker, M.L. 1977. *Natural Resources of British Columbia and the Yukon.* Vancouver: Douglas, David and Charles.

Bone, R.M. 2002. *The Regional Geography of Canada.* 2nd ed. Don Mills, ON: Oxford University Press.

Cannings, S., and R. Cannings. 1999. *Geology of British Columbia: A Journey through Time.* Vancouver: Douglas and McIntyre.

–. 2004. *British Columbia: A Natural History.* 2nd ed. Vancouver: Greystone Books.

Christopherson, R., and M. Byrne. 2009. *Geosystems: An Introduction to Physical Geography.* Toronto: Pearson.

Dalichow, F. 1972. *Agricultural Geography of British Columbia.* Vancouver: Versatile.

Energy, Mines and Resources Canada. n.d. "Earthquakes in Southwest British Columbia." *Geofacts.* Pamphlet. Sidney, BC: Geological Survey of Canada, Pacific Geoscience Centre.

Environment Canada. 2019. "Canadian Climate Normals: 1981–2010 Climate Normals and Averages." National Climate Data and Information Archive. climate.weatheroffice.gc.ca/climate_normals/index_e.html.

Foster Learning. 1997–2004. "An Overview: The Shaping of Western Canada." OTS Heavy Oil Science Centre. lloydminsterheavyoil.com/geooverview.htm.

Gore, A. 2006. *An Inconvenient Truth: The Planetary Emergency of Global Warming and What We Can Do about It.* Emmaus, PA: Rodale Press.

Hare, K., and J. Thomas. 1974. *Climate Canada.* Toronto: John Wiley.

Hernandez, J., and T. Lovgreen. 2017. "'It's Alarming': Wildfire Emissions Grow to Triple B.C.'s Annual Carbon Footprint." *CBC News,* Aug 24. cbc.ca/news/canada/british-columbia/it-s-alarming-wildfire-emissions-grow-to-triple-b-c-s-annual-carbon-footprint-1.4259306.

Hickson, C.J., and M. Ulmi. 2006. "Volcanoes of Canada." Natural Resources Canada. mineralsed.ca/site/assets/files/3451/bc_volcanoes_jan_03_06_v4.pdf.

Intergovernmental Panel on Climate Change. 2019. *2019 Refinement to the 2006 IPCC Guidelines for Greenhouse Gas Inventories.* ipcc.ch/report/2019-refinement-to-the-2006-ipcc-guidelines-for-national-greenhouse-gas-inventories/.

Jones, R.K., and R. Annas. 1978. "Vegetation." Sec. 1.4 in *The Soil Landscapes of British Columbia,* ed. K.W.G. Valentine, P.N. Sprout, T.E. Baker, and L.M. Lavkulich. BC Ministry of Environment: Soils. env.gov.bc.ca/esd/distdata/ecosystems/Soils_Reports/Soil_Landscapes_of_BC_1986.pdf.

Lavkulich, L.M., and K.W.G. Valentine. 1978. "Soil and Soil Processes." Sec. 2.2 in *The Soil Landscapes of British Columbia,* ed. K.W.G. Valentine, P.N. Sprout, T.E. Baker, and L.M. Lavkulich. BC Ministry of Environment: Soils. env.gov.bc.ca/esd/distdata/ecosystems/Soils_Reports/Soil_Landscapes_of_BC_1986.pdf.

Monger, J. 1990. "Continent-Ocean Interactions Built Vancouver's Foundations." *Geos* 4: 7–13.

Natural Resources Canada. 2007. "Cordilleran Geoscience: The Five Belt Framework of the Canadian Cordillera." Accessed 2011. gsc.nrcan.gc.ca/cordgeo/belts_e.php.

–. 2010. "Sea to Sky Story: Volcanoes." publications.gc.ca/collections/collection_2010/nrcan/M4-83-5-2010-eng.pdf.

–. 2019. "Where Are Canada's Volcanoes?" chis.nrcan.gc.ca/volcano-volcan/can-vol-en.php.

Pojar, J. 2010. *A New Climate for Conservation Nature, Carbon and Climate Change in British Columbia.* Commissioned by the Working Group on Biodiversity, Forests and Climate. davidsuzuki.org/wp-content/uploads/2010/01/new-climate-conservation-nature-carbon-climate-change-british-columbia.pdf.

Ryder, J.M. 1978. "Geology, Landforms, and Surficial Materials." Sec 1.3 in *The Soil Landscapes of British Columbia,* ed. K.W.G. Valentine, P.N. Sprout, T.E. Baker, and L.M.

Lavkulich. BC Ministry of Environment: Soils. env.gov.bc.ca/esd/distdata/ecosystems/Soils_Reports/Soil_Landscapes_of_BC_1986.pdf.

Valentine, K.W.G., and L.M. Lavkulich. 1978. "The Soil Orders of British Columbia." Sec. 2.4 in *The Soil Landscapes of British Columbia,* ed. K.W.G. Valentine, P.N. Sprout, T.E. Baker, and L.M. Lavkulich. BC Ministry of Environment: Soils. env.gov.bc.ca/esd/distdata/ecosystems/Soils_Reports/Soil_Landscapes_of_BC_1986.pdf.

Walker, I., and R. Sydneysmith. 2008. "British Columbia." Chapter 8 in *From Impacts to Adaptation: Canada in a Changing Climate, 2007,* ed. D.S. Lemmen, F.J. Warren, J. Lacroix, and E. Bush, 329–86. Ottawa: Natural Resources Canada.

Welsted, J., J. Everitt, and C. Stadel. 1996. *The Geography of Manitoba: Its Land and Its People.* Winnipeg: University of Manitoba Press.

Wilson S.J., and R.J. Hebda. 2008. *Mitigating and Adapting to Climate Change through the Conservation of Nature.* Land Trust Alliance of British Columbia. ltabc.ca/wp-content/uploads/2012/02/LTA_ClimateChangePrint.pdf.

Witherick, M., S. Ross, and J. Small. 2001. *A Modern Dictionary of Geography.* 4th ed. London: Arnold.

Wynn, G., and T. Oke, eds. 1992. *Vancouver and Its Region.* Vancouver: UBC Press.

Geophysical Hazards and Their Risks

A geophysical hazard, or natural hazard, is an extreme natural event in the crust of the earth that poses a threat to life and property. As defined, geophysical hazards have two components: (1) the natural forces at work and (2) jeopardy to people and their property (2). In hazards research, unless humans or their property are involved in the geophysical event, there is no hazard.

The natural forces that give rise to geophysical hazards can be placed into three categories:

- "Tectonic hazard" refers to the spreading of the seafloor and the consequent collision, or subduction, of continental and oceanic plates. The risk here is from earthquake activity, volcanic eruption, and tsunamis.
- "Gravitational hazard" refers to the discharge, under the force of gravity, of surface material such as rock, earth, snow, or all manner of debris down a slope. Snow avalanches, rockslides and mudslides, debris flows, and debris torrents are all examples of gravitational hazards.
- "Climatic hazard" refers to unusual weather conditions that result from extreme temperatures, lack of moisture (drought), excess moisture (floods), lightning, hail, and violent winds (hurricanes, typhoons, and tornadoes).

Although categorized separately, hazards can overlap. An earthquake (a tectonic hazard), for example, may trigger a landslide (a gravitational hazard). A volcanic eruption, such as that of Mount St. Helens in 1980, may cause debris torrents, floods, and hurricane-force winds, and it may even affect the global climate because of the amount of ash and debris dumped into the atmosphere. The eruption of Mount St. Helens was a catastrophic event.

Because the conditions and technologies that shape the human landscape change over time, so does the potential for disaster or catastrophe. Recognizing risks and developing measures to reduce or eliminate them require decision making at various levels of society, from the individual through to all levels of government. Furthermore, institutions such as universities, legal firms, banks, insurance companies, and engineering firms must be part of the solution.

As you'll learn in this chapter, British Columbia – with its rugged, mountainous landscape, tectonic activity, turbulent rivers, and exposure to multiple climate regimes – has many geophysical hazards. Unfortunately, lack of knowledge about risks and a tendency to simply ignore them has led to a great deal of death and destruction. As the province's population grows, the risk of future disasters will only increase. Land-use planning in this context is an absolute necessity. In this chapter, you'll learn how a natural hazards model can help ensure that decision makers seriously consider the physical processes that can affect the landscape and thus reduce or eliminate risks when they develop a location. In particular, you'll learn about the four types of geophysical events that are primarily responsible for loss of life and property in British Columbia – floods, wildfires, avalanches, and earthquakes. And, most importantly, you'll learn how individuals can gain knowledge of geophysical hazards and avoid or modify risk to themselves and their property.

A NATURAL HAZARDS MODEL

To assess risk and ensure the reduction of risk from natural hazards, it's useful to employ a model that separates the two components or systems (see Figure 3.1). The term "system" implies that complex processes are involved, and the potential for conflict between the two is reflected by the term "versus" in the model. Each system requires further investigation.

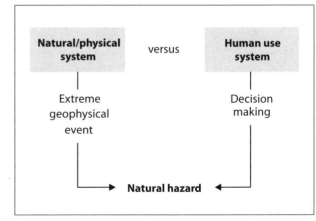

Figure 3.1 Hazards model

The natural/physical system includes all the physical processes (outlined in the three categories above) that affect the landscape of British Columbia. In Chapter 2, you learned about many of the processes responsible for geophysical hazards: tectonic activity, the gravitational activity of a vertical landscape, and the forces responsible for weather and climate variations. The concept of extreme geophysical events recognizes that the forces of nature can be hazardous. Earthquakes, typhoons, floods, and so forth may be infrequent, but they are natural. The challenge is to understand when, where, and why these physical processes occur.

Because people or property have to be involved for an event to be categorized as a hazard, the many earthquakes recorded off the west coast of Vancouver Island each year are not considered hazards. Similarly, although avalanches occur throughout the mountainous terrain of the province in winter, only a few affect people and consequently fit into the hazards model.

The human use system includes homes, commercial structures, institutional buildings, industrial parks, railway lines, highway systems, and recreational facilities, and the location of each of them is governed by decision making. Housing subdivisions, for example, are frequently located on flat land as a result of political decisions governing zoning, density, and services. Economics play a role in decision making by developers, who in turn involve architects and financial institutions as they decide on the type of buildings and how many will be built in any location.

Decisions about location can be understood from a historical perspective. In British Columbia, river systems served as an effective means of transportation and irrigation, and river flood plains provided fertile soil for agriculture and flat land for housing. Subsequent decisions to build and expand on these initial locations have resulted in whole communities being situated on flood plains. Unfortunately, the decision-making process of the human use system often shows little respect for the natural/physical system until a flood occurs.

When the two systems are in conflict during any hazard, it might be tempting to ask, Which system is to "blame"? In other words, when a flood causes loss of life or property damage, is it the fault of the extreme geophysical event or of the human use system in which decisions were made to build in a risky location? The literature on hazards suggests that the natural/physical system is passive, implying that it is absolutely necessary to gain as much knowledge about our physical environment as possible (Burton, Kates, and White 1978; Whittow 1980). For example, in studying river systems, we've come to understand the processes for water discharge, how a river creates a flood plain, and where the risk of locating is greatest. In other words, "flooding is a hazard only because humans have chosen to occupy flood-vulnerable areas" (de Loë 2000, 355).

The human use system is active and therefore must take the responsibility for damage and loss of life. Location decisions are usually made from an economic and political perspective, with little consideration of the physical environment. When a flood occurs or an avalanche hits a community, however, there is a tendency to view the extreme geophysical event as the cause of the catastrophe. In fact, these disasters are commonly referred to as "acts of God," and insurance companies normally do not insure against them. In this simplistic way, extreme geophysical events in nature are put into a realm that suggests we know nothing about physical processes and have no responsibility for our location decisions. Governments usually come to the rescue when a catastrophe occurs, compensating individuals for losses. The negative aspect of this process is that compensation is viewed as a right rather than a gift, and responsibility is rarely taken at the individual level. The importance of the hazards model is its ability to help us understand the processes operating in each system, so that decision making will seriously consider the natural/physical system and thus reduce or eliminate risks.

The story of Walhachin, on the South Thompson River between Cache Creek and Kamloops, illustrates the potential for conflict between the two systems. In the early 1900s, American owners of these bench lands, or terraces, sold off ten-acre farms, mainly to British remittance men from England. (The men had commissions in the military, and their families had money.) Land in the Thompson River Valley, which was known as Walhachin, was promoted as comparable to that of the Okanagan in British Columbia and the Wenatchee Valley in Washington. By the beginning of the twentieth century, both areas had excellent reputations for fruit farming. Immigration to Walhachin began in 1910, and the remittance men soon discovered that the area was no Okanagan Valley. The area was a desert and the struggle to obtain water for irrigation never-ending. In those days, it was not possible to pump water from the Thompson River. Over 200 kilometres of flumes were built to carry water from dammed-up streams. Even with these expensive efforts, the community rarely had enough water, and the soil was alkaline and not conducive to tree fruit agriculture. To make matters worse, the region was subject to frosts severe enough to kill fruit trees. When the First World War broke out in 1914, nearly all the men from Walhachin enlisted in the armed forces. Most never returned, and the community was abandoned in the 1920s (Riis 1973).

The town of Sandon, in the West Kootenay region, had a more disastrous story. It was a fairly typical mining town, hastily constructed following the discovery of rich veins of silver in the 1890s. The town was located in the Selkirk Mountains, a region where peaks can exceed 2,700 metres, where six metres of snow in winter is not uncommon, and where stream valleys are narrow. When the snow melts in the spring, Carpenter Creek, which winds through the community, becomes a raging torrent. Many of the trees surrounding the community were cut, leaving Sandon vulnerable to avalanches and mudslides.

The hazards were many for Sandon. The town was made of haphazard frame structures built too close together. When fire broke out, the town burned to the ground – the upper town in 1900 and the lower town in 1906. After the fire of 1900, the town was rebuilt, and much of Carpenter Creek was encased with a culvert shaped like an upside-down U, so that a road could be put over the top. The encasement required constant maintenance to stop debris from nearby logging from building up at its front end. Over time, as the price of silver declined, so too did the mining. People abandoned the town, and the maintenance of the Carpenter Creek encasement was not kept up. The town was eventually destroyed as the creek sought its own course.

Another mining community, Britannia Beach, just north of Vancouver on the way to Squamish, faced its share of natural hazards. Unlike Sandon, Britannia Beach was a planned **company town**. Though the main community was located on the shores of Howe Sound, a number of self-contained mining camps, including Jane Camp, were established in the mountains above the main town. The physical setting included steep, mountainous terrain; heavy rainfall in the winter months; and snow at higher elevations. Geologists had warned of the instability of the slopes above Jane Camp. Unfortunately, no action was taken, and a major landslide occurred in 1915, claiming over fifty lives and destroying much of the camp.

The main community of Britannia Beach was also unwisely situated on the flood plain of Britannia Creek. The accumulation of debris in the creek from logging and mining caused water to pond behind debris dams. The dams broke with the rains of 1921, and the combined flooding and debris torrent washed many of the homes located on the flood plain into Howe Sound. With little warning, great destruction had occurred, and many lives were lost.

These examples are just a few of the many tragic stories that colour the history of British Columbia. Common to each is how the human use system and the natural/physical system can combine to create catastrophes. Had people or corporations made decisions based on the physical characteristics of the region, the risks could have been reduced or eliminated. Part of the calamity is that knowledge about the natural/physical system was available, but it was ignored.

MEASURING AND RESPONDING TO EXTREME GEOPHYSICAL EVENTS

There are a number of ways to measure extreme geophysical events, and these methods are usually the basis of risk assessments for any location. "Magnitude" refers to the intensity of an event and thus describes the potential for damage. A common measure of magnitude for earthquakes is a seismometer that measures the shock waves. These waves are calculated by a logarithmic calibration, upon which each number represents a tenfold increase in the intensity of shaking. A reading of 4.0 or less means some degree of shaking but usually little damage. A reading of 7.0 or higher indicates a catastrophic event. Among the climatic hazards, high winds are measured by various categories of speed. Hurricanes, for example, begin at 121 kilometres per hour and are feared for their destructive power. Floods, another climatic event, are measured by the rising volume of water, usually past a base level. All extreme geophysical events are measured by magnitude in some way.

"Frequency" refers to how often an event occurs in any location. Lightning-induced forest fires are fairly common in the interior of the province during July and August but much less so in coastal locations, and they are very rare in either location during the winter. Some river systems, such as the Fraser, have a high potential for flooding each spring, while others are considerably less likely to do so. As can be observed from these examples, seasons influence the frequency of some extreme geophysical events, but this is not the case for tectonic hazards.

"Speed of onset" describes how rapidly an event occurs and, consequently, how much warning is likely. Earthquakes occur with little or no warning, whereas snowmelt floods may take weeks to raise water levels enough to inundate areas.

"Duration" refers to the length of time the event lasts. Earthquakes may be over within seconds. When a flood occurs, it may take weeks before water levels recede. Windstorms often end within hours.

"Spatial pattern" may be the most important descriptive measure of all because the term refers to whether any pattern can be discerned on the landscape where an extreme geophysical event occurs. In other words, can it be mapped? Spatial patterns remind us that the various measures of assessment should not be isolated. It is of fundamental importance to recognize that most geophysical events have a spatial pattern for which risk can be estimated. Clearly, some locations are at greater risk than others for particular types of natural hazards. Earthquake risk in British Columbia, for example, is much higher in coastal locations than in the interior. If you want zero risk of earthquake, move to Saskatoon. Of course, you'd then be exposed to the risk of frigid temperatures in the winter.

By combining all these types of measurement, we can predict extreme geophysical events. We have yet to refine prediction to the point of being able to state the exact day and time that a particular magnitude of earthquake will occur, although research is being conducted towards that end. The accumulation of magnitude, frequency, and location measures over time gives us statistical data about how often floods, earthquakes, and other events

have occurred, and at what intensity, in various locations. Statistical probability then allows us to assess the risk of any location. Understanding risk is therefore essential to the decision-making process when choices about location are made.

Because people are often unaware of the range of hazards, they simply react to extreme geophysical events when they occur. Knowing that a hazard can happen is an important first step towards a proper response. Knowing how to respond is the crucial next step. How can we protect our lives and property from natural hazards? Can we prevent disasters from happening?

As an individual, here are some factors you should consider when thinking about potential hazards:

1. Hazard experience: Have you experienced the hazard before, and will that experience make a difference?
2. The probability of a hazard's occurrence: If the likelihood of a hazard occurring is high, will it make a difference to your response?
3. The extent of economic investment: How much do you have to lose in terms of property, and will it make a difference in how you perceive the threat? (Sims and Baumann 1974, 26)

These three factors appear to be reasonable, but the underlying assumption is that we all act rationally. Unfortunately, that is not the case: "Before the lava cools or the flood waters fully recede, people are back on the volcano's slope and the river's edge, rebuilding" (Sims and Baumann 1974, 26). It is very difficult to respond in a rational manner when you are involved personally. Moreover, we often have few choices about where we live. As the flood water recedes, so too do memories of the nasty experience.

If it is valid to ask how the individual perceives hazards, it is equally valid to ask how neighbourhoods, communities, regional governments, provincial governments, and the federal government perceive them. To reduce and eliminate risks, each level must be involved in the decision-making process. In the remainder of this chapter, we'll look at the four types of geophysical events that are primarily responsible for loss of life and property in British Columbia: floods, wildfires, avalanches, and earthquakes. Each will be discussed in terms of their physical

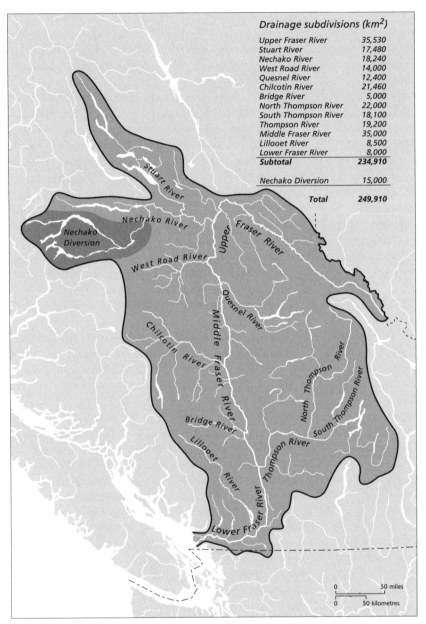

Drainage subdivisions (km²)	
Upper Fraser River	35,530
Stuart River	17,480
Nechako River	18,240
West Road River	14,000
Quesnel River	12,400
Chilcotin River	21,460
Bridge River	5,000
North Thompson River	22,000
South Thompson River	18,100
Thompson River	19,200
Middle Fraser River	35,000
Lillooet River	8,500
Lower Fraser River	8,000
Subtotal	**234,910**
Nechako Diversion	15,000
Total	**249,910**

Figure 3.2 Fraser River drainage basin
Source: Modified from Fraser Basin Management Program (1994).

storms, and volcanic eruptions have occurred in the province throughout its history. The Tseax cone eruption of 1775 killed an estimated two thousand Nisga'a (Hickson and Ulmi 2006). The Hope Slide of 1965 was one of the largest recorded in Canadian history and killed four people (Geography Open Textbook Collective 2014). Debris flows and torrents took out the M Creek Bridge on the Sea-to-Sky Highway in 1981, killing nine people (Taylor 2003). These are but a few examples. All of these hazards require corrective and preventative measures.

FLOODS

The human need for water means that housing, industry, and transportation systems are often built in close proximity to rivers or larger bodies of water. It is therefore unsurprising that floods have caused the greatest amount of property damage in the province (Foster 1987, 48). Three types of flooding are associated with the river systems of British Columbia: **snow-melt flooding**, **flash flooding**, and **ice jam flooding**.

Snow-melt flooding, often referred to as spring runoff flooding, occurs on river systems such as the Fraser, Columbia, Peace, and Skeena, which drain the interior of the province. The volume of water, or discharge, through these systems depends on the size of the drainage basin, the amount of snowpack or accumulation, and spring weather conditions. The Fraser and all its tributaries encompass nearly one-quarter of the land base of British Columbia (Figure 3.2). The system has a large drainage basin and represents a potentially huge volume of water. The amount of water

characteristics, the destruction they cause, and the responses by individuals and governments. Keep in mind, though, that other extreme geophysical events such as debris flows and torrents, landslides, windstorms, hail-

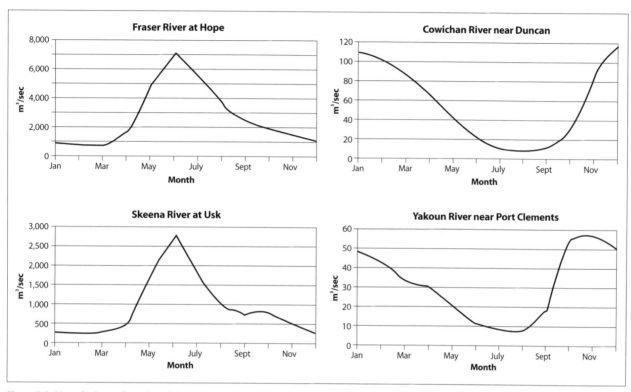

Figure 3.3 River discharge for selected rivers
Source: Data from Environment Canada (1991).

that runs off depends on the amount of snow that has accumulated over the winter and how quickly it melts.

The smaller river systems of Vancouver Island, Haida Gwaii, and the coastal region can produce conditions that result in flash flooding. For these river systems, the rapid rise of water is directly related to intense amounts of rainfall, which can cause the rivers to overflow their banks.

Most northern rivers in British Columbia freeze over in winter, and this can result in ice jam floods. These floods can occur during freeze-up or unusual midwinter warming, but they most frequently occur with the spring breakup of river ice. In all of these instances, ice fragments break free only to jam into downstream ice still stuck firmly to the shore or piled against bridge piers. The jam then causes water in the channel to back up and flood the regions behind the dam of ice. This temporary dam may release suddenly, shifting the potential flooding and damage downstream.

Figure 3.3 shows the range of water discharge for selected rivers in British Columbia. There is a great difference between the volumes discharged by the large drainage basin of the Fraser River and the small drainage basins of the Yakoun and Cowichan River systems. Although volume of water is important, it is the pattern of discharge that determines flood potential at different times of year. Peak periods of discharge occur in the late spring for both the Fraser and Liard River systems, for example, producing snow-melt flood conditions in these interior drainage basins. Because large river systems drain so much of the interior of British Columbia, the potential for flooding during spring runoff affects most of the province. As the graphs for the Yakoun River in Haida Gwaii and the Cowichan River on Vancouver Island show, the greatest rates of discharge occur in the winter months, when flash flooding usually happens. For example, in 2009, three hundred households in Duncan and North

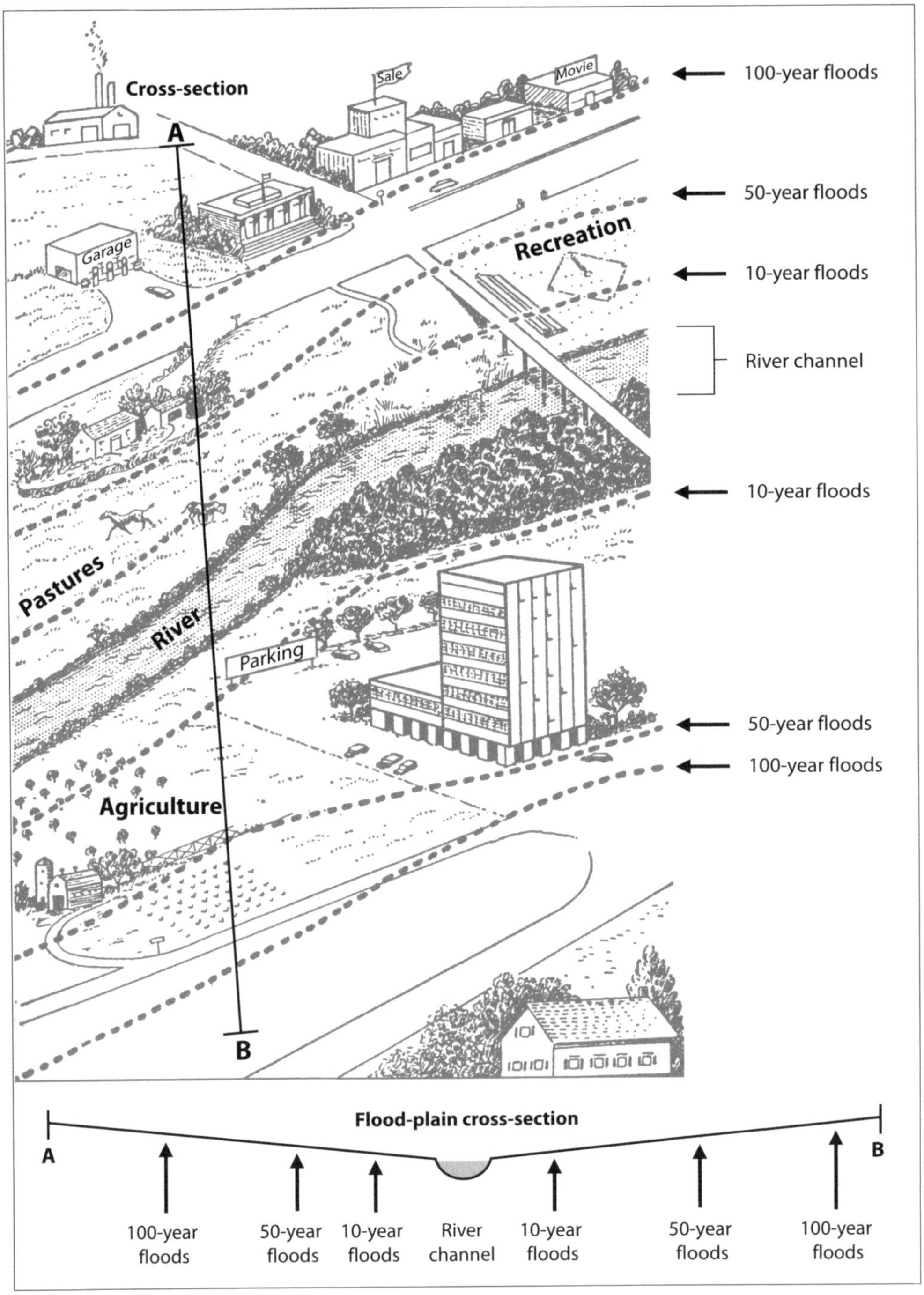

Figure 3.4 Flood plains and land uses
Source: Modified from Tufty (1969), with permission.

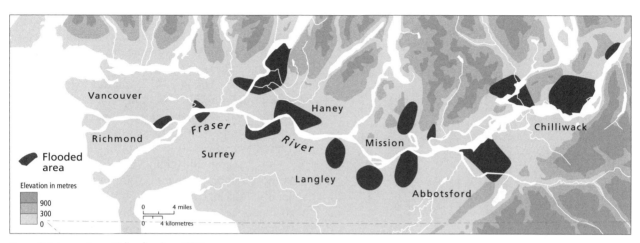

Figure 3.5 Lower Fraser Valley flooding, 1948

Cowichan were forced to evacuate when heavy rains, melting snow, and a high tide caused two rivers to overflow their dikes and flood low-lying neighbourhoods (CBC 2009).

Documenting flood events is essential to forming a proper response. To do this, the volume of water discharged in any given year is measured in relation to benchmarks for the community. The flood plains affecting most flood-prone communities in British Columbia and the rest of Canada have been mapped. The extent of flooding depends on the level to which the river rises. The high-water "contours" can be mapped to show the areas that will be inundated. By accumulating information about both the magnitude and frequency of flooding, a map of statistical probability, or what is referred to in Figure 3.4 as a 10-, 50-, or 100-year flood plain, can be produced. The 50-year flood plain shows the potential of the water rising to that level, or contour, once in fifty years. The flood plain mapping program financed by both the federal and provincial governments includes the 200-year flood plain. It is interesting to note that the "Great Flood" in Manitoba in 1997 reached the 500-year flood plain level. In terms of decision making, it is obvious from Figure 3.4 that the risk from flooding is reduced substantially by locating higher in elevation from the river's edge.

The 1948 flood in the Fraser Valley was catastrophic: "This flooding of 22,260 hectares (55,000 acres) left 200 families homeless, caused 10 fatalities and washed out

82 bridges. Together the federal and provincial governments provided approximately 20 million dollars to rehabilitate flood victims and repair and strengthen the diking system" (Foster 1987, 49). Figure 3.5 gives an idea of the extent of flooding in the Lower Fraser Valley alone. Flooding also occurred throughout the Fraser River drainage system as well as in the Columbia River system in the Kootenays and south into Washington and Oregon States.

The unusual condition most responsible for the 1948 flood was the prolonged spring season. Snow and cold weather continued well into April and early May. Then, as many farmers commented, "summer came." The hot weather melted the snow rapidly, and the flooding began. Figure 3.6 shows the Fraser River discharge at Hope, measured in cubic metres per second, for 1948 as well as for several other high-water years. Although the 1948 flood caused a great deal of damage and loss of life, considerably higher flood waters were recorded in 1894. But there is little record of the extent of the damage in the earlier flood, because few people lived in the Lower Fraser Valley then. As the number of people on a flood plain increases, so does the risk. The population of the Lower Fraser Valley has exploded since 1948, and many more homes have been built in the areas most susceptible to flooding. Are we prepared for flood levels that would match or exceed those of 1894? A flood of the same magnitude has a statistical probability of recurring at Mission

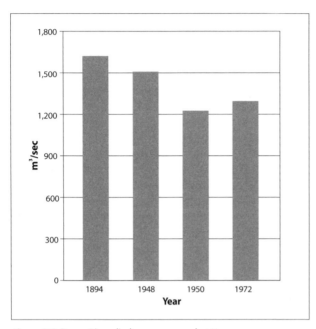

Figure 3.6 Fraser River discharge measured at Hope
Sources: Data from Fraser Basin Management Program (1994), Environment Canada (1991).

once every 140 years (Fraser Basin Management Program 1994, 27).

People sometimes get romantic notions about floods and imagine rowing to work or school as an adventure. This myth needs to be dispelled. One of the greatest concerns related to flooding is the spread of diseases. Flood waters carry many pathogens, and each house must be inspected by a health officer after the waters have receded. Floods also cause an enormous amount of damage to home furnishings, and the fine silt deposited by flood water usually renders motors useless. Moreover, no price tag can be put on the pets, photo albums, and many other items lost or destroyed. In the 1948 flood, the Fraser Valley was cut off from the rest of the province for most of June because the waters washed away the railways and roads. The army had to be called in to prevent looting, and the price of staple food products such as bread and milk had to be fixed because of shortages. Floods have a huge cost, and their risk needs to be reduced or avoided.

The most common way to guard against flooding has been to build diking systems. In fact, too much reliance

has been placed on this form of technology. Subject to erosion, dikes must be continuously maintained to remain viable, particularly given that river drainage systems can change course. In some cases, dikes are partly responsible for the change. As communities in a drainage system grow in population, range land is overgrazed, forests are clear-cut, and more pavement is added. The cumulative impact of all these human activities is an increased rate of runoff in the spring, bringing greater pressure to bear on the diking system.

As more people have moved into the Lower Fraser Valley and interior communities, the diking systems has been expanded. As more and more kilometres of dikes have been built, more and more of the flood waters have been contained, resulting in ever greater volumes of water and pressure on downstream dikes. Rivers carry enormous amounts of silt (the Fraser, in particular, has a "muddy" reputation), which either is deposited on the land during a flood or builds up on the riverbed and the delta as the river slows. When a diking system is increased, more sediment stays in stream and builds up the bottom of the river floor, thus raising the level of the water and again increasing pressure on the downstream dikes. The UBC Department of Geography has been monitoring the Lower Fraser River and is concerned "that the dikes might be too low in some locations between Hope and Mission as gravel moving from the mountains is deposited on the diked and confined channel" (Department of Geography, University of British Columbia 2004). There is no such thing as a permanent dike. They are subject to erosion and must constantly be reinforced and maintained to reflect the changing dynamics of river systems.

Frequency and magnitude recordings reveal that ice jam floods also have a spatial pattern. Environment Canada recognizes that features such as bridge piers, islands, bends, shallows, slope reductions, and constrictions enhance the probability of ice jams (Environment Canada 2009). As in other flood situations, possible corrective and preventive measures include monitoring, warnings, evacuation, diking, and flood proofing, along with measures specific to the event. Prior to breakup, flow can be enhanced and jamming minimized by cutting or blasting the ice, drilling holes in it, covering it with dark material, or injecting it with warm water (Beltaos, Pomerleau, and Halliday 2000).

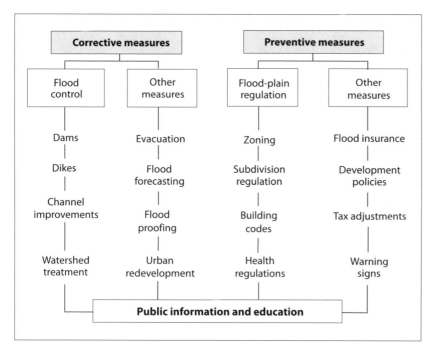

Figure 3.7 Corrective and preventive measures for flood hazards
Sources: Modified from Sewell (1965) and Whittow (1980).

Figure 3.7 shows the variety of potential corrective and preventive measures to reduce or eliminate the risk from flooding. Corrective options include the many short- and long-term responses available to people affected by flooding. Quite simply, how can people who live in a flood-prone area protect themselves and their property? Immediate responses range from warning systems to sand bagging and evacuation. More long-range approaches involve various means of controlling water (including dams, dikes, and channel improvements such as dredging) and the recognition that human activities in the watershed, such as clear-cutting, affect water runoff and flooding. Homeowners in the flood plain can invest in flood proofing, thus keeping flood waters from entering basements or ground floors. Urban redevelopment may result in an entire high-risk section of the community being flood-proofed or built on pillars.

Preventive measures keep human activity out of flood plains. Many of these options require various government institutions to establish regulations and barriers.

Zoning can be effective in keeping high-risk areas as pasture lands, parks, or other uses that exclude housing. Regulation can achieve the same end by not allowing the subdivision of flood plains. Building codes and health regulations may also stipulate rules that prevent construction.

There are other ways to prevent building in flood plains. A rather subtle form of prevention, for example, is the posting of warning signs. Flood insurance and tax adjustments can also be scaled to risk as a preventive measure: the higher the risk, the higher the insurance premium, or the taxes, will be. Development policies may discriminate against construction within a flood plain or, if structures are to be built, require the developer to build up the land to an elevation above the 200-year flood level.

It is an interesting exercise to assess the theoretical range of options any community has with respect to reducing and preventing floods and to match these options to decisions that have already been made. Public information and education, a component of both correction and prevention, is essential and arguably the most important element in all levels of decision making. Individual homeowners, investors, politicians, and others must be aware of the options if they are to be implemented. As Rob de Loë, a professor of water policy and governance, argues, "Floods are considered hazards only in cases where human beings occupy floodplains and shorelands. Therefore, the problem is one of human behaviour rather than the vagaries of the hydrological system" (de Loë 2000, 357).

Although Figure 3.7 was designed to examine the range of options for flood conditions, keep the framework in mind as other extreme geophysical events are discussed. Reducing or eliminating risk from avalanches, debris flows, and earthquakes involves similar considerations.

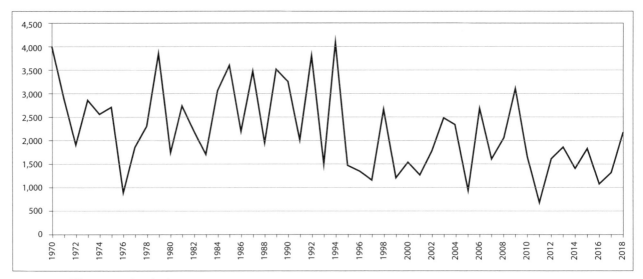

Figure 3.8 Wildfires in British Columbia, 1970–2018

Note: Data for 2018 are preliminary estimates.

Sources: BC Wildlife Management Branch (2009a, 2009b), de Loë (2000), and BC Ministry of Forests, Lands and Natural Resources Operations (n.d.-c).

WILDFIRE AND INTERFACE FIRE

Forest fires are a common occurrence in British Columbia, especially in the dry interior, although no parts of the province are immune from this hazard. Figure 3.8 indicates the wide fluctuation in the number of fires from year to year. Keep in mind that although these fluctuations are important, there is little relationship between the number of fires and the area destroyed by wildfire or the cost in terms of fighting fires or damage inflicted (Table 3.1). For example, there were a moderate number of fires in 2003, but records were set for the area burned and the cost of damages. Likewise, in 2017 the number of fires was not record-breaking, but the area destroyed and the cost of fighting the fires was one of the highest on record.

Firestorm 2003 was the worst on record in the province for interface fires – forest fires that affect personal property. The fires destroyed over three hundred homes and many businesses, and they led to the evacuation of over forty-five thousand people. The firestorm was particularly catastrophic for the Kelowna area, where 238 homes burned, and for the communities of Louis Creek and Barrier north of Kamloops. The total cost of the firestorm was estimated at $700 million, but the greatest cost was borne by the families of the three pilots who died in the line of duty (Filmon 2003). Many of the impacts were immediate and short-term, such as dealing with evacuees or those who had lost their homes and had to rebuild.

Others are more long-term. The loss of the sawmill, the main employment base in Louis Creek, was a major economic blow for the whole community, as was the destruction of many of the historic Kettle Valley Rail bridges at Myra Canyon near Kelowna, which devastated a growing tourist industry in the area built around this popular site for walking and cycling tours. (Through provincial and federal funding, the twelve trestle bridges were restored and back in service by 2008.) The intense wildfires also lead to short- and long-term ecosystem instability. Fire pollutes the air, adds carbon dioxide to the atmosphere, destroys forests, and kills wildlife. Destabilization of soil can lead to mud slides and even flooding during rainstorms, and it takes considerable time for forests and vegetation to regenerate.

Table 3.1

Area destroyed and cost of wildfires in British Columbia, 1970–2018

Year	Area (ha)	Cost of fighting ($ millions)	Year	Area (ha)	Cost of fighting ($ millions)
1970	49,742	12.6	1995	26,888	38.5
1971	128,354	11.4	1996	2,670	43.0
1972	5,974	2.2	1997	286	19.0
1973	20,554	10.7	1998	76,574	153.9
1974	9,993	6.0	1999	11,581	21.1
1975	8,526	7.7	2000	17,673	52.7
1976	17,721	1.1	2001	9,677	53.8
1977	1,582	9.2	2002	8,581	37.5
1978	35,728	17.7	2003	255,466	371.2
1979	18,461	23.2	2004	230,000	156.0
1980	32,743	18.0	2005	34,588	47.2
1981	57,277	39.2	2006	139,265	159.0
1982	280,676	42.3	2007	29,440	98.8
1983	32,848	24.8	2008	13,233	82.1
1984	12,227	37.5	2009	247,419	382.1
1985	54,231	101.5	2010	337,149	212.0
1986	9,474	21.5	2011	12,604	53.5
1987	22,308	35.8	2012	102,122	133.6
1988	3,284	53.0	2013	18,298	122.0
1989	11,089	64.0	2014	369,168	298.0
1990	52,575	60.2	2015	280,605	277.0
1991	11,249	32.2	2016	100,366	129.0
1992	17,212	69.9	2017	1,216,053	649.0
1993	1,376	25.2	2018	1,354,284	615.0
1994	20,737	90.9			

Note: Cost does not include property losses. Data for 2018 are preliminary estimates.
Sources: Filmon (2003), BC Wildfire Management Branch (2009a, 2009b), Brown (2017), and BC Ministry of Forests, Lands and Natural Resource Operations (n.d.-c).

Of course, the conditions that give birth to wildfires do not go away, meaning that they can happen again, as they did in the summers of 2009, 2017, and 2018. Wildfires are a natural phenomenon, and several physical characteristics are responsible for their occurrence. Prolonged periods of dry weather and the presence of fuel are key factors. A fire can be ignited by lightning or by human activity, which causes 40 percent of wildfires (Filmon 2003). Once a fire is in progress, wind is an unpredictable and sometimes deadly factor. As well, several relatively recent factors have increased the risk of wildfire overall and interface fires in particular. Mountain pine beetle has plagued British Columbia, mainly in the lodgepole pine forests of the southern and central interior. The combination of mild winters and mature forests has made many parts of the province vulnerable to this infestation, resulting in large areas of standing dead trees, which represent a significant increase in fuel.

Past forest fire–fighting practices included the burning of slash piles left over from logging activities. However, this activity has increasingly been postponed or neglected on account of local opposition to smoke, but the consequence has been more fuel. Furthermore, although the province has been effective in fire suppression, aggressive fire-fighting methods have led to a buildup of forest fuels. Forests have spread onto grasslands, and open spaces in the dry forests of the southern interior and other areas have filled in. These changes have increased the risk of devastating wildfires and "negatively impacted biodiversity and forest health" (Province of British Columbia 2017b).

One of the equations that applies to flood hazards also applies to wildfires: as the population living in a hazard area increases, so too does the risk. More and more people in British Columbia are opting to live in rural settings either adjacent to forests or on larger lots with plenty of trees. Whole subdivisions on the outskirts of cities project an image of being closer to nature and include cedar-sided homes nestled in trees. During an interface fire, these homes are extremely vulnerable, as the Kelowna experience has shown.

The available information on wildfires and interface fires confirms a spatial pattern based on prolonged dry weather and fuel buildup. The speed of onset can be exceedingly rapid, particularly under conditions of strong winds, which can cause fires to leapfrog a kilometre or more at a time. Once the fire has engulfed an area, its duration may not be long. Witnesses in Kelowna reported that from the time a fire reached their home, it took twelve minutes to reduce it to ashes (BC Ministry of Forests, Lands and Natural Resource Operations n.d.-b).

The severity of Firestorm 2003 resulted in a commission, headed by Gary Filmon, past premier of Manitoba, to investigate the factors responsible for the firestorm and make recommendations for corrective and preventive measures (CBC 2004). Filmon advised that fuel could be reduced by burning slash from logging activities and deadfall in parks in the wet season. He also recommended that the standing fuel from beetle kill be reduced by encouraging forest companies to harvest these trees. Fire insurance companies, he advised, should be encouraged to adjust rates for risk-reduction practices, and the provincial and municipal or regional district levels of government needed to consider land-use practices such as the installation of sprinklers on houses. Another recommendation included ongoing education for adults and schoolchildren. Filmon believed funding for these measures should be provided by both the provincial and federal governments because "this investment in prevention will undoubtedly result in a reduction in future damage costs" (Filmon 2003).

As a consequence of the Filmon Commission, Natural Resources Canada has developed a "Forest Fire Danger Rating System" that gives six regions throughout British Columbia a danger class report each day (Filmon 2003). Furthermore, the province, through Natural Resources Canada, has also developed a downloadable *FireSmart Homeowner's Manual*, which includes detailed illustrations and easy-to-read instructions to make one's home less susceptible to wildfire (Province of British Columbia 2017c; BC Ministry of Forests, Lands and Natural Resource Operations n.d.-a). FireSmart has also developed a much larger text, *Protecting Your Community from Wildfire* (Natural Resources Canada 2003). These manuals detail the many corrective and preventive measures that must be undertaken by individual homeowners and those responsible for designing subdivisions. They include a warning that, during an interface fire, firefighters will concentrate on properties that have implemented FireSmart principles, such as removing woodpiles and debris within ten metres and installing fire-resistant roofing and siding.

Wildfires will continue to occur in the future, and it is important to have warning systems for early detection and the means of early suppression, especially if the fire is in proximity to homes. Water bombers, trained firefighters, fire guards, back burning, spraying fire retardants on buildings, and evacuation procedures are some of the reactive measures taken once fire threatens a community, but banning campfires and even back-country travel are critical measures that can reduce human-caused wildfires.

AVALANCHES

Of all the hazards, avalanches have resulted in the greatest loss of life in the province, primarily because railways have been built through steep mountain passes. Although many transportation systems and a few communities are still threatened, the greatest danger today is in the skiing and snowboarding and snowmobile industries.

There are five interrelated factors that produce avalanches:

1 *Type of snow.* When the type varies from hard pellets to the typical "star" shape, pellets do not bind well, creating unstable snow conditions.
2 *Rate of snowfall.* Avalanches are more likely if snow accumulates at 2.5 centimetres per hour or more.
3 *Terrain, or slope.* Because slopes greater than sixty degrees are usually too steep for large accumulations of snow, they experience frequent, small avalanches instead. Slopes below thirty degrees rarely pose any threat of an avalanche. Snow can build up on slopes between thirty and sixty degrees, and the terrain is steep enough to be unstable.
4 *Change in temperature.* A change of temperature of six degrees Celsius per hour or more, either warmer or colder, can create the unstable conditions that result in avalanche.
5 *Vibration.* Vibration caused by thunder, gunshots, explosions, or other loud noises can trigger avalanches. (Fraser 1966; Natural Disasters Association 2017a)

Because these factors must be taken into account, predicting avalanches is not a simple process. Figure 3.9 classifies and illustrates the various forms of avalanche conditions. Many single-point avalanches are relatively small, but when they are confined to valleys, they can be extremely hazardous. On the other hand, slab avalanches, the most common winter avalanche hazard, can be dangerous and unpredictable and are the primary target of preventative or control measures (LaChapelle

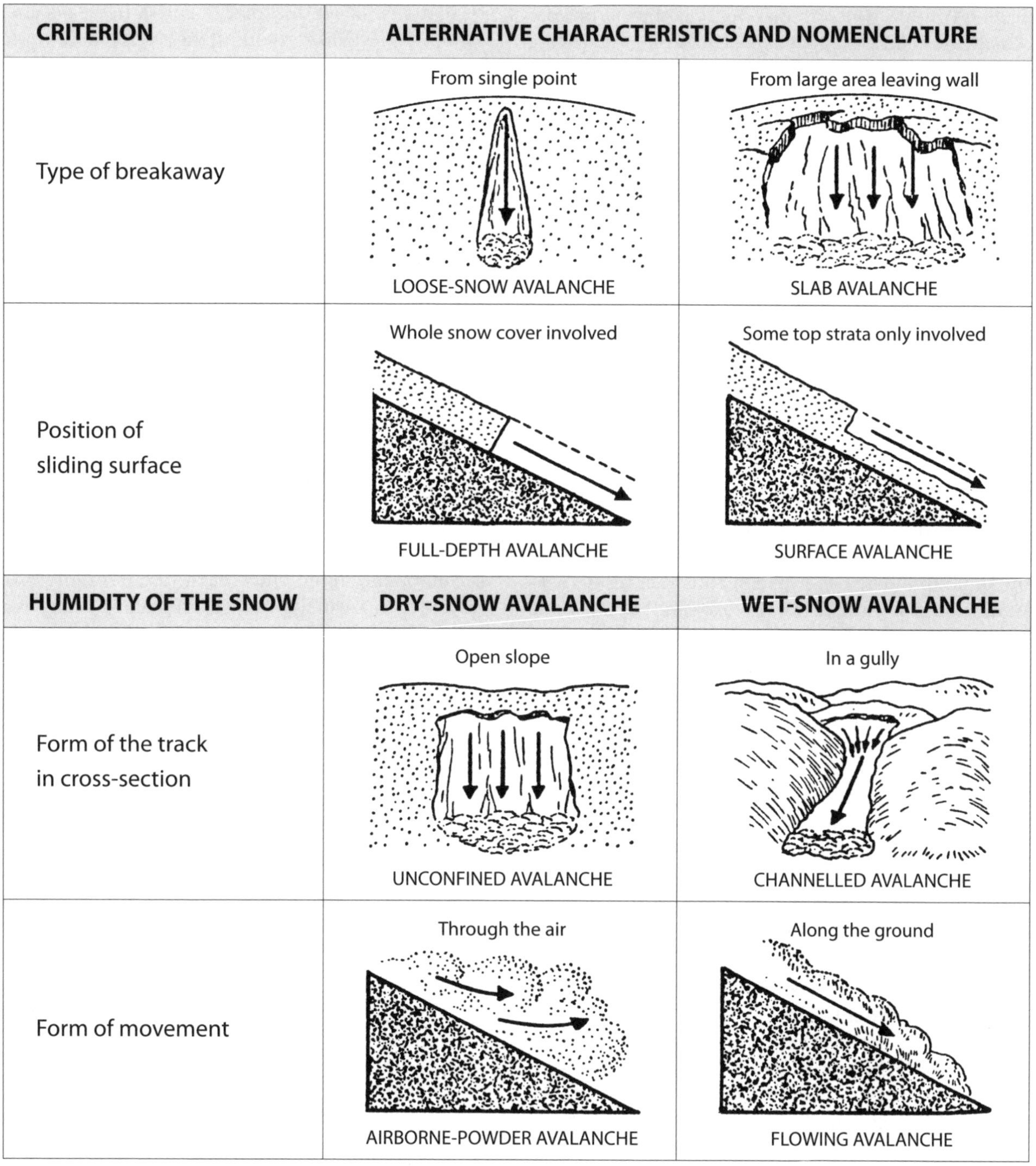

CRITERION	ALTERNATIVE CHARACTERISTICS AND NOMENCLATURE	
Type of breakaway	From single point LOOSE-SNOW AVALANCHE	From large area leaving wall SLAB AVALANCHE
Position of sliding surface	Whole snow cover involved FULL-DEPTH AVALANCHE	Some top strata only involved SURFACE AVALANCHE
HUMIDITY OF THE SNOW	**DRY-SNOW AVALANCHE**	**WET-SNOW AVALANCHE**
Form of the track in cross-section	Open slope UNCONFINED AVALANCHE	In a gully CHANNELLED AVALANCHE
Form of movement	Through the air AIRBORNE-POWDER AVALANCHE	Along the ground FLOWING AVALANCHE

Figure 3.9 General types of avalanche
Source: Fraser (1966).

1998–2001). And when there is a dry, airborne-powder avalanche, it can reach speeds of up to 325 kilometres per hour, destroying practically everything in its path.

The first steps towards corrective and preventive measures are understanding the underlying physical processes and gathering statistics. In British Columbia, mapping the spatial pattern of the highest-risk areas is also essential. The province's road network was mapped for risk after a tragic avalanche destroyed the North Route Café west of Terrace on 22 January 1974, claiming seven lives. The provincial government then commissioned a task force to assess all highways and recommend measures to reduce the risk from this extreme geophysical event (Province of British Columbia 2017a). The provincial government's "Avalanche Safety Program" includes advice for travellers of BC's roads and the "DriveBC" website, which maps road conditions throughout the province (Avalanche Canada 2017).

Other measures have also been used in British Columbia's road and rail systems, including erecting roadside warning signs; using artillery fire and dynamite to prevent the buildup of large snowpacks; constructing tunnels, snow fences, or snowsheds over roads and rail lines; and building berms or catchment basins, which form a semi-circle at the bottom of an avalanche path to prevent it from continuing on to a highway. Terrain management, another measure, involves planting trees to stabilize snow conditions and building deflecting barriers. Where buildings are involved, regulatory mechanisms include building codes, restrictive zoning in high-risk areas, tax incentives, and prohibitive insurance programs.

As mentioned earlier, the highest avalanche risk today is for skiers and those involved in outdoor snow activities more generally. With the growing popularity of back-country skiing, heli-skiing, and snowmobiling into untracked and often untested snow conditions, the risks are high. Even on managed ski slopes, adventurers who ski out of the prescribed boundaries run a serious risk. Avalanche Canada's website has considerable information on avalanche risk, including an up-to-date interactive map for British Columbia (and the rest of Canada) for skiers, snowmobilers, and those going into the backcountry (Canadian Association of Geophysical Contractors 2010). In 2009, WorkSafeBC introduced new regulations for all workplaces at risk from avalanche. These new rules likely reduced the risk for those working in avalanche-prone regions, but recreational users of the backcountry continue to be at risk.

EARTHQUAKES AND TSUNAMIS

The coastal area of British Columbia is an extremely complex and active earthquake zone. From northern California to just north of Vancouver Island, the Juan de Fuca Plate subducts under the North American Plate. To the north of the Juan de Fuca Plate and in close proximity to Haida Gwaii, the Pacific Plate slips past the North American Plate, moving in a northerly direction until it reaches Alaska, where it subducts under the North American Plate. (See Chapter 2 for an explanation of plate tectonics.) These movements produce earthquakes – in some cases of high magnitude. The whole coast of British Columbia is one of the highest-risk earthquake zones in Canada because of these tectonic processes.

The Canadian National Seismograph Network measures the magnitude of earthquakes in Canada using a seismometer, upon which each number represents a tenfold increase in the intensity of shaking. Many would hardly notice a magnitude 2.0 earthquake, and even a magnitude 5.0 earthquake would not likely cause much damage. However, an earthquake of magnitude 7.0 and above could cause considerable damage. Any structural damage related to magnitude should be read with some caution, because the effects of an earthquake depend on many variables, including the slope of the land, the depth and type of soil, the nature of the bedrock, the materials, height, and design of a building, and so on. Brick and stone buildings are less flexible than frame structures, for example, and high-rise complexes may not be able to withstand the horizontal motion of an earthquake. Structures located on landfill may be subject to soil liquefaction and lose their support. This information can be mapped for any community, clearly showing areas of higher and lower risk.

Earthquakes occur frequently off the west coast of Vancouver Island and Haida Gwaii, although some occur inland as well; for example, a magnitude 7.3 earthquake occurred between Powell River and Comox in 1946. Fortunately, most of them are of small magnitude and pose little risk. Some are of considerable magnitude; for example, there was a 6.8 quake on 28 June 2004 north of Haida Gwaii, and a 7.7 quake hit Haida Gwaii on 27 October 2012. The west coast has a record of earthquakes of catastrophic magnitude, and the potential for a megathrust earthquake along the Cascadia subduction zone is of great concern (Natural Resources Canada 2018). Megathrust earthquakes occur when the oceanic Juan de Fuca Plate builds up tremendous pressure while subducting under the continental North American Plate. A megathrust is the rapid release of this pressure, resulting in an earthquake that could exceed 9.0 in magnitude. Indigenous Oral Tradition, Japanese records of tsunamis, and coastal subsidence tell us that a megathrust earthquake of approximately 9.0 occurred on 26 January 1700 (Meissner 2018). *Quake Hunters* (Thompson 1998), a well-researched documentary, traces the evidence of this megathrust earthquake. A follow-up documentary, titled *Shock Wave* (Thompson 2010), focuses on the Cascadia subduction zone and assesses the impact of a tsunami on the west coast. Natural Resources Canada (2018) informs us: "In the Cascadia subduction zone 13 megathrust events have been identified in the last 6,000 years, an average [of] one every 500 to 600 years. However, they have not happened regularly. Some have been as close together as 200 years and some have been as far apart as 800 years. The last one was 300 years ago."

Seismic activity can produce landslides, slumping, building damage, and destruction of telephone lines, hydro lines, roads, rails, and pipelines. These related hazards can bring their own damage and threat to life. The great earthquake of 1964 in Alaska caused enormous destruction to the port of Valdez and the city of Anchorage, and the ensuing tsunami damaged port communities all the way to California. One of those hardest hit by the tsunami was Port Alberni on Vancouver Island, which suffered millions of dollars in damage.

Tsunamis stem mainly from subduction earthquakes when the seafloor is violently thrust vertically upward and/or downward (see Figure 3.10). They can also be caused by volcanic eruptions and coastal and underwater landslides (BC Earthquake Alliance n.d.). Tsunami waves are unlike normal waves in that the whole column of water rushes towards the shore and can cause extensive

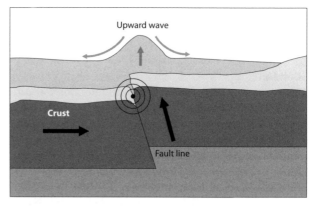

Figure 3.10 Tsunami from subduction earthquake

damage. The megathrust earthquake and tsunami that hit Japan in 2011 left some 16,000 people dead. There is considerable tsunami risk on the west coast of British Columbia, especially from a megathrust along the Cascadia subduction zone, an event often referred to as the "big one." The 1964 Alaska earthquake and subsequent tsunami mentioned above is only one example of the destruction that can occur. Legends from First Nations have been passed down through generations and include accounts of whole villages being washed away on the west side of Vancouver Island because of the megathrust earthquake of 1700, which was followed by a tsunami (Natural Disasters Association 2017b).

The frequency, magnitude, and spatial pattern of earthquakes and tsunamis in BC's coastal region are grounds for concern for the people who live there. The BC Earthquake Alliance offers considerable information on earthquake preparedness, including drill manuals and planning documents, as does Natural Resources Canada, which produces an *Earthquake and Tsunami Smart Manual* (Emergency Management BC 2015). At the policy level, important measures are being taken to reduce the risk through the regulatory mechanism of building codes, which building materials and structural design that are resistant to seismic activity (Natural Resources Canada 2016). These initiatives often require older buildings; schools and other institutional, commercial, and residential sites; and dams to be retrofitted. Building codes are important because whether individuals live or die during an earthquakes usually depends directly upon land use and, especially, the siting and design of buildings (Hewitt 2000, 333).

SUMMARY

The study of natural hazards is an interesting and important area of research that combines the skills of the physical and the human geographer. A lot of information is required to understand the physical processes that produce extreme geophysical events and to produce higher degrees of accuracy in predicting where and at what magnitude these events will occur. One of the most important elements in risk reduction is public information and education to ensure that informed location decisions are made, from the individual to the governmental level. As the population of British Columbia expands, so too does the risk. Hazards research offers considerable hope that many of the risks can be avoided.

REFERENCES

Avalanche Canada. 2017. Interactive map. avalanche.ca/map.

BC Earthquake Alliance. n.d. "What Is a Tsunami?" shakeoutbc.ca/downloads/Tsunami_FAQ.pdf.

BC Ministry of Forests, Lands and Natural Resource Operations. Emergency Management BC. BC Wildfire Service. n.d-a. *FireSmart Homeowner's Manual: FireSmart Begins at Home*. gov.bc.ca/assets/gov/public-safety-and-emergency-services/wildfire-status/prevention/prevention-home-community/bcws_homeowner_firesmart_manual.pdf

–. n.d.–b. "Prescribed Burning." gov.bc.ca/gov/content/safety/wildfire-status/prevention/vegetation-and-fuel-management/prescribed-burning.

–. n.d.-c. "Wildfire Averages." gov.bc.ca/gov/content/safety/wildfire-status/wildfire-statistics/wildfire-averages.

BC Wildfire Management Branch. 2009a. "Current Statistics." bcwildfire.ca/hprScripts/WildfireNews/Statistics.asp.

–. 2009b. "Fire Averages." bcwildfire.ca/History/average.htm.

Beltaos, S., R. Pomerleau, and R.A. Halliday. 2000. "Ice-Jam Effects on Red River Flooding and Possible Mitigating Methods." Report prepared for International Red River Basin Task Force, International Joint Commission. researchgate.net/publication/308520236_Ice-jam_effects_on_Red_River_flooding_and_possible_mitigation_methods.

Brown, S. 2017. "2017 Is British Columbia's Worst Wildfire Season on Record." *Vancouver Sun,* 16 August.

Burton, I., R.W. Kates, and G.F. White. 1978. *The Environment as Hazard.* New York: Oxford.

Canadian Association of Geophysical Contractors. 2010. "Snow Avalanche Assessment (British Columbia)." worksafebc.com/en/resources/law-policy/ppcc-meeting-minutes/g411-snow-avalanche-assessment?lang=en.

CBC. 2004. *Almanac.* 14 July. radio.cbc.ca.

–. 2009. "Vancouver Island Residents Question Flood Plan." *CBC.ca,* 24 November. cbc.ca/canada/british-columbia/story/2009/11/24/bc-duncan-cowichan-flood-damage.html.

de Loë, R. 2000. "Floodplain Management in Canada: Overview and Prospects." *Canadian Geographer* 44 (4): 355–68.

Department of Geography, University of British Columbia. 2004. "Fraser River Gravel Reach Studies – Overview." geog.ubc.ca/fraserriver/overview.html.

Emergency Management BC. 2015. *Earthquake and Tsunami Smart Manual.* gov.bc.ca/assets/gov/public-safety-and-emergency-services/emergency-preparedness-response-recovery/embc/preparedbc/2015_earthquake_tsunami_smartmanual_v07.pdf.

Environment Canada. 1991. *Historical Streamflow Summary, British Columbia.* Ottawa: Inland Water Directorate, Water Resources Branch, Water Survey of Canada.

–. 2009. "Causes of Flooding – Ice Jams." ec.gc.ca/eau-water/default.asp?lang=En&n=E7EF8E56-1#icejams.

Filmon, G. 2003. *Firestorm 2003 Provincial Review.* llbc.leg.bc.ca/public/PubDocs/bcdocs/367694/Firestorm_Report.pdf.

Foster, H.D. 1987. "Landforms and Natural Hazards." In *British Columbia: Its Resources and People,* ed. C.N. Forward, 43–63. Western Geographical Series, vol. 22. Victoria: University of Victoria.

Fraser, C. 1966. *The Avalanche Enigma.* London: Murray.

Fraser Basin Management Program. 1994. *Review of the Fraser River Flood Control Program.* Vancouver: Fraser Basin Management Board.

Geography Open Textbook Collective. 2014. *British Columbia in a Global Context.* Version 1.5. Victoria: BCcampus. opentextbc.ca/geography/.

Hewitt, K. 2000. "Safe Place or 'Catastrophic Society'? Perspectives on Hazards and Disasters in Canada." *Canadian Geographer* 44 (4): 325–41.

Hickson, C.J., and M. Ulmi. 2006. "Volcanoes of Canada." Natural Resources Canada. mineralsed.ca/site/assets/files/3451/bc_volcanoes_jan_03_06_v4.pdf.

LaChapelle, E.R. 1998–2001. "Snow Avalanche: Their Characteristics, Forecasting and Control." people.uwec.edu/jolhm/EH3/Group4/references/snow%20avalanche.htm.

Meissner, D. 2018. "Next Megathrust Quake Will Imperil B.C.'s Pachena Bay, Experts Say." *Globe and Mail*, 12 May.

Natural Disasters Association. 2017a. "Natural Hazards: Avalanches." n-d-a.org/avalanche.php.

–. 2017b. "Natural Hazards: Tsunami." n-d-a.org/tsunami.php.

Natural Resources Canada. 2003. *FireSmart: Protecting Your Community from Wildfire.* 2nd ed. firesmartcanada.ca/images/uploads/resources/FireSmart-Protecting-Your-Community.pdf.

–. 2016. "2015 National Building Code of Canada Seismic Hazard Maps." earthquakescanada.nrcan.gc.ca/hazard-alea/zoning-zonage/NBCC2015maps-en.php.

–. 2018. "Questions and Answers on Megathrust Earthquakes." http://earthquakescanada.nrcan.gc.ca/zones/cascadia/qa-en.php..

Province of British Columbia. 2017a. "Avalanche Safety Program." gov.bc.ca/gov/content/transportation/transportation-environment/avalanche-safety-program.

–. 2017b. "Causes of Wildfires." gov.bc.ca/gov/content/industry/forestry/managing-our-forest-resources/wildfire-management/wildfire-response/fire-characteristics/causes.

–. 2017c. "Fire Danger." gov.bc.ca/gov/content/safety/wildfire-status/wildfire-situation/fire-danger.

Riis, N. 1973. "The Walhachin Myth: A Study of Settlement Abandonment." *BC Studies* 17 (Spring): 3–25.

Sewell, W.R.D. 1965. "Water Management and Floods in the Fraser River Basin." Research paper no. 100. Chicago: University of Chicago.

Sims, J.H., and D.D. Baumann. 1974. "Human Response to the Hurricane." In *Natural Hazards: Local, National, Global,* ed. G.F. White, 25–30. New York: Oxford.

Taylor, Alison. 2003. "Renewed Calls for Bridge Warning System. *Pique*, 24 October. https://www.piquenewsmagazine.com/whistler/renewed-calls-for-bridge-warning-system/Content?oid=2145647.

Thompson, Jerry, dir. 1998. *Quake Hunters: Tracking a Monster in the Subduction Zone.* Produced by Terence McKeown. Ottawa: Omni Films/Canadian Broadcasting Corporation.

–. 2010. *Shock Wave.* Produced by Omni Films in association with CBC-TV. Ottawa: Raincoast Storylines/Canadian Broadcasting Corporation.

Tufty, B., 1969. *1001 Questions Answered about Natural Land Disasters.* New York: Dodd, Mead.

Whittow, J. 1980. *Disasters: The Anatomy of Environmental Hazards.* Markham: Penguin.

Resource Development and Management

4

Resources have often been the engine of change globally and a major influence in the settlement and development of British Columbia. Because the value of resources often fluctuates and new resources are discovered, reliance on them brings change and sometimes conflict to regional and world economies. Resource control and management has political, social, and economic dimensions. How resources are used, or valued, has changed over time and is influenced by the interrelated and shifting roles of labour, technological innovation, energy, capital, land, and other regional factors affecting the production of goods and services. Adding to this complexity are the unforeseen and often outside influences of wars, recessions or depressions, natural hazards, and many political decisions.

All of these factors have led to an enduring pattern of boom and bust for resource-dependent economies, such as British Columbia's. However, from the 1970s on, the province's economy and landscape have undergone a major transformation, including a shift from being predominantly rural and dependent on resources (the **goods-producing sector**) to being predominantly urban and dependent on the service sector. The shift to a service economy led the Lower Mainland's population to boom while the outlying regions of central, north, and east British Columbia have lost population, often because of a declining resource-based employment.

To understand these processes and shifts, in this chapter you'll learn how resources are defined, how they can be categorized to better understand their use and management, how theorists have tried to explain how some resources have given advantage to the industrialization of Canada, and whether their theories are still relevant in the age of globalization. In the process, you'll come away with a basic outline of the history of resource development, management, and mismanagement in British Columbia, which will lay the foundation for the discussions in Part 2 of this book.

DEFINING RESOURCES AND CATEGORIZING THEIR USES

A simple definition of a resource is any naturally occurring substance that is of value to society. The key to this definition is not that it names an enormous number of substances but that it centres on their function in society.

It implies that resources are culturally defined. Herring roe, for example, is a significant resource in this province, bringing in several million dollars to BC fishers, even though very little roe is consumed in the province. It is highly valued in Japan, however, where it is considered a delicacy. Different cultural groups value materials, or resources, in different ways.

This definition also implies that the value of a resource can change. Consider the value of coal from the beginning of the twentieth century to now. Coal used to be the most important energy source for cooking and heating homes, running locomotives, and firing steam-driven engines in industry. By the 1950s and 1960s, however, petroleum, in the form of diesel fuel and gasoline, was replacing the transportation uses of coal, while electricity and natural gas were replacing its residential and industrial uses. These decades were not good times for coal miners in the Fernie area of southeastern British Columbia. Then new technologies and demand from new markets increased the value of coal in the 1970s. Japan initially became the main purchaser of BC coal, using coking (or metallurgical) coal in the iron and steel industry and thermal coal to create electricity. The new communities of Elkford and Sparwood in the Fernie area and Tumbler Ridge in the Peace River area symbolized new interests, urban patterns, and employment related to the metamorphosis of this resource.

The same resource may have more than one value. Fresh water, for example, has numerous values. Anglers, swimmers, boaters, environmentalists, and others value fresh water in its natural setting from a variety of perspectives that are mainly complementary. The use of fresh water for hydroelectric energy production or as a system to discharge effluent from industry or domestic sewage is more contentious. These competing values explain the need for resource management.

Resources can be categorized in a number of ways to help us understand the dynamics of their use and management. One way is to separate renewable from nonrenewable resources. Renewable resources are living, or biotic, resources, which can be divided further into plants and animals. Trees, agricultural crops, domestic and wild species of animals, and humans are all examples of renewable resources. But these resources are renewable only if they are properly managed. The extinction of plants and animals is a concern throughout the world. Noted Canadian geneticist David Suzuki has enlightened Canadians about the threat to biodiversity caused by decimation of species and a reduction in resource options.

Nonrenewable resources are nonliving, or abiotic. Minerals and fossil fuels are two examples. The term "nonrenewable" implies that there is a fixed or finite amount of these resources. In practical terms, it is difficult to determine the fixed amount of a nonrenewable resource such as oil or natural gas, much less gold or copper. More often, we view these resources within a more specific, regional context and make projections about how long a particular mine or oil well will produce until it is exhausted. Even then, these estimates are often good only for a particular time and a particular technology. The concept of a finite amount of any resource is important to the management process.

Figure 4.1 represents an ecosystem model of resource use. It shows how our use of resources affects the environment and the use of other resources. The term "fundamental resources" refers to air (atmosphere), water (hydrosphere), and land (lithosphere). The mix of these three essential spheres, along with energy from the sun, produces photosynthesis, in which plants absorb carbon dioxide and give off oxygen. These resources are thus fundamental to life and are therefore known as life-giving resources.

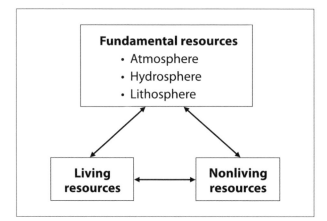

Figure 4.1 Ecosystem model of resource use

The dynamics of the ecosystem model can be illustrated further if we consider what happens when people rely on and use more and more fossil fuels for energy and materials. Using these nonliving resources results in ever greater amounts of waste (pollution) such as sulphur dioxide and carbon, which is emitted into to the atmosphere, one of the fundamental resources. In the atmosphere, sulphur dioxide (SO_2) combines with water vapour (H_2O) and falls as acid precipitation (H_2SO_4), which affects the land (lithosphere) and the resources of the land, including living resources. Increased acidity from acid rain can have a serious impact on drinking water, forests, agriculture, and fish living in the streams and lakes (hydrosphere). Similarly, the increased production of carbon, primarily from the ever-increasing burning of fossil fuels, has largely contributed to greenhouse gases and climate change, which is already causing catastrophic effects on this province (e.g., through the mountain pine beetle epidemic, wildfires, and so on) and the rest of the world.

The ecosystem model demonstrates the cyclical nature of resource use, thereby recognizing environmental impact as a "cost" in the production of any given commodity. It incorporates the real cost of commodities and showcases the concept of the **sustainability** of resources. To manage resources, it is essential to view the whole process of how they are developed into products. As a society, we've continually struggled to recognize and assess the impact and costs associated with resource development, whether they be an oil pipeline rupture, the failure of a tailing pond containing toxic materials, the discharge of effluents from factories into our water and land, or the side effects of herbicide and pesticide use. What level of effluent or discharge is "safe," and who pays the costs of cleanup? Considering that many of the consequences to human health and the environment take a great deal of time to assess, it is not surprising that these negative factors have been labelled as externalities to the production process. Corporations have therefore not had to include environmental repercussions among the costs of production.

Political pressures have altered this view. Concern over the risk of environmental catastrophe has led to demands (usually by those potentially affected) for measures such as the **precautionary principle**. For example, environmentalists, stakeholders in the commercial salmon-fishing industry, a number of First Nations, and others believe that fish farms that raise salmon in the ocean (**aquaculture**) cause harm to wild salmon populations. The precautionary principle, if applied, would mean that the **burden of proof** would lie with industry and the federal government, not the public. Moreover, if it is affirmed that wild runs of salmon have been affected, then the fish farm industry should pay (the polluter pays principle). In the case of other environmental catastrophes (e.g., oil spills), environmentalists advocate for the polluter to set money aside for potential disasters and pay for legal representation of those impacted. They argue that compensation must take into account the long-term degradation of the environment and the impacts on those affected.

The concept of sustainability, or sustainable development, has only been in use since the 1980s. Fundamentally, sustainable development means that resources should not be exploited to a level where they will not be available for future generations. This concept rests on recognizing that the destruction or extinction of species brings ruin to an economy and a way of life and that, in some cases, this destruction occurs because of government policy. Because the concept of sustainability recognizes ecological relationships, it is not easy to put into practice, particularly when increasing populations create a corresponding increase in demand for resources and products. The debate continues as to where the emphasis on sustainability should be placed: on development (i.e., sustaining a certain standard of living) or on ecological processes (McManus 2000, 813). The message of this book is that we need to gauge resource use and its impacts for each region of the province. Resource use and sustainability, however, must also be viewed in both a national and global context.

STAPLES THEORY

In the 1930s, the Canadian economic historian Harold Innis used **staples theory** to describe Canada's national economic development (Watkins 1963; Barnes and Hayter 1997). His theory focuses on the exploitation of resources, or staples. Its main assumption is the existence of an external demand for these resources. Innis identified five resources in the development of Canada: fish, furs, timber, wheat, and minerals. These five resources represent both a historical and an east-to-west pattern of development (but minerals, the last resource on the list, are the exception to the directional trend). The first resource, cod, was taken off the coast of Newfoundland beginning in the late 1400s. Then continental resources were developed from east to west during the fur trade, followed by timber and agriculture. Minerals tended to be discovered sporadically from region to region. Through these resources and the economic activities tied to them, the Canadian economy developed.

The terms "backward linkage," "forward linkage," and "final demand linkage" are used in staples theory to describe the types of economic activity, including employment, associated with each staple. "Backward linkage" refers to all the conditions necessary to export a resource. The most important backward linkage for any resource is transportation systems because they can influence so many other economic activities. Development in Canada meant building and running port facilities with warehouses, boat repairs, and all the employment related to loading resources onto ships for export. Over time, ports connected to canals, railways, and road systems. Backward linkages also include the employment created from building these facilities, as well as the construction and manufacturing of rails, boats, trains, trucks, and any of the inputs to export a resource.

Forward linkage is the process of adding value to any resource through processing or manufacturing prior to export. In British Columbia, there has always been concern about the export of raw logs versus the forward linkage of milling the logs into dimension lumber. Obviously, if the sawmilling activity occurs in the province, then more jobs are created. Even higher value can be added to wood if it is manufactured into furniture, doors, or even musical instruments, with greater benefits to employment and government revenues. Once a resource is tagged for export, a community is usually involved, whether at the port or at the location of resource extraction. Over time, and with the accumulation of backward and forward linkages, some of these communities become major centres.

Final demand linkages are defined as the demand for production of goods and services for the local or domestic market. In smaller communities, consumer goods may have to be imported. Nevertheless, as the population increases, the ability to reach thresholds for local production also increases.

The accumulation of all these linkages is referred to as the multiplier effect. For example, a pulp mill or mine locating in a community will add five hundred workers (forward linkage). These workers, in turn, need housing, food, and many other necessities and luxuries, and so the local economy grows to fulfill this increased demand (final demand linkages). One more pulp mill or mine in the province may be all it takes to stimulate the manufacture of pulp or mine machine components (backward linkages). All this economic activity can mean considerably more than five hundred people working.

Keep in mind, though, that the main assumption of this theory is that there is an external demand for the resource. Historical periods of recession and depression show that external demand has not always been sustained. When the mill or mine shuts down, the multiplier effect works in reverse, and many more people, apart from those employed in the mill or mine, face unemployment. When a mill or mine is the only industry, its closure may jeopardize an entire community.

Is staples theory relevant today? Certainly, but Innis could not have accounted for the **digital revolution** or for the fact that technologies would shrink time and space, facilitating the movement of goods via air cargo carriers and supertankers or the flow of information and capital in an instant. Globalization has fundamentally altered the various linkages that Innis set down. Moreover, the emerging dominant resource – human resources – are much more intangible than natural resources. Revolutionary technologies enabled a new global division of labour, with labour-intensive goods and services being produced in the cheap labour regions of the world. As a consequence, fewer and fewer workers were required for resource industry employment, whereas the global demand for services such in medicine, pharmacy, education, real estate, and high-tech industries (e.g., gaming) accelerated. In other words, Innis's traditional resources are no longer the driving force of the economy.

In addition, by the 1970s, traditional resource industries in British Columbia (e.g., lumber, pulp, paper, concentrates of metals) faced many other challenges such as foreign ownership, contracting out, and capital-intensive technologies that reduced the need for labour. And resource-development issues became considerably more complex as companies had to take environmental impacts into consideration and consult with and compensate First Nations when a project impacted unceded Crown lands. Vacillating government policies that strengthened or weakened environmental stewardship and produced uncertainty, or ignored Indigenous Title, also began to play a significant role. Staples theory remains relevant to the experiences of those involved in the traditional resource-extraction and resource-processing industries, but the theory is not as relevant to British Columbia's (and Canada's) economy today as it was in the 1930s. Some evidence of the decline in the importance of resources and the whole goods-producing sector can be gained by examining British Columbia's **gross domestic product (GDP),** which measures the total production of all goods and services (Table 4.1).

THE CHALLENGE OF FORDISM AND GLOBALIZATION
One of the most fundamental changes to resource processing occurred early in the twentieth century, when Henry Ford invented the assembly line process. Since then, the term **"Fordism"** has been used to define mass production of standardized goods, usually at a centralized assembly plant. Assembly line technology transforms resources into consumer goods. In British Columbia, Fordist methods were employed by the resource industries in the canning of salmon, the

Table 4.1

BC's gross domestic product, by industry, in millions of 2012 chained dollars, 2018

	$ (millions)	%
All industries	246,506	
Goods-producing sector	60,281	24.4
Agriculture, forestry, fishing, and hunting	6,157	2.5
Mining, quarrying, and oil and gas extraction	10,208	4.1
Utilities	4,940	2.0
Construction	21,110	8.6
Manufacturing	17,711	7.2
Service-producing sector	186,236	75.6

Source: Statistics Canada (2019b).

concentration of minerals, the manufacture of two-by-fours and other dimension lumber, and the production of pulp, paper, plywood, and oriented strand board (particle board or OSB).

After the Second World War, a period of prosperity created the greatest ever demand for consumer goods. The principles of Fordism were refined, and an increased demand for the forest and mineral resources of British Columbia resulted in a major expansion of these industries. Investments in manufacturing plants, however, tended to be made in central Canada and other global metropolitan locations. Many single-resource communities, dependent on one particular resource, were created.

The 1960s and 1970s saw the rise of oligopoly conditions, in which large corporations with little competition began to control the resource sectors of British Columbia and elsewhere. These corporations employed new scientific management principles, including capital-intensive technologies to produce ever greater amounts of goods from resources, but these new machines and techniques began to replace labour. The old, labour-intensive ways of processing resources, from sawmilling to mineral extraction, were fast disappearing.

The struggle for provincial governments, and to a lesser degree for the federal government, was to encourage – usually by providing the infrastructure – corporate investment in the resource sector to produce more employment. So long as new plants were being opened, the potential multiplier effect of new employment would result, and greater revenues from resource taxation would accrue to the government. Once the plant and equipment were in place, however, new capital-intensive technologies eventually reduced the number of workers required. More and more, the government's role was to establish a social safety net, along with educational retraining for displaced workers. Keep in mind, though, that until the early 1970s almost the whole country had experienced good economic times; there were plenty of job opportunities, and governments had money to spend.

Unions faced new challenges. The new technologies often necessitated increased productivity to remain competitive in an industry, making it difficult for unions to oppose the equipment that reduced their rank-and-file

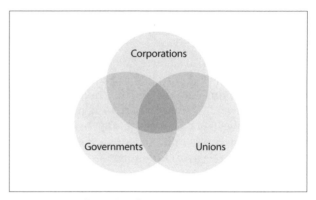

Figure 4.2 Interrelationship of corporations, governments, and unions

members. Instead, their job became to ensure better working conditions and wages – for those workers who remained. Figure 4.2 illustrates the interlocking relationship of various groups within British Columbia – change in one sphere influences change in the others. For example, the introduction of an automated green chain (an assembly line of freshly cut dimension lumber) at a sawmill immediately increases the productivity of the mill but leaves many workers seeking Employment Insurance or retraining and fewer rank-and-file members in the Industrial, Wood and Allied Workers of Canada.

Figure 4.3 illustrates the condition of British Columbia's resource-based economy in the 1970s and 1980s, decades when the province's resources achieved only low value-added processing. The corporations that owned the resource industries resided in heartland areas beyond the province's borders such as in southern Ontario, the United States, Britain, and Japan. The corporations decided where to locate their head offices and research facilities and where to develop new products and implement new technologies. The high-paying and often professional jobs were not located in the hinterland (that is, British Columbia). Similarly, the production of high value-added products (mainly consumer goods produced from semiprocessed resources) and the suppliers of transportation, goods, and services were also located outside of British Columbia. Moreover, because resource industries within the province were controlled extraprovincially, the profits generated were exported, and the province had to import goods, services, and high value-added consumer goods.

At the time, there was serious concern about encouraging large foreign corporations to invest in the resource sector. In his 1982 assessment of resources and provincial policy, Tom Gunton wrote: "Overall ... a staple region such as B.C. which is dominated by externally controlled firms interested in obtaining a secure supply of resources for externally located operations will not develop strong regional linkages. Instead the economy will develop only lower order processing and service activities tied to the regional market. It will lose out on much of the 'footloose' employment such as head office management, research and development and higher order processing which will be located outside the region simply because the firm is externally controlled" (Gunton 1982, 10). Gunton suggested that British Columbia should develop many more linkages with its resource base.

Gunton's prediction came true, and BC's economy developed in four unique ways in comparison to the rest of Canada (Barnes et al. 1992). First, industries employing Fordism were decentralized in the resource frontier. Second, BC's Fordist industries gained the fewest forward and backward linkages of any province in Canada. Third, the provincial government played the most facilitative role of any government in Canada in providing infrastructure and attracting corporations. Fourth, British Columbia experienced the largest boom and bust cycles of any region in Canada.

In good economic times – for BC, from the end of the Second World War to the early 1970s – wealthy federal and provincial government programs could handle cyclical changes and their impact on workers. They established community colleges with training programs and expanded the social safety net. The economic engine, not just for British Columbia but also for the Western world, was based on the manufacturing industry, and this **industrial economy** led to increased urbanization.

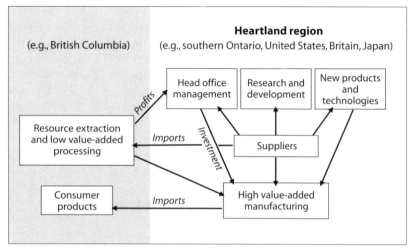

Figure 4.3 Staple dependency and external control
Source: Gunton (1982); reproduced with permission.

POST-FORDISM AND RESTRUCTURING IN A NEW GLOBAL SYSTEM

Global conditions changed, however, and these changes had serious implications for regional economies that relied on resource development and manufacturing. By the 1970s, the world was moving into **post-Fordism** – meaning that the industrial economy was being surpassed by the service economy in terms of providing employment. Early in the decade, currencies that had been pegged to fixed gold rates became destabilized, and the energy crisis created global unease. Increased pressure for global trade – along with major economic recessions in 1981–86, the early 1990s, the late 1990s, the early 2000s, and 2008–09 – indicated that the traditional resource economy was not working.

Two trends transformed the production process and caused a major restructuring of resource-based economies: time-space convergence and the fragmentation of the production process. The advent of supertankers, jet air cargo carriers, coaxial cables, satellites, telephones, fax machines, computers (also referred to as the digital revolution), the internet, email, and other technologies conquered distance, permitting a global scale of organization in the movement of goods, services, information, and capital. In this context, the fragmentation of the production process meant the manufacture of components could be assigned to various locations around the world to reduce costs. Corporations had to ask: What are the components of any good or service? Where are the cheapest places in the world to produce these components?

Time-space convergence and process fragmentation gave birth to the multinational or transnational corporation. For example, Sears, which declared bankruptcy in 2017, used the Kenmore product name, but Sears manufactured no products itself (O'Neill 2017). Transnational corporations contracted well-known and lesser known corporations such as Whirlpool, Daewoo, Corning, GE, Electrolux, Samsung, and Inglis to produce a large array of products. These products, when taking into account all their inputs, make up a map of the world. For example, Samsung "televisions sets are made in Europe, in the Czech Republic, Hungary, Romania, Slovakia and Slovenia," and their "phones are produced in South Korea, Vietnam and China" (IAC Publishing 2017).

Post-Fordism put the global marketplace in a state of uncertainty, particularly with respect to traditional production processes for standardized, one-size-fits-all materials. Post-Fordism brings with it the concept of flexible specialization, by which goods can be produced to fit individual consumer demands, often by contracting out to small firms throughout the world. The term "restructuring" refers to the ways in which resources and technologies are combined to produce goods and services in the global marketplace. Restructuring has a somewhat negative connotation: when corporations, or whole economies, restructure, it implies major changes, including employment loss and uncertainty for many.

A serious recession from 1981 to 1986 caused many traditional Fordist operations to restructure in British Columbia. MacMillan Bloedel's sawmill in Chemainus on Vancouver Island is a typical example of the impact of restructuring. This fairly large mill produced dimension lumber mainly for the American housing market. When that market collapsed in the early 1980s, the mill closed in 1982, laying off 642 workers. The mill closure was a major blow to the town's economy. A new mill opened in 1985, but it was organized according to the principle of flexible specialization. It employed specialized equipment to cut lumber to fit the market demands of Japan,

Europe, and the United States, and it only required 145 employees. Meanwhile, the company contracted out a number of the functions that it had previously run itself under the old Fordist system, such as a dry kiln operation (Barnes et al. 1992; Rees and Hayter 1996).

For the community of Chemainus, these were tough times marked by high unemployment and outmigration as people left in search of jobs. In response, the town turned to the tourism industry, promoting itself as the "Little Town That Did!" The history of Chemainus was depicted in large murals on the walls of downtown stores, successfully attracting tourists. New service-oriented jobs, often at minimum wage, along with contract employment in fields that paid much less than working in the resource industry, were part of the restructuring.

The service sector was also restructured in this period. Consider banking. Once automated banking machines were introduced, the need for branch employees declined dramatically. Both large and small firms experienced layoffs, or what the media commonly referred to as "downsizing" or "re-engineering." All communities, whether small resource-based ones or large metropolitan centres such as Vancouver, were affected by these global economic systems (Norcliffe 1994).

Because British Columbia and Canada are part of a global system of trade and commerce, they are affected by worldwide recessions, such as the one that followed the collapse of Asian economies in the late 1990s and the Great Recession of the late 2000s and early 2010s. However, some of these events were offset by other conditions. For example, the effects of the Great Recession were buffered by construction and preparation for the 2010 Winter Olympics. And not all the changes have been negative. The twenty-first century has been referred to as the "Pacific century" in reference to the fact that Asian countries are becoming the new leaders in industrial growth, development, and investment. The Pacific Rim nations include over 50 percent of the world's population, and British Columbia is well positioned to engage with these markets. Metro Vancouver is a major gateway to the rest of North America; it is home to Canada's largest port, which supports trade with more than 170 economies around the world (Port of Vancouver 2017). As well, Vancouver's International Airport (YVR), which is one of the largest in Canada, offers many direct flights throughout the Asia Pacific region. To a considerably lesser degree, Prince Rupert, with which has an international port at Ridley Island, is another important link to Asia. However, it is not resource development that is the driving force of investment from Asia – it is mainly real-estate investment. BC's experience in resource production and in infrastructure and urban development has itself become a potential knowledge-based export. British Columbia is politically stable and comes complete with modern banking and financial institutions, excellent educational facilities, and a highly educated population, attributes that attract immigration and investment, especially in the service industries. In addition, the province still has resource-based commodities required throughout the world.

THE NEW REALITY OF TWO GEOGRAPHIES

Today, BC's economy is spatially differentiated between two geographies, one based on the service sector, the other on the goods-producing sector. The service economy, which is knowledge- and information-based, is located in the larger urban centres of the province, but Metro Vancouver, with a population of nearly 2.7 million out of a total provincial population of nearly 5 million, dominates. The growth of the Lower Mainland represents increasing economic opportunities and development in manufacturing, tourism, the movie industry, the high-tech industry, and many other services.

Table 4.2 details the various employment categories for the province and Metro Vancouver as of 2017. Clearly, Metro Vancouver, with nearly 57 percent of the jobs in the province, is the most important centre. Only 20 percent of jobs are in the goods-producing sector, and employment is concentrated in construction and manufacturing. Well over 80 percent of the jobs in this highly

Table 4.2

BC and Metro Vancouver, employment by industry, annual average, 2017

	British Columbia		Metro Vancouver		% of BC
	N (000s)	%	N (000s)	%	
All industries	2,466.8		1,400.9		56.8
Goods-producing sector	491.7	19.9	239.0	17.1	9.7
Agriculture	26.2	1.1	5.6	0.4	0.2
Forestry, fishing, mining, quarrying, oil and gas	49.8	2.0	7.4	0.5	0.3
Utilities	12.9	0.5	7.6	0.5	0.3
Construction	228.6	9.3	118.3	8.4	4.8
Manufacturing	174.2	7.1	99.9	7.1	4.0
Service sector	1,975.1	80.1	1,161.9	82.9	47.1
Trade	374.0	15.2	217.2	15.5	8.8
Transportation and warehousing	139.4	5.7	85.9	6.1	3.5
Finance, insurance, real estate, and leasing	156.2	6.3	109.0	7.8	4.4
Professional, scientific, and technical services	197.2	8.0	134.4	9.6	5.4
Business, building, and other support services	102.6	4.2	61.6	4.4	2.5
Educational services	166.6	6.8	95.8	6.8	3.9
Health care and social assistance	303.5	12.3	158.5	11.3	6.4
Information, culture, and recreation	136.8	5.5	83.1	5.9	3.4
Accommodation and food services	182.6	7.4	97.8	7.0	4.0
Other services	114.4	4.6	61.2	4.4	2.5
Public administration	101.9	4.1	47.4	3.4	1.9

Source: Statistics Canada (2019a).

urban centre are in the service sector, and the statistics do not indicate employment categories such as the film, television, and high-tech industries, most of which are located in the Vancouver area. In 2016 alone, employment in the high-tech sector increased 4.1 percent to 106,430, and this sector now employs 5 percent of British Columbia's workforce, more than the mining, oil and gas, and forestry sectors combined (BC Stats 2017).

As Table 4.2 indicates, less than 20 percent of provincial employment is in the goods-producing sector. Still, these are important jobs, particularly for the less populated regions of the province. As Table 4.3 shows, the Lower Mainland–Southwest region dominates in terms of population, so much so that the other four economic regions – Kootenay, Cariboo, North Coast and Nechako, and Northeast – now have less than 9 percent of the population, even though the regions make up 75 percent

Table 4.3

BC population, by economic region, 2001 and 2017

	2001		2017	
	N	*%*	*N*	*%*
Vancouver Island and Coast	717,546	17.6	815,997	16.9
Lower Mainland–Southwest	2,383,643	58.5	2,968,592	61.6
Thompson-Okanagan	484,927	11.9	564,166	11.7
Kootenay	151,130	3.7	148,162	3.1
Cariboo	167,365	4.1	153,485	3.2
North Coast and Nechako	65,617	1.6	53,112	1.1
Total	4,076,881		4,817,160	

Source: Statistics Canada (2018).

of the geographic area of British Columbia. These relatively rural regions have considerably more employment in the goods-producing sector, and they are the regions most affected by catastrophes such as the mountain pine beetle epidemic and sawmill closures. These are also the regions where population is declining.

What remains of BC's traditional resource economy in these regions is a combination of older Fordist production systems and the new post-Fordist ways of doing business. The resource economy's structure is much more complex than it used to be, and there are many more variables responsible for economic growth. Globalization has also changed the role of government as a manager of resources. Faced with pressure to privatize and deregulate, governments "increasingly find themselves with diminished control over the processes and strategies of production, except insofar as they can represent the vested interests of their corporations in the development of international rules of corporate behavior and/or privilege" (Wilkinson 1997, 131).

SUMMARY

The early settlement and development of British Columbia was very much tied to resources such as furs, coal, forests, fish, energy, gold and other minerals, and some agriculture – in other words, commodities produced for an external market. The value and use of these resources have changed over time, and so have production processes, corporations, governments, and markets. Categorizing resources as renewable or nonrenewable and

using an ecosystem model of resource interaction are two ways to recognize the need for and potential limitations of resource management.

Staples theory helps to unravel the economic links between resources and related development. But staples theory is built on the assumption that traditional resources such as fur, forests, and minerals are the driving force of the economy. This was true of British Columbia up until the 1970s, but no longer. These resources have been eclipsed as a revolution in technologies has facilitated the dominance of human resources in a service-oriented economy based on ideas and information.

The global economic system is not static, nor are the technologies used to transform resources. Fordism was adopted by industrial firms in British Columbia following the war, and employment and development expanded into the 1970s. Changing global economic conditions from the 1970s onward, however, led to the formation of multinational corporations that could organize the production of goods and services on a global scale with a global division of labour. Post-Fordism is marked by uncertainty in resource production, restructuring, further globalization, and a shift to a service-based economy.

In this context, new opportunities have arisen for BC, particularly with respect to the economies of the Pacific Rim. The province still has traditional resources, but agriculture, forestry, fishing, mining, quarrying, and oil and gas amount to only 3.1 percent of overall employment in the province. However, resource development continues to be significant for the lives and fortunes of the 9 percent of the population who live in the interior and north end of Vancouver Island. In general, though, British Columbia has transitioned into an urban, service-oriented province as Metro Vancouver and its surrounding region have captured many of the new economic and employment opportunities.

REFERENCES

Barnes, T.J., D.W. Edgington, K.G. Denike, and T.G. McGee. 1992. "Vancouver, the Province, and the Pacific Rim." In *Vancouver and Its Region*, ed. G. Wynn and T. Oke, 171–99. Vancouver: UBC Press.

Barnes, T.J., and R. Hayter, eds. 1997. *Troubles in the Rainforest: British Columbia's Forest Economy in Transition.*

Canadian Western Geographical Series, vol. 33. Victoria: Western Geographical Press.

BC Stats. 2017. "Technology Sector Profile: 2017 Edition," *Infoline* (17–149): gov.bc.ca/gov/content/data/statistics/infoline/infoline-2017/17–149-bc-tech-sector-profile-2017.

Gunton, T. 1982. *Resources, Regional Development and Provincial Policy: A Case Study of British Columbia.* No. 7. Ottawa: Canadian Centre for Policy Alternatives.

IAC Publishing. 2017. "Where Are Samsung Products Made?" *Reference.com.* reference.com/business-finance/samsung-products-made-891e3f5391f0ef1a#.

McManus, P. 2000. "Sustainable Development." In *The Dictionary of Human Geography,* 4th ed., ed. R.J. Johnston, D. Gregory, G. Pratt, and M. Watts, 812–16. Oxford: Blackwell.

Norcliffe, G. 1994. "Regional Labour Market Adjustments in a Period of Structural Transformation: An Assessment of the Canadian Case." *Canadian Geographer* 38 (1): 2–17.

O'Neill, D. 2017. "The Purchase: Who Make's What? Kenmore." *Appliance 411: Information.* appliance411.ca/purchase/sears.shtml.

Port of Vancouver. 2017. "About Us." portvancouver.com/about-us/.

Rees, K., and R. Hayter. 1996. "Enterprise Strategies in Wood Manufacturing, Vancouver." *Canadian Geographer* 40 (3): 203–19.

Statistics Canada. 2018. "Annual Demographic by Economic Region, Age and Sex Based on the Standard Geographical Classification (SGC) 2011." CANSIM Table 051–0059. statcan.gc.ca/cansim/a26?lang=eng&id=510059.

–. 2019a. "Employment by Industry, Annual, Census Metropolitan Arres (× 1,000)." CANSIM Table 282–0131. statcan.gc.ca/t1/tbl1/en/tv.action?pid=1410009801.

–. 2019b. "Gross Domestic Product (GDP) at Basic Prices, by Industry, Provinces and Territories (× 1,000,000)." Table 36-10-0402-01. www150.statcan.gc.ca/t1/tbl1/entv.action?pid=3610040201.

Watkins, M.H. 1963. "A Staple Theory of Economic Growth." *Canadian Journal of Economics and Political Science* 29 (2): 141–58.

Wilkinson, B.W. 1997. "Globalization of Canada's Resource Sector: An Innisian Perspective." In *Troubles in the Rainforest: British Columbia's Forest Economy in Transition,* ed. T.J. Barnes and R. Hayter, 131–47. Canadian Western Geographical Series, vol. 33. Victoria: Western Geographical Press.

Part 2
The Economic Geography
of British Columbia

"Discovering" Indigenous Lands and Shaping a Colonial Landscape

5

There were many names – Indigenous names – for the region that eventually came to be known as British Columbia. For example, much of the Lower Mainland was known as S'ólh Téméxw in Stó:lō language (Carlson 2001, 6). The lands on the Pacific side of Canada were "discovered" by Europeans who had never been to this part of the world before but believed so-called discovery entitled them to ownership. But the land was not vacant. Indigenous Peoples thrived throughout British Columbia, and their ways of life, values, and relationship to the land were critical to the region's history. In this chapter, you'll learn about the complex processes, developments, and events that went into the construction of the place names, borders, and boundaries we now take for granted. You'll learn about the historical developments and events that led to the exploration and colonization of the region, the creation of the British colonies of Vancouver Island and British Columbia, their eventual entry into Canadian Confederation as the province of British Columbia, and how these developments in turn affected First Nations. You will also learn how the values and beliefs of colonizers led to racism and discrimination.

To the Europeans who believed they were entitled to plunder other lands (along with their resources and peoples), colonialism may have been fuelled by idealism. But it involved disruption, displacement, dependency, and, in some instances, slavery, and it often led to death for Indigenous Peoples. In British Columbia, Indigenous Peoples and natural resources were spared the colonial assault until the latter 1700s simply because geography placed them at the end of a seafaring world. However, before then, First Nations suffered the deadly results of smallpox, which arrived from the south before the arrival of the Spanish and the British by ship in the mid-1770s. Nor would were they spared subsequent epidemics in the 1800s.

The accidental discovery of sea otter pelts by the British in the 1770s led to incursions by fur traders and conflicting territorial claims over the region. It didn't take long, however, for sea otters to become "commercially extinct," the first documented tragedy of the commons in the region. Another fur trade quickly took its place. Within British Columbia, the Hudson's Bay Company (HBC) competed fiercely with the North West Company (NWC) for territory and furs, inflicting forts, flags, and new rules on the land. Conflict with the Americans also led to the creation of a border between British North America and the United States at the forty-ninth parallel.

The overland fur trade began to wane by the time coal was discovered on Vancouver Island, which became an official British colony, but under the authority of the HBC. The non-Indigenous population grew slowly until the discovery of gold. It was the Fraser River–Cariboo Gold Rush that opened up British Columbia to development, shaped its boundaries, and allowed it to become influential in Confederation.

THE LAND BEFORE COLONIZATION

The generally accepted theory of North American settlement is that the first humans came from Siberia some 11,000 to 12,000 years ago, when glaciers from the last ice age were receding. The last ice age was at its peak about 20,000 years ago, when vast sheets of ice covered polar regions of the world, including most of Canada and northern portions of the United States (see Chapter 2). These ice sheets gained their moisture from oceans; as a result, the "sea level was about 120 m lower than it is now, so that a land bridge existed between Siberia and Alaska" (Aguado and Burt 2001, 441). These physical conditions were responsible for a relatively flat plain that extended from Siberia to the Mackenzie delta, known as Beringia. The present-day Bering Strait became the land bridge, and Beringia, because of its dry climate, was free of ice. Archaeologists suggest that early Indigenous groups followed the herds of animals from northern Asia across to North America and travelled in one of two directions: a coastal migration and an inland migration via an ice-free corridor. The evidence of a coastal migration is much more difficult to obtain as the historical coastline is now covered by 120 metres of water.

In most of Canada, the glaciers receded to their present location in the alpine Cordilleran region and the high Arctic some seven thousand to six thousand years ago. Consequently, the boreal forest migrated even farther north than its present location, salmon swam in the rivers flowing to the Pacific, and other plants and animals established habitats throughout the province. For the most part, the original Indigenous Peoples were nomadic hunting-gathering societies that had to adapt to the many changes. They were, however, in sufficient numbers and possessed hunting technologies that resulted in the extinction of a number of species, including mastodon, several birds, and giant beaver (Wynn 2007, 22–23).

As the ice melted and receded, the land stabilized, and these groups became seminomadic. For example, coastal peoples settled adjacent to shellfish bays or beside rivers when salmon migrated upstream; at other times of the year, some would move upland, where berries and other goods were harvested. They used the same sites for hundreds and sometimes thousands of years. Claiming a territory allowed a nation to gather intimate knowledge of the region and to develop tools and technologies suited to the region. On the other hand, claiming a territory also meant they needed to defend it. Wars did occur, and boundaries were remoulded through these conflicts. But nations also shared their territories and resources through trade and other agreements, and the inclusion of outside nations made defining boundaries even more complex. The population and density of Indigenous peoples varied with geographic location. The highest and most densely populated regions were along the coast and along salmon-bearing streams in the interior. Population density was considerably lower in the rest of the interior.

Indigenous societies were based on oral tradition; history, legends, stories, and wisdom were held by the Elders and passed on to the next generation. Stories, dances, and songs were important possessions in defining the resources of the land and the territorial limits of the land base. These communally based societies functioned through an elaborate division of labour in which each individual had a role in sustaining the group.

First Nations had complex social, spiritual, and political organizations. The potlatch was an extremely important political institution practised almost exclusively by coastal First Nations. It was a form of government that organized and legitimized the decisions each nation made: "Potlatch government business is conducted at feasts. The house invites members of other clans to its feasts to witness the decisions taken, and to receive gifts that reinforce cooperation between the various houses. Potlatch feasts are held to commemorate deaths, to pay debts, to resolve disputes, and to confirm positions within the house leadership. They also provide a forum for discussing and resolving community problems. Potlatch

feasts are organized according to strict procedures as to invitations, seating, speaking, gift giving, payments, dress and conduct" (Brown 1992, 4).

Indigenous spiritual beliefs, specifically animism, were intimately tied to the land, as was the social and spatial organization of the various peoples. Territorial boundaries based on language group were further defined in terms of ownership by individual houses or clans. British Columbia, and particularly coastal British Columbia, had far more linguistic groups and clans, living in much higher densities, than anywhere else in Canada. Figure 5.1 shows the Traditional Territories of First Nations throughout British Columbia. There are 203 First Nations with 34 unique languages in the province (First Peoples' Cultural Council 2014).

For these First Nation societies that made their living through hunting, fishing, gathering, and trading, land was their culture. However, their ways of life, their policies, and their organization of land use could not prepare them for European colonization.

EUROPEAN COLONIALISM AND FIRST CONTACTS

European colonialism of lands and peoples around the world depended on the ability to sail into the wind, the dry compass, weapons such as cannons and flintlock rifles, and Christianity, which put God on the side of colonizers (as far as the colonizers were concerned). Wealth drove colonialism, and the main "treasure" was gaining access to the riches of the East. Chinese silk, tea, and porcelain. Indian pepper and cloth. Ceylon cinnamon and pearls. Indonesian nutmeg, cloves, and mace. All of these treasures were well known in Europe.

By the end of the 1400s, the Portuguese had sailed south and rounded the Cape of Good Hope (South Africa). They used their powerful cannons to suppress the Arab traders of India and gain control of the spice trade. In the early 1500s, one Portuguese sailor, Ferdinand Magellan, sailed in the opposite direction – southeast to South America, through the Magellan Strait, and on to the Moluccas (islands in Indonesia). In the era of sailing vessels, these two routes – either around South Africa or around South America – were the only two ways to gain access to the riches of the East.

Coastal British Columbia was the farthest destination from these two routes. To Europeans, it was the end of

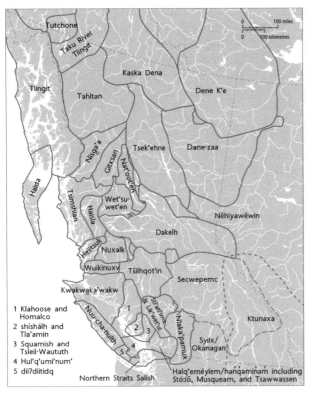

Figure 5.1 Traditional Territories of First Nations in British Columbia
Sources: Muckle (1998), First Peoples' Culture Council (2019).

the world. But the European quest for colonies and resources eventually resulted in other "riches" being plundered on the east coast of North and South America as well as the coastal regions of much of the rest of the world. As colonizers explored North America for potential sources of wealth, the idea of another passage to Asia became part of popular mythology. A great many attempts were made to discover the Northwest Passage, with no successes and the loss of a great many ships and lives.

To the north of British Columbia, in what would become Alaska, the Russians began to extract furs in the 1740s. The sea otter was highly valued in China, especially after Steller's sea cow (a marine mammal that was already an endangered species and resided only in Alaska) became extinct because of the early fur trade. The Russian fur trade had also devastated Indigenous Peoples in the region. For instance, within fifty years after first contact, the Unangax̂ population was reduced by about 80 percent, to about 2,500 people, as warfare with Russians, forced labour, and the introduction of diseases decimated the people and their culture (Veltre n.d.).

Claiming foreign territory and holding that territory often resulted in conflict and wars among the colonizers. In some instances, they engaged in negotiations to delineate territorial boundaries. The Spanish, who laid claim to much of South and Central America as well as Mexico, believed that they also had control of the west coast of North America. They were not pleased with Russian claims to Alaska and in 1774 sent explorer Juan José Pérez Hernández to spy on Russian traders and take possession of the land as far north as the sixty-first parallel (Center for the Study of the Pacific Northwest n.d.). On the journey, Pérez made contact with Haida and Nuu-chah-nulth, including Chief Maquinna (Mowachaht) at Nootka Sound on the west side of Vancouver Island.

Another navigational event – the arrival of Captain James Cook in 1778 – made the international situation even more complex. Cook represented British interests, and his third voyage had been publicized as a journey to return an Indigenous man who had been picked up from one of the Islands of Tahiti during Cook's second voyage and then put on display as a "novelty" in England. The man was returned to his homeland, but Cook had another more surreptitious quest: to discover the Northwest Passage from the Pacific side of North America. The discovery of this elusive route would come with a prize of £20,000.

Cook arrived in North America in 1778 from Hawaii at approximately the forty-third parallel (present-day Oregon State) and named the place "Cape Foul Weather." He then turned north and landed at Nootka Sound (Vancouver Island), where he stayed a month at Resolution Bay, on what would be named Bligh Island, to repair his ships, stock up on water, and trade with Chief Maquinna (Bligh was a surveyor and navigator with Cook who later became famous for a mutiny). These events were the beginning of European place-name geography in British Columbia.

For Cook, this was just one more meeting with Indigenous Peoples, although he and other Europeans were curious about "foreign" people and lands. The artist he brought with him sketched local landscapes, including Indigenous settings. But the relationships that developed were unequal. Europeans arrived with superior weapons and a view that the new land and its resources and people belonged to them by right of their being the first Europeans to arrive. The colonizers also came with a rigid set of values based on class, status, and religion. A strong belief in the superiority of Christianity and Western culture caused them to judge Indigenous Peoples as "heathens" or "pagans." Because the British believed in trade, though, which often required animal harvesting by Indigenous Peoples, they avoided conflict.

From Nootka Sound, Cook sailed north in search of the Northwest Passage. He sailed through the Bering Strait but did not collect the £20,000. A wall of ice forced his return to the southern shores of Alaska, where he ran into Russian traders, who informed him of the value of sea otter furs in China. Eventually, Cook returned to Hawaii, where he was killed by Indigenous Peoples. But his ships continued on to Canton, China, where the value of sea otter pelts was realized. His journal, which stated that he was the first European in the area, made its way back to England. Many more vessels would set sail for the Pacific Northwest to engage in the maritime sea otter trade.

By the 1780s, the region that would become British Columbia generally and Nootka Sound on Vancouver Island in particular were about to become embroiled in a European quest for sea otter pelts and the next wave of global colonial claims over land and resources. British, Spanish, Russian, and American fur traders were in the process of immersing themselves into a sophisticated, complex, and fairly densely populated Indigenous landscape.

CONFLICTING TERRITORIAL CLAIMS IN THE MARITIME FUR TRADE

The pursuit of sea otters was complicated because the furs were intended for China, not Europe. The Chinese had not been particularly interested in European goods but were forced at gun-point, initially by Britain, to trade. China established rules for the trade: trade could only take place in Canton, and Chinese goods could only be traded for silver or gold. European nations, however, were reluctant to part with their gold or silver and were constantly on the lookout for other items that the Chinese would take in exchange – sea otter pelts became one of those commodities.

Following Captain Cook's third voyage, Britain made a more formal claim to territory in the Pacific Northwest through Captain James Strange, who sailed to Nootka Sound in 1786: "He hoisted the colours and turned the soil and pronounced the name of His Britannic Majesty" (Bowering 1996, 51). Three years later, in 1789, John Meares established a trading fort in Nootka Sound; however, that same year, Spanish war ships arrived to build a fort, Fort San Miguel, at the entrance to Nootka Sound and to claim the region for Spain. The Nootka Incident of 1789 nearly escalated into war when conflicting claims caused the Spanish to seize four British ships and take two to San Blas, in Mexico.

In an attempt to resolve the conflict, the first Nootka Convention occurred in 1790. Britain, with its military power, prevailed, and Spain agreed that the two countries could share the territory for navigation and trading. The argument presented by the British was that a Spanish ship might have been at Nootka Sound before James Cook arrived, but the Spanish had not come ashore, as both Cook and Captain Strange had. The argument set a new precedent in colonial claims of entitlement.

A second round of negotiations occurred in 1792. This time, Captain George Vancouver represented the British and Captain Bodega y Quadra Spain. They did not find agreement as to a dividing line for the territory. Both Spain and Britain completed extensive surveys of the coast that resulted in the addition of many more Spanish and British place names, many of which are still in place to this day. During the survey, Captain Vancouver noted another discovery: "The skull, limbs, ribs and backbones, or some other vestiges of the human body, were found in many places, promiscuously scattered about the beach in great numbers" (Hopper 2017). Although he made this observation in what is now Washington State, there is evidence that smallpox had arrived in the Pacific Northwest. Between 1769 and 1780, it had spread from Mexico overland to the Prairies and then over the Rockies to the west coast via the Columbia River system and on up the coast (Boyd 1994).

During the third round of negotiations in 1794, both the British and Spanish agreed to abandon Nootka Sound. Although the Spanish left the area the following year, the British, with their naval power, took control of the Pacific Northwest. The region became a British possession, but no formal boundaries were created.

Hostilities in British Columbia were not confined to the British and Spanish. Contact between First Nations and European and American fur traders resulted in numerous conflicts, particularly because of the lack of respect shown for Indigenous Peoples and aggressive trading practices – particularly seizing furs by force and the random shooting of First Nations. Accounts by American, Spanish, British, and Russian fur traders include stories of thievery by First Nations and often harsh retaliation by traders, who in many cases shot them. In some instances, ships were attacked by First Nations, as was the case when the Haida sunk the sea otter–trading vessels *Ino* and *Resolution* in 1794 and when Chief Maquinna had the *Boston* sunk in Nootka Sound in 1803 and the *Tonquin* in Clayoquot Sound in 1811. However, overall, far more First Nations suffered loss of life than did foreign fur traders.

One of the most detailed accounts of Chief Maquinna and of Mowachaht life in Nootka Sound comes from John R. Jewitt, an armourer who wrote about being held captive by Maquinna, whom he referred to as a king. Jewitt was on the ship *Boston* in 1803 when Maquinna, who had been badly insulted by Captain Salter, showed up, pretending to trade. All were killed except for Jewitt, who could build and fix metal implements such as guns and was thus viewed as useful by Maquinna. A second man, Thompson, who hid during the skirmish, was later found. Jewitt pleaded for his life to be spared, claiming that Thompson (who was older) was his father.

Jewitt and Thompson were held for two years until the *Lydia* arrived in 1805 and took Maquinna hostage to negotiate for the release of the two men, which caused even greater humiliation for Maquinna. Eventually, Jewitt made it back to Boston, where he wrote a book titled *A Narrative of the Adventures and Sufferings of John R. Jewitt: Only Survivor of the Crew of the Ship Boston, during a Captivity of Nearly 3 Years among the Savages of Nootka Sound: With an Account of the Manners, Mode of Living, and Religious Opinions of the Natives.* During his stay with Maquinna, Jewitt had accompanied the Mowachaht on their seasonal rounds, learned the language, had a wife and child, and recorded aspects of their way of life, such as the importance, and abundance, of salmon. Chief Maquinna told him of the many injustices done to him and his people by the European traders – taking furs by

force, killing his people for no reason, and displacing their summer homes to build a fort.

At the time, massive numbers of sea otter pelts were flooding the Canton market, depressing prices and leading to boom-and-bust cycles. The final bust occurred in the early 1800s with the **commercial extinction** of sea otters. Maquinna and many other coastal First Nations involved in the sea otter trade suffered. For example, at first Maquinna had become "rich" by trading sea otter pelts for Western goods, which he then doled out through the Potlatch system of redistribution. By the mid-1790s, however, traders had exterminated most of the sea otters in the region. With few boats arriving, Maquinna's status as a provider of luxury goods diminished, which likely contributed to his attack on the *Boston*.

Although the maritime sea otter trade waned among coastal First Nations, it had only begun for First Nations in the interior.

THE OVERLAND FUR TRADE

In British North America, the overland fur trade was considerably older than the Pacific maritime fur trade. It had been sparked in the late 1500s by an increased demand for beaver fur for felt hats. In fact, all furs from North America were in demand. The fur trade took off in the 1600s among the French, who were located in the St. Lawrence River region and in the Maritimes, and among the British, who were located in the Thirteen Colonies to the south. Conflict was inevitable, because the British wanted a greater share of the valuable trade. A series of wars between France and Britain resulted in the Maritimes being taken over by the British and in the British awarding Rupert's Land to the Hudson's Bay Company in 1667. Rupert's Land was a vast territory defined by the river basins that flowed into Hudson and James Bays. Wars continued, and the French were eventually forced out of North America in 1763.

Precedence is a central concept in British civil society, and the conquest of New France set an important precedent that would have major implications for all of Canada, including British Columbia. With the French removed from the North American landscape, the British held a monopoly on the fur trade. First Nations who recognized that this monopoly was not in their best interests launched Pontiac's War. The British, tired of war, established the

Royal Proclamation of 1763. It recognized that Indigenous Peoples had been in North America first, that they had forms of self-government that should continue, that they should not be abused, and, most important, that they had Indigenous Title (a right to their Traditional Lands). The British also recognized that, over time, this right would have to be extinguished (ceded or surrendered), and the means to do so was through a treaty with the Crown.

Another historical event that had far-reaching implications was the American Revolution, which saw the British forced out of the Thirteen Colonies and the establishment of the United States in 1783. Following the Conquest and the American Revolution, the area that would eventually become Canada gradually became British North America, and it was shaped by the quest for furs. When furs were depleted in eastern and central British North America, trappers and hunters moved west and north. The HBC controlled most of this trade until 1783, when stiff competition came with the amalgamation of several smaller fur trade companies into the North West Company (NWC), centred in Montreal.

The NWC was much more aggressive than the HBC. It established a string of fur trade forts across the Prairies (through Rupert's Land), thus intercepting furs destined for HBC forts located on Hudson Bay. The rivalry became intense as the NWC sent company employees farther and farther west in search of new fur sources in lands previously unexplored by Europeans. During his first voyage for the NWC, in 1789, the Scottish explorer Alexander Mackenzie followed the river that would later take his name all the way to the Arctic. However, since he had wanted to get to the Pacific, he named the river the "River of Disappointment."

His second trip was more successful. In 1792, he set out and crossed the Rockies via the Peace River then headed south to the Fraser River, where he branched out west overland and down the Bella Coola River to the Pacific, which he reached in 1793. It was the first overland journey from sea to sea by a European (see Figure 5.2). Mackenzie was stopped, however, by the Heiltsuk First Nation (from Bella Bella) from reaching the open Pacific. The Heiltsuk were somewhat hostile to Europeans because they had experienced conflict with Captain George Vancouver only six weeks earlier. The notation

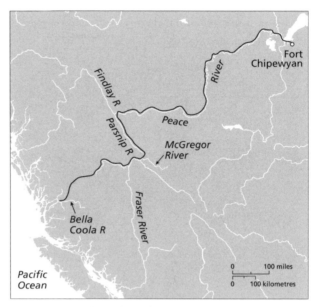

Figure 5.2 Alexander Mackenzie's route to the Pacific

that Mackenzie made in his journal betrayed deep-rooted colonial attitudes of ownership over resources and land: "I gravely added that the salmon, which was not only their favourite food but absolutely essential to their existence, came from the sea which belonged to us white men; and that as, at the entrance to the river, we could prevent those fish from coming up it, we possessed the power to starve them and their children" (Swenerton 1993).

Other explorers supported by the NWC followed, including Simon Fraser, who travelled the Fraser River in 1808. As he got closer to the mouth of the river, he noted in his journal that he had witnessed the devastating consequences of smallpox, including whole villages that had succumbed to the disease (Hume 2008, 233). David Thompson, who was active at the same time as Fraser, surveyed and mapped the Columbia River system by 1811. Although these men worked for the NWC, the company was not alone in exploring the region. The Lewis and Clark Expedition (1804–06) ended in present-day Oregon and Washington State, and when Thompson finally made it to the mouth of the Columbia River in 1811, he discovered that the American Pacific Fur Company had recently erected Fort Astoria, establishing an American presence in the fur trade.

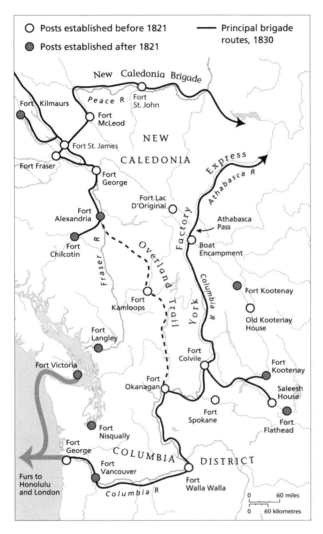

Figure 5.3 Fur trade forts and routes, 1805–46
Source: Modified from Harris (1997, 37).

and isolated environments. Although they carried the British flag, the forts in reality symbolized a corporate landscape where profit from furs was the sole justification for their existence.

A number of adverse conditions set British Columbia's fur trade apart from that of eastern and central Canada. Although the birch bark canoe was a mainstay in transporting people and furs, some regions in British Columbia had no birch, so the vessels couldn't be made or repaired. Because turbulent river systems made navigation impossible, overland trails were constructed, and pack horses were used to carry goods. Forts became small ranches (see Figure 5.3).

Reliance on First Nations to gather furs became another issue, largely because of the sheer number of First Nations in the region. When learning multiple languages and negotiating with hundreds of Hereditary Chiefs became a burden, the traders devised the fur brigade. These highly mobile, well-armed units were made up of many ethnic groups – English, Scots, Irish, Hawaiians (Kanakas), Iroquois, and Métis (all of whom frequently viewed one another with suspicion) – along with the various First Nations in whose territory they were trespassing. From the fur trade company's perspective, First Nations were largely an obstacle; they did, however, continue to trade with them. For First Nations, the fur trade was the beginning of assimilation.

The fur trade landscape in British Columbia extended beyond the trade relationship between First Nations and traders. As the number of forts increased in the early 1800s and became more permanent, the forts expanded to include vegetable gardens and farm animals. The forts also represented Britain's claim to the territory: the head of the forts, the factors, mainly of British extraction, represented "law and order" and often employed violence to create a "safe" trading landscape, even as smallpox and other diseases remained an overriding concern for First Nations. Adding to the complexity was the fact that traders, when they were in the territory, were encouraged to take Indigenous wives, and their wives and children, who were Métis, became a part of the fur trade landscape. Missionaries frequently established missions adjacent to forts in their fervour to convert First Nations and Métis to Christianity.

The overland fur trade differed fundamentally from the maritime trade. While the maritime trade was transitory, the land-based trade required the permanent presence of colonial interests – in other words, forts. Furthermore, the overland fur trade also had its own set of rules and relationships that had been established over two hundred years in eastern Canada. The forts served as fortified establishments (*entrepôts*) in often violent, hostile,

THE FORTY-NINTH PARALLEL, COAL, AND PERMANENT SETTLEMENT

In the early fur trade years, the British had little interest in actually settling the territory. But by the mid-1830s the discovery of a new resource had altered this position. By then, British territory in the Pacific Northwest included Vancouver Island, Haida Gwaii, and the mainland. But none of these territories had been proclaimed an official British colony. Moreover, the British claim to the mainland was not well defined. The War of 1812 between the United States and Britain had resulted, in 1818, in the creation of the forty-ninth parallel as the boundary between British and American territory west of Lake of the Woods in western Ontario. However, the line extended only as far as the Rockies.

Beyond the Rockies was fur-trading territory, claimed by both the British and Americans through the Anglo-American Convention of 1818 but governed mainly by the Hudson's Bay Company, which merged with the NWC in 1821 and had an exclusive licence to trade. The British referred to the northern portion of this territory (extending to latitude 54°40') as **New Caledonia** and the southern portion as the Columbia District, but the whole territory was also referred to as Oregon Territory, especially by the Americans. In 1824, the Russians claimed the Alaskan Panhandle down to latitude 54°40'.

The discovery of coal at the north end of Vancouver Island, albeit small and tentative, led to the beginning of permanent settlement in the region. The discovery was made at the mouth of Suquash Creek in 1835, and the HBC erected Fort Rupert nearby to house a group of coal miners. A shaft was sunk, and coal was mined intermittently, but there were many problems: the coal seams ran under the Queen Charlotte Strait; the quality varied, as did the width of the seams; and Kwakwa ka'wakw were hostile to trespasses on their land.

More serious settlement, and even a territorial challenge over Oregon Territory, occurred south of the border. Although both the British and the Americans viewed the region as a fur-trading area, furs and fur prices were on the decline after the 1830s. Silk hats were replacing fur and felt, and the territory was becoming over-trapped. And there was new interest in Oregon Territory, which was being opened to agricultural settlement, particularly in the Willamette Valley of present-day Oregon State.

The concept of **manifest destiny** had gained popularity in the United States by the 1840s after an essay by the same name proclaimed that the nation's westward expansion was a God-given right. For some Americans, it meant entitlement to the whole of North America. Essentially, manifest destiny was the belief that God wanted the United States to spread from east to west and then north and south to take over much of North America. The showdown for Oregon Territory was ignited during the 1844 presidential election with the slogan "54-40 or Fight." The Americans wanted the whole of Oregon Territory. War did not occur, but negotiations did, and the Oregon Treaty of 1846 extended the forty-ninth parallel from the east to form the border we know today. The exception was Vancouver Island, which dipped below that latitude. British Columbia's borders were becoming defined, although it was still referred to as New Caledonia (see Figure 5.4).

In 1844, during the negotiations, Robert Greenhow, an American geographer, had written a report for the US government in which he argued that the Oregon Territory would be marginal to settlement and of little value in terms of resources:

> It produces no precious metals, no opium, no cotton, no rice, no sugar no coffee; nor is it, like India, inhabited by a numerous population, who may easily be forced to labor for the benefit of the few. With regard to commerce, it offers no great advantages, present or immediately prospective. It contains no harbor in which articles of merchandise from other countries will probably at any future period be deposited for re-exportation; while the extreme irregularity of its surface, and the obstruction to the navigation of its rivers, the removal of which is hopeless, forbid all expectation that the productions of China, or any other country bordering on the Pacific, will ever be transported across Oregon to the Atlantic regions at the continent. (Greenhow 1844, 399)

Clearly, Greenhow saw only a rugged land not likely to attract a great deal of colonization from an agricultural

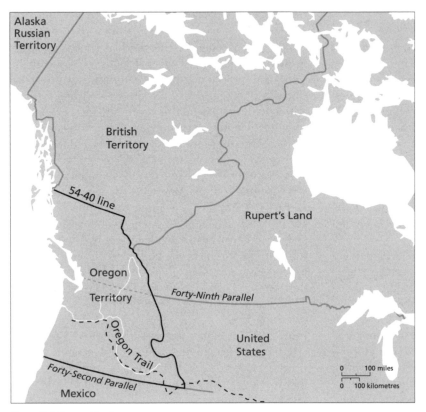

Figure 5.4 Oregon Territory

perspective. His report exposed entrenched colonial values with respect to "foreign" lands and Indigenous Peoples. Small wonder the American government settled for the continuation of the forty-ninth parallel.

As a result of the Oregon boundary dispute, Fort Vancouver, the Hudson's Bay Company's main administrative fort at the mouth of the Columbia River, was abandoned and turned over to the Americans. In 1843, the new British centre was located on the south end of Vancouver Island and named Fort Camosun, which was then changed to "Fort Albert" and, finally, to "Fort Victoria" in 1846.

The creation of a fixed international southern boundary and permanent settlements caused problems for the HBC. Permanent British settlement, including farms (albeit small) on Indigenous Lands, was beginning to occur. Conflict with the United States had led the British to set up a naval base at Esquimalt, near Victoria, in 1846. Many First Nations were not pleased with more

permanent settlement on their territories. For example, they attacked Fort Camosun in 1844 just as it was being built. The new boundary had also failed to take into consideration that many First Nations territories crossed the forty-ninth parallel.

The new border also disrupted the HBC's fur trade route from the southern interior on the Columbia River to Fort Vancouver. Negotiations allowed the route to continue, but the Americans imposed taxes on all goods coming into or leaving Fort Vancouver. The HBC formed a new route in 1847. It followed the Thompson River then extended overland to the newly created Fort Hope, where the Fraser River could then be navigated.

None of these developments encouraged major settlements, but everything changed with the HBC's discovery, in 1849, of coal on Vancouver Island at Nanaimo (called Coleville until 1860), which tipped the balance in terms of British settlement. The British Colonial Office, recognizing that permanent settlement was inevitable, designated Vancouver Island a Crown Colony in 1849. However, the new colony did not come under the direct control of the British; rather, it was handed over to the HBC, which held a ten-year grant. Granting a lease was expedient to the Colonial Office, and it was considerably less expensive to have a private company (with plenty of experience dealing with Indigenous Peoples) manage the new colony. However, the Colonial Office retained the right to appoint the governor, Richard Blanchard. And it also stipulated that the HBC had to promote settlement. Colonizing the land close to Victoria was in the company's interest since it owned the land and surveyed and subdivided properties for sale. Moreover, the HBC also had a monopoly on the import, export, and sale of all consumer goods. Finally, it owned the coal deposits at Fort Rupert (although the Suquash Mine was abandoned in 1851) and Nanaimo.

The production of coal on the island brought it into the new industrial era occurring in Europe, the United States, and parts of British North America. These regions were embracing the technology changes that came with the Industrial Revolution. The steam engine altered and improved the production of goods by centralizing plants, equipment, and labour; it also drove trains along steel tracks and powered vessels over the ocean, thus shrinking time and space and improving the standard of living – for some.

Few of these industrial, mechanized conditions applied to British Columbia in the late 1840s, but some did, primarily because of another unpredictable event – the California Gold Rush, which began in 1848. Over 100,000 gold seekers swarmed into the Bay area, creating a demand for goods, particularly coal, which was used to fuel steam-driven paddlewheelers or steamers, which carried people destined for the goldfields and reduced the time required to move people, mail, and other goods. Steamers left New York and docked at the Isthmus of Panama, where the journey between the Atlantic and Pacific (approximately 97 kilometres) was then undertaken by horse and canoe until 1855, when a railway was built. Steamers then continued on to San Francisco. These innovations reduced the trip to less than one month (plus another four days to Victoria).

In the new colony of Vancouver Island, Governor Blanchard, disappointed in the HBC's extensive control, resigned in 1851. James Douglas was then appointed governor. Douglas was chief factor of the HBC from 1849 to 1858 and governor from 1851 to 1864. Well aware of the conflict that European settlement could elicit with local First Nations, who were by far in the majority, he engaged in treaty making. Between 1850 and 1854, fourteen small treaties were signed on Vancouver Island in and around Victoria, southwest to Sooke, north of Victoria on the Saanich Peninsula, and in the coal-mining regions of Fort Rupert and Nanaimo. First Nations received a small amount of money, HBC blankets, and small reserves, where they could live. But the **Douglas Treaties** also confirmed that they could continue to hunt, fish, and gather on unoccupied lands. The transition from a fur trade landscape to one with settlers, communities, coal mining, and some logging and farming was a profound change.

Despite the treaties, First Nations on Vancouver Island remained concerned that their lands were being taken over. For example, in 1853, a Hudson's Bay Company man was killed in the Cowichan Valley (near Nanaimo) by First Nations in that region. Douglas assembled a significant military force to quell further conflicts and hanged two First Nations men charged with the murder. The Douglas Treaties went a ways towards resolving land-use conflicts, but, after 1854, Douglas was no longer able to negotiate treaties because of a lack of colonial funding. He did, however, continue to allocate reserves to First Nations, reserves that First Nations selected themselves.

Colonization and settlement followed a prescribed set of rules in terms of how the land was organized. The introduction of treaties and reserves, which segregated First Nations, who were not considered to be "using" the land, was the first step. Then British institutions of law and order, democracy, and education all needed to be established. One of the most fundamental changes, however, was that land became a **private property resource** to be bought, sold, and developed, whereas it had previously been held communally by First Nations. Enforcing British civil society involved imposing other values on the land – the superiority of the English language and Christianity and a prevailing belief in a hierarchical class and race system that rendered all who were not white and upper-class inferior. Entrepreneurial leadership was regarded as essential to the colony's economy and investment in resource development its engine. "Good" governments struck policies to encourage investment. "Good" government established policies to ensure that all conformed to this definition of civil society. For First Nations, the grip of colonialism tightened as settlement led to the imposition of assimilation.

But attracting new settlers from Britain to remote, sparsely inhabited Vancouver Island was not easy. The journey from Britain took six months, and the land was not cheap – one pound per acre, with a minimum purchase of twenty acres (Bowering 1996, 102). The west coast of the United States held much greater appeal because of the lure of the California Gold Rush, and free homesteads of 640 acres were available in Oregon. Still, some came, and the non-Indigenous population increased on Vancouver Island. The 1855 census (actually conducted in 1854) indicated that there were 774 white settlers in the whole region; Victoria had 232 and Nanaimo, 151.

THE GOLD RUSH AND NEW TENSIONS

Unlike furs, fish, and forests, minerals are a nonrenewable resource destined to run out, leading to boom-and-bust cycles and ghost towns. But there was a vast difference between coal and gold. Coal mining in the 1800s depended on the discovery of broad coal seams with quality coal (the higher the carbon content, the more efficient the burn). A significant capital outlay was then required to invest in equipment to sink shafts, hire labour, supply housing for workers, and transport the coal to a port facility. The miners were hired by corporations that distributed most of the profits to owners, and they worked in high-risk underground coal seams where they were in danger of suffering methane explosions and cave-ins.

The California Gold Rush sparked a gold rush throughout the Pacific Northwest. It brought 100,000 gold seekers to San Francisco within a year, and the population continued to escalate. The gold seekers engaged in **placer mining**. Gold is one of the few elements that can come in its pure form – gold nuggets and fine dust. It is relatively heavy and can accumulate in river and stream beds, and even small quantities are valuable. As a consequence, gold fills the imagination of those who hope to become rich. Miners need only stake a claim and invest in a shovel and gold pan. With considerable work, they can potentially find quantities of gold. In other words, there was no need for large corporations and vast capital outlays.

Small amounts of gold were discovered on Haida Gwaii in the early 1850s, and similar to the coal deposits, the HBC attempted to gain an exclusive right to it, but they were prevented by the Haida, who captured and ransomed a ship, the *Georgiana,* in 1851. In addition, a number of Americans wanted to take part in the gold finds. Britain sent gun boats to the region to secure its interests, and the Queen Charlottes (Haida Gwaii) were formerly made a colony in 1853, under Governor Douglas. However, little gold materialized.

At this time, in the mainland, the HBC was mainly engaged in trapping for furs and the barrelling of salmon. The region was only marginally involved in industrial resource-extraction activities. But by the mid-1850s, HBC forts in the central interior, such as the one at Kamloops, had begun to acquire gold from First Nations, who gathered it from the Fraser and Thompson Rivers. Douglas, however, did not want this discovery known, especially since New Caledonia was not officially a colony and there was fear that the United States would annex the region if it flooded with gold prospectors.

By 1857, the rumours of gold in the region had spread to Oregon and California, and the migration north began. In 1858, some 25,000 to 30,000 miners made their way into the lower reaches of the Fraser River. Many came from eastern Canada, Britain, other parts of Europe, and China, but many more miners – Americans, Chinese, and African American – came from California. War on the Fraser River seemed inevitable as armed Americans mining for gold began displacing First Nations and instilled fear that placer mining would destroy life-giving salmon runs. The introduction of alcohol did not improve the situation. There were plenty of skirmishes and murders, and the majority of the victims were First Nations. All-out war was avoided when Douglas became involved. The British took direct control of the colonies and cancelled the HBC's grant. Douglas was named governor of both Vancouver Island and the new mainland colony of British Columbia in 1858.

As thousands of gold seekers rushed to Victoria and then on to the interior, the region became less isolated, the landscape even more altered. Victoria, the main centre for incoming gold seekers, boomed overnight, initially as a large tent town but then as a full-fledged town with wood-frame hotels and shops. Miners stocked up on provisions before they chased their dreams up the Fraser River. Many other groups were attracted to Victoria, including the Chinese and several thousand First Nations. They included members of the local Songhees First Nation who had signed one of the Douglas Treaties, and they came from other parts of coastal British Columbia – Haida from Haida Gwaii, Tsimshian from the Skeena River region, Tahltan from the Stikine River, Tlingit from Atlin, Heiltsuk from Bella Bella, and Kwaguʼł from Fort Rupert. Each of these First Nations carved out their own camps in proximity to the town. Tragically, smallpox arrived on two steamers from San Francisco in 1862, leading to fear, panic, the inoculation of some First Nations and the quarantining of others, and forced relocation by gunboat of many northern First Nations (a good example of spatial diffusion). As the deadly disease made its way inland, First Nations of the interior were also affected.

Within two years, the pandemic had killed approximately 20,000 people – mostly First Nations. The Haida First Nation, which had an estimated population of 18,000 to 20,000 before European contact, had been reduced to only 6,000 by 1835 and 800 by 1885, reaching a low of 588 in 1915 (Duff 1965, 32). All First Nations experienced reductions in population, although not as severe as the Haida. First Nations were becoming a minority population in their own land, ravaged by diseases and the loss of land and resources, which severely altered their communal and seminomadic way of life.

With the creation of the colony of British Columbia, Douglas wanted the new capital to be Fort Langley (which was temporarily renamed Derby), but the sloping banks behind New Westminster were chosen instead in 1859, because the location offered better defence if the Americans attacked. The British became more serious in fortifying the region. The "hanging judge," Matthew Baillie Begbie, was appointed chief justice, and the Royal Engineers (also responsible for surveying) were to help establish law and order. Historical concerns about the effects of belief in US manifest destiny, the discovery of gold on the Queen Charlottes, and the massive influx of Americans up the Fraser had produced enormous concern about retaining British sovereignty.

Gold seekers made their way up the Fraser by paddle-wheelers that served new communities along the lower reaches of the Fraser: New Westminster, Fort Langley, Mission, Hope, and the last stop, Yale. Initially, small tributaries of the Fraser River such as Emory Creek near Yale yielded sufficient gold to result in a small community being hastily constructed, but these communities were soon abandoned as gold yields diminished. Some communities, such as Hope and Yale, became service centres. As the miners continued on into the Cariboo, steamers also operated on Harrison Lake and the upper reaches of the Fraser River. Many small communities sprang up along the Fraser only to become ghost towns. In the Cariboo, where some of the greatest yields of gold were

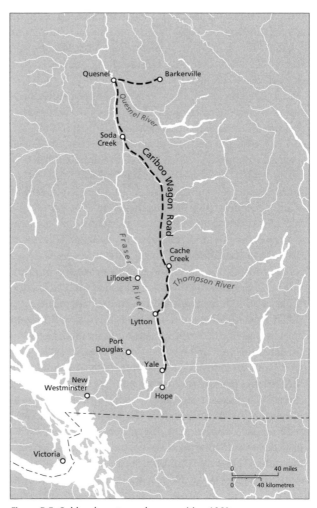

Figure 5.5 Gold rush routes and communities, 1860s
Source: Yardley (1998).

discovered by the early 1860s, the "large" town of Barkerville was erected (see Figure 5.5). The town contained approximately five thousand people, and a significant proportion of the population was Chinese men, who were segregated to the north end of the community.

As prospectors moved up the Fraser River, there was more demand for roads and trails to the interior. Initially, the favoured route was up Harrison Lake to Port Douglas, from which prospectors followed a series of roads and even a tramway to Lillooet (originally named Cayoosh).

North of Lillooet, a road was built to Clinton and then on up into the Cariboo. Lillooet became "Mile 0" of the route. It was a difficult and time-consuming route that required off-and-onloading multiple times. As a consequence, a second route was demanded.

In 1862, the Royal Engineers designed a new route that began at Yale and followed the Fraser River, then the Thompson River, before branching off north to the Cariboo. By 1865, the Cariboo Wagon Road (a toll road) had been completed from Yale to Barkerville, a distance of 380 miles (612 kilometres), which could be covered in four days by a regular stagecoach. Building the road was a major undertaking that depended on the contractual labour of Chinese and First Nations. Many First Nations contracted smallpox and died in 1863. Because the new route bypassed Lillooet, Lillooet's advantage as a service centre diminished. The new road also resulted in communities such as 100 Mile House, located on the Cariboo Road, no longer having a "mile zero" starting point.

Had it been completed, the Waddington Trail Project would have significantly reduced the distance between Victoria and the Cariboo. Prospectors would have travelled by boat from Victoria up the coast to the head of Bute Inlet, where a road, running adjacent to the Homathko River, would then take them up into the Chilcotin region and on to Barkerville. The project was initiated in 1862, but conflict with the Tŝilhqot'in First Nations resulted in the death of nineteen workers in 1864. In 1864, Alfred Waddington wrote about some of the causes of the "massacre," including the military insulting a Tŝilhqot'in man's son and the introduction of smallpox. From the Tŝilhqot'in perspective, this was not a reactive skirmish – it was a war. They were defending their land from white intruders – the insults, trespass, and smallpox were genuine cause. Following the war, believing they were going to attend peace talks, six Tŝilhqot'in Chiefs were hanged under false pretences.

Other gold-mining sites likewise attracted gold miners and more conflict. In 1859, gold was discovered at Rock Creek, not far from the US border in the West Kootenays. The majority of miners were American and did not pay the British miners' tax. They ran the gold commissioner from the region and had little use for Chinese in the area. Murders occurred, but conflict was settled with the arrival of Douglas and Begbie.

Douglas recognized that allocating reserves – some of significant size – to First Nations reduced tensions over the incursion of foreigners into their Traditional Lands. Unfortunately for First Nations, Douglas retired in 1864, and his successor, Governor Frederick Seymour, did not recognize a number of the new reserves that Douglas had surveyed. Moreover, the new colonial administration appointed Joseph Trutch as chief commissioner of lands and works, and he had a different attitude: "The Indians have really no right to the lands they claim, nor are they of any actual value or utility to them, and I cannot see why they should either retain these lands to the prejudice of the general interests of the Colony, or to be allowed to make a market of them either to the Government or to Individuals" (Cumming and Mickenberg 1981, 193). From Trutch's perspective, First Nations stood in the way of settlement, and he reduced the size of the large reserves created under Douglas. As large reserves were pared down, the colonial administration was adamant that reserve allocation should be limited to ten acres per family; if First Nations were farming or ranching, an allocation of twenty acres was justified. Discontent and bitterness by First Nations over these policies of discrimination resulted in a prolonged struggle for justice.

By the early 1860s, most claims in the Cariboo had been staked, and some adventurous miners moved beyond the northern borders of British Columbia (then defined at 54°40' north latitude) to seek gold in the Stikine River. Some gold was discovered, attracting a few hundred prospectors and Britain's interest. In 1862, Britain created Stikeen Territory – which had an arbitrary boundary that extended north to the sixty-second parallel and west to the 125th line of longitude. As soon as this boundary was in place, gold was discovered on the Peace River, but east of the 125th line of longitude. The British solution, in 1863, was to amalgamate the boundaries into the province's present configuration, which is north to the 60th line of latitude and east to the 120th line of longitude until it intersects with the divide in the Rockies (see Figure 5.6). Other gold discoveries within these boundaries, such as in the Kootenays, first at Rock Creek and then a Wild Horse Creek, resulted in other patterns of development, including the building of the Dewdney Trail, which linked these communities to Hope. In 1866, the mainland's new northern and eastern regions were united with the separate

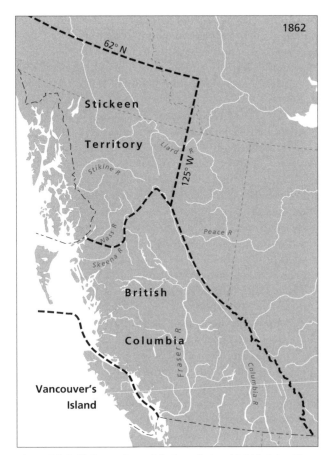

Figure 5.6 Gold discoveries and the boundaries of British Columbia
Source: Modified from Nicholson (1954).

colonies of Vancouver Island and the Queen Charlottes (Haida Gwaii). Victoria served as the capital.

A LAND TRANSFORMED

The gold rush and technological advances in mining transformed the land and the region's river and stream systems. Trees were logged in the vicinity of new communities to make way for roads and produce lumber for frame structures, pre-emptions (essentially free land for doing improvements) for farming and ranching was encouraged, and streams were diverted over sluices in the search for gold.

Sluices allowed miners to use water to wash and sort out gold from the gravel in which it was embedded. Over time, the sluices became larger and larger, creating demand for more and more water. Before long, whole streams were dammed and diverted. Miners also burrowed and tunnelled down into former stream beds to access placer deposits, and they used dredges anchored in rivers and hydraulic systems to wash down river banks and divert gravel over sluices. By these means, streams and rivers were turned inside out, and fish habitats were destroyed, especially salmon, a main source of food for First Nations. There were no policies in place to curb these activities.

A major food staple for miners was beef, and the growth of cattle ranches and cattle drives (often from the United States) through the Okanagan and Thompson

River Valleys and up to communities such as Barkerville brought other environmental changes. Cattle survived the long trek by eating the dominant vegetation, bunch grass, which grew naturally in these semiarid valleys. This grass was so abundant that the colonial government had created a land act in 1860 to make a quarter section of land (160 acres or 73 hectares) available for purchase to British subjects. Ranches soon dotted these lush valleys. Tragically, with no policies governing overgrazing, the bunch grass was depleted by the 1870s. Today, the dominant vegetation is sage brush, which cattle do not eat.

In addition to developments connected to mining, unpredictable events also fostered development and transformation of the land. In the 1860s, British Columbia was well placed geographically when the US Collins Overland Telegraph, which was backed financially by the Western Union Telegraph Company, was built to connect North America to Europe via Russia. By 1866, California (along with much of the United States and central and eastern Canada) had been linked to New Westminster, and the telegraph continued through British Columbia to Hazelton on the Skeena River. A branch line through the San Juan Islands connected Victoria. Although the project died abruptly because of the successful Atlantic cable connection in 1866, the surveying for it left names on maps, including Telegraph Creek on the Stikine River. As British Columbia considered joining Confederation, it was already part of a global communication system. J.S. Helmcken of Victoria put this development in perspective: "In 1850 one year was consumed in sending a message to England and receiving an answer. A few months ago a message was sent by telegraph to Manchester in England and an answer received in nineteen seconds, the distance travelled being 13,000 miles! Electricity knows neither time nor distance" (Helmcken 1895, 6).

The American Civil War, which raged from 1861 to 1865, also affected British North America, particularly when the South, which Britain supported, lost. There was little discussion of manifest destiny or annexation while the war was being waged, but when it ended there was interest in punishing the British for their interference and a renewed push for US territorial expansion. In this context, the Fenian Raids in the Maritimes and central British North America persuaded Britain to set the colonies on a path to becoming an independent nation.

Several other political events occurred in the mid-1860s that would shape the future direction of British Columbia and Canada. In 1866, the colonies of Vancouver Island and British Columbia united, with the capital remaining at Victoria. The following year, on 30 March, the Americans purchased Alaska for $7.2 million, re-establishing concerns about annexation of British Columbia and the Prairies. Canadian Confederation – the coming together of Ontario, Quebec, Nova Scotia, and New Brunswick into one federation – occurred on 1 July 1867, and by 1869 Rupert's Land had been purchased from the HBC.

By this time, BC's gold rush was waning, which meant the population of the colony was declining as disillusioned prospectors left the region. But some remained and attempted to turn what they perceived as a "wild" landscape into a tamed British one. At the same time, the colony's British administrators were concerned about mounting debt due to road and trail building, keeping the peace, and increasing administrative expenses.

British Columbia was no longer an isolated colony of the British Empire. To the south, the Americans completed the first transcontinental railway in 1869, thus reducing the New York to San Francisco trip to one week and renewing interest in taking control of the territory that stretched from California to Alaska. To the east, in Canada, there was conflict over westward expansion, especially in the Red River Valley area of Manitoba. The Riel Resistance of 1869 resulted in the recognition of the Métis, and the new province of Manitoba was added to the map of Canada in 1870.

There was considerable debate in British Columbia, with its declining economy, whether it should join the rest of Canada, join the United States, or become a separate nation-state. The BC delegation that went to Ottawa was persuaded by promises that Canada would absorb the colony's mounting debt along with the promise of a railway. Before the colony could join Confederation, however, a referendum was necessary, but not everyone was allowed to vote. London had already established voting rights for the colony by 1870: only males of at least twenty-one years of age who were British subjects and could read and write English were granted the franchise. By design, these qualifications excluded women, First Nations, Chinese, and Americans. With a Confederation referendum imminent and given concern about American annexation, further restrictions were placed on who could vote. The ruse worked, and British Columbia became the largest province (by size) in the Dominion. Sixty-two percent of the province's population was of First Nations or Chinese origin, while people of British origin accounted for 30 percent of residents (Elections Canada 2007).

During the negotiations, nobody mentioned the small parcels of land that had been set aside as reserves for First Nations or the lack of treaties in most of the region. Instead, the Crown's **fiduciary trust** to First Nations was emphasized – that is, it was agreed that the federal government had a legal obligation to protect the interests of First Nations. Many believed First Nations were heading for extinction. The federal government gave the colony's land and resources to the new provincial government without consulting First Nations. Unlike in the Numbered Treaties on the Prairies, the federal government did not uphold its fiduciary trust in British Columbia.

SUMMARY

During the colonial era, there were three distinct phases in the transition of the people, resources, and landscapes. Between the arrival of Captain Cook in 1778 and the discovery of coal in 1835, the maritime and overland fur trades led to the extinction of Steller's sea cow and the commercial extinction of the sea otter. In the process, Britain established a territorial claim to Vancouver Island, Haida Gwaii (the Queen Charlottes), and the mainland (New Caledonia). Fur trade forts were erected, but there was little intention to create permanent British settlements. Indigenous Peoples wanted European goods but often learned that the price was too high and included abuse, the loss of their Traditional Territories, and diseases or death.

The second phase opened with the discovery of coal in northern Vancouver Island and the establishment of permanent settlements. American annexation of the Oregon Territory prompted negotiations and settlement of the forty-ninth parallel in 1846. When richer deposits of coal were discovered in the Nanaimo area in 1849, Vancouver Island became a colony managed by the HBC.

Conflicts with First Nations erupted when the company encouraged permanent settlement and led to the Douglas Treaties. The discovery of gold on Haida Gwaii was the catalyst to establish the region as a colony in 1853.

The third phase began when gold was discovered in New Caledonia in 1858 and some thirty thousand gold seekers advanced up the Fraser River and into the Cariboo. The British government took direct control and turned New Caledonia into a colony named British Columbia. The gold rush created new tensions, including fear of American annexation, anti-Asian sentiment, conflicts between miners and First Nations, and concern over the effects of placer mining and ranching on salmon habitats and grasslands. As roads, trails, communities, and telegraphs started to dot the landscape and more gold discoveries were made in the north, which reshaped political boundaries, Britain created the amalgamated colony of British Columbia in 1866. As the gold rush waned and Alaska was purchased by the Americans, British Columbia joined Canadian Confederation in 1871 on the promise that the federal government would absorb its debts and build a transcontinental railway. First Nations were not part of Confederation. It was widely believed that they, too, would become extinct.

REFERENCES

Aguado, E., and J.E. Burt. 2001. *Understanding Weather and Climate.* 2nd ed. Upper Saddle River, NJ: Prentice Hall.

Bowering, G. 1996. *Bowering's BC: A Swashbuckling History.* Toronto: Penguin.

Boyd, R. 1994. "Smallpox in the Pacific Northwest: The First Epidemics." *BC Studies* 101: 5–40.

Brown, D. 1992. "Aboriginal Rights." Educational report for the Carrier-Sekani Tribal Council, Prince George.

Carlson, K.T., ed. 2001. *A Stó:lō Coast Salish Historical Atlas.* Vancouver: Douglas and McIntyre.

Center for the Study of the Pacific Northwest, University of Washington. n.d. "Indians and Europeans on the Northwest Coast: Historical Context." washington.edu/uwired/outreach/cspn/Website/Classroom%20Materials/Curriculum%20Packets/Indians%20&%20Europeans/II.html.

Cumming, P.A., and N.H. Mickenberg. 1981. "Native Rights in Canada: British Columbia." In *British Columbia: Historical*

Readings, ed. W.P. Ward and R.A.J. McDonald, 184–211. Vancouver: Douglas and McIntyre.

Duff, W. 1965. *The Indian History of British Columbia.* Victoria: Royal British Columbia Museum.

Elections Canada. 2007. "Modernization, 1920–1981." Chapter 3 in *A History of the Vote in Canada.* elections.ca/content.aspx?section=res&dir=his&document=chap3&lang=e.

First Peoples' Cultural Council. 2014. *Report on the Status of B.C. First Nations Languages 2014.* 2nd ed. Brentwood Bay, BC: First Peoples' Cultural Council. fpcc.ca/files/PDF/Language/FPCC-LanguageReport-141016-WEB.pdf.

First Peoples' Culture Council. 2019. "Language Map of British Columbia. maps.fpcc.ca/map.

Greenhow, R. 1844. *The History of Oregon and California and the Other Territories on the North-West Coast of North America.* London: John Murray.

Harris, R.C. 1997. *The Resettlement of British Columbia: Essays on Colonialism and Geographic Change.* Vancouver: UBC Press.

Helmcken, J.S. 1895. *Victoria Daily Colonist,* March 31.

Hopper, T. 2017. "Everyone Was Dead: When Europeans First Came to B.C., They Confronted the Aftermath of a Holocaust." *Vancouver Sun,* 21 February.

Hume, S. 2008. *Simon Fraser: In Search of Modern British Columbia.* Madeira Park: Harbour.

Muckle, R.J. 1998. *The First Nations of British Columbia.* Vancouver: UBC Press.

Nicholson, N.L. 1954. *Boundaries of Canada, Its Provinces and Territories.* Ottawa: Queen's Printer, Department of Mines and Technical Surveys.

Swenerton, D.M. 1993. *A History of Pacific Fisheries Policy.* Ottawa: Department of Fisheries and Oceans. dfo-mpo.gc.ca/Library/165966.pdf.

Veltre, D.W. n.d. "Unangax̂: Coastal People of Far South-Western Alaska." Aleutian and Pribilof Islands Association. apiai.org/culture-history/.

Wynn, G. 2007. *Canada and Arctic North America: An Environmental History.* Santa Barbara: ABC-CLIO.

Yardley, C. 1998. "The Cariboo Gold Rush Roadhouses." cariboogoldrush.com/wagonroa/wagon.htm.

Boom and Bust from Confederation to the Early 1900s

6

When British Columbia entered Confederation in 1871, colonial values, including racism and class privilege, did not go away. The new province now had the ability to develop provincial policies to exploit the land and resources for revenue, and the catalyst for its vision of growth and prosperity was the federal government's promise of a transcontinental railway. But the world was on the brink of experiencing a global recession, and the federal government failed to make good on its promise until the 1880s. In this chapter, you'll learn how the province's resource dependence led to boom-and-bust cycles, how governments used revenues and Crown lands to promote railway and resource developments, and how economic developments combined with colonial-inspired policies of exclusion and assimilation to reshape the spatial organization of the land and its people.

COLONIAL ATTITUDES, RESOURCE DEVELOPMENT, AND THE 1870S RECESSION

British Columbia was the largest province in Canada by area and rich in natural resources when it joined Confederation, but its population was small, composed of only 26,000 Indigenous People and 11,000 Europeans (Dunn and West 2011). The rules governing jurisdiction over people, the land, and its resources were spelled out in section 92 of the Constitution Act, 1867. The provincial government gained control of lands and resources, except for a swath twenty miles (32 kilometres) wide on each side of the proposed railway. In return, the province handed over 3.5 million acres (1.4 million hectares) known as the Peace River Block to the federal government, along with the telegraph lines (Seager 1996, 208). The Pacific Ocean, including the fish and the ports, remained under federal jurisdiction, as did First Nations and their reserves.

Colonial values and attitudes did not change when British Columbia entered Confederation. The new province's ladder of privilege continued to be rigid in its structure. At the top were those with an education such as doctors, lawyers, and judges, as well as politicians, ship captains, military leaders, successful business entrepreneurs, and individuals with wealth and family connections to British royalty. Teachers, nurses, small business owners, ranchers, and large farm owners were perhaps a

rung lower but fit in well with so-called civil society. And then there were white labourers who worked in the mines and mills, on commercial fishing vessels, or in service industries in communities such as Victoria. Considerably lower down the social ladder were non-British subjects – First Nations, Chinese, and all those not white – who were paid lower wages than white workers. Where one fit into the social hierarchy was not absolutely static. Being convicted of corruption, bribery, or some other crime could lower your standing in society, while gaining wealth through, for example, the discovery of gold would allow you to "jump the social cue."

The "rules" as to who was acceptable shaped everything from the province's new voting rules to residential patterns. In both provincial and federal elections, only British male subjects who were twenty-one years of age and who had resided in the province for twelve months could vote. These rules, of course, disenfranchised women, First Nations, Chinese immigrants, recent British arrivals, and those who were not British subjects. **Residential segregation** created another spatial pattern of social acceptance. As towns and cities emerged, the wealthiest lived in the largest houses, apart from those with considerably less wealth and influence. Chinese were confined to their own enclaves in Chinatowns, whereas First Nations were placed on reserves or, if they moved to centres such as Victoria, separate camps.

As in the colonial period, the provincial elite and leaders continued to believe that resources and the land were endless, that they existed to be exploited, even though economic conditions in the province were fairly dismal by 1871 because of a combination of global recession and local events. The gold rush had waned, many were leaving the region, and the colony had racked up considerable debt largely because of the costs of building the Cariboo Wagon Road. In this context, both levels of government continued to implement policies put in place during the colonial era to facilitate and promote economic growth and development. The provincial government encouraged white settlers to take up farmland through pre-emption or, after 1872, as homesteads on federal lands. Other provincial policies encouraged the private sector to harvest forests and stake mining claims. The federal government also began formulating policies to regulate and manage commercial fisheries.

While these policies were relatively straightforward, policies pertaining to public spending on infrastructure – ranging from schools, hospitals, and court houses to transportation improvements – were more complicated. Transportation schemes promised the latest, most progressive technologies to shrink time and space. They promoted an image of modernization, improvements to the economy, and rising standards of living, and transportation infrastructure was intimately linked to resource exploitation. By collapsing time and space, railways and steamers made it profitable to move relatively low-value bulk commodities in British Columbia (such as metals, wheat, coal, lumber, or tins of fish) over long distances to markets in the Prairies, central and eastern Canada, and the United States. Public and private investment in railways was viewed as the main conduit for opening up territories for resource harvesting, and it was recognized that railways had a multiplier effect. For example, a rail line could be built for a single use – for example, from a mine to a smelter to a port facility. However, other industries – such as forestry, commercial agriculture, or even tourism – could then use the rail line. Moreover, in the mind of promoters, railways translated into construction jobs, opened up regions to farming, and attracted settlers, who, in turn, bought land from the rail line. This money, in turn, assisted in the cost of building the railway and expanding the economy. Railways also lowered the price of goods and travel, thereby improving standards of living. They represented progress and modernity and were popular with politicians and the voting public.

Although there was certainly hope of better economic times for British Columbia with Confederation, little changed. Surveyors searched out prospective routes for the CPR – and there were many proposals. Sanford Fleming was in charge of surveying routes and a terminus. One proposal involved the rail passing through Pine Pass in the Peace River region and ending at Port Simpson, near present-day Prince Rupert. The federal government, however, favoured a route down the Homathco River (the former proposed Waddington Trail) to Bute Inlet and ending at Esquimalt (Leonard 2002, 166). This decision certainly pleased Victoria. However, other terminuses were also proposed, and the Burrard Inlet route, which would terminate at Port Moody, was recommended by Fleming as the most cost-effective. All of

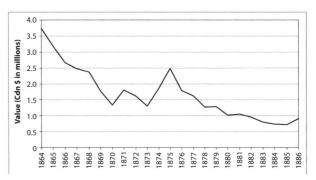

Figure 6.1 Value of gold production, 1864–86
Source: Statistics Canada (1888).

these proposed routes encouraged considerable land speculation.

Although the route was chosen and surveyed, the Pacific Scandal of 1873, in which Prime Minister John A. Macdonald was charged for taking bribes to build the railway, brought the federal government down. The new Liberal government put the railway on hold, which was a major setback for BC's economy. Although resource industries not reliant on rail transportation (such as coal, lumber, and salmon) saw somewhat improved employment numbers in the 1870s, they were all coastal activities. In the interior, the Cariboo Gold Rush had slowed down considerably, but miners continued to prospect in the Kootenays and in remote rivers and streams in the north. Some gold was discovered in the Omineca and Cassiar regions, but the overall amount of gold declined steadily after 1875, even though newer technologies such as dredges and hydraulics were being employed (see Figure 6.1).

Gold mining may not have been a growth industry for the new province, but coal mining on Vancouver Island certainly was (see Figure 6.2). Coal was in demand to fuel the increasing number of steam engines (e.g., at Victoria's Albion Iron Works) being built, especially in the United States, where the completion of the transcontinental railway was followed up by the emergence of numerous regional rail lines in the Bay area. The Nanaimo coal fields, along with new coal seams discovered at Wellington and Cumberland, required coal miners to be imported from throughout Europe. They lived in company towns and worked in high-risk mines that claimed

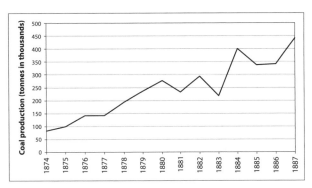

Figure 6.2 Coal production, 1874–87
Source: Statistics Canada (2014).

many lives. The infamous Dunsmuir mines rejected unionization, did nothing to improve poor living conditions in their company towns, kept wages low, and failed to make safety a priority. Strikes became common.

Forestry also emerged as an important export industry. Large sawmills on Burrard Inlet (such as Hastings Mill in Vancouver and Sue Moody's Mill in North Vancouver) and on Vancouver Island (Chemainus and Port Alberni) began to tap into reserves of enormous old-growth Douglas fir, hemlock, and western redcedar, trees that grew in abundance along the coast and were milled for markets throughout the Pacific, including San Francisco, Hawaii, and Japan. These logs were drawn by oxen on corduroy roads (small logs placed horizontal to logs skidded to the foreshore, also referred to as skid roads) to the water and then towed by boat to the sawmill, creating a linear logging pattern along the coast. The line rarely moved inland. Logging and lumber production was minimal in the interior, confined to serving local demand in mining communities, which declined along with the number of gold discoveries.

Although the resources of the ocean such as fish, sea otters, whales, and fur seals are renewable resources, they are renewable only within their biological limits. However, these limits were of little concern so long as the colonial myth of inexhaustible abundance continued to be believed. Unfortunately, this myth was perpetuated after Confederation (even after the sea otter became commercially extinct in the early 1800s), when the federal government, which had jurisdiction over the oceans, established policies for the salmon industry that failed

to address the concept of biological limits. Historically, the Hudson's Bay Company had barrelled and exported small quantities of salmon from Fort Langley, but it was the technology of canning and the building of canneries, especially at the mouth of the Fraser River, that allowed the industry to take off. However, because canneries were a significant investment, as well as being labour-intensive, they had a cautious beginning. There were only three canneries at the mouth of Fraser River in 1876, when the federal government began officially recording commercial fishing on the West Coast.

From the beginning, the government's annual reports mentioned issues with salmon harvesting. Initially, because the main catch was sockeye salmon, the industry only operated five to six weeks per year. From the federal government's perspective, there was little need to regulate this part-time industry, but huge conflicts were brewing, conflicts that touched on multiple aspects of BC society, including white value systems and a sense of entitlement to determine when to fish, how to fish, and who could fish.

Salmon are a mobile resource and are difficult to see except in their spawning channels. Salmon are also anadromous, meaning that they are born in fresh water, migrate to the ocean, and return to the same stream in which they were born to spawn and die (although some 10 percent may select a different river). There are five species of Pacific salmon (sockeye, Chinook or spring, coho, pink, and chum), and each species has its own migration path within the ocean environment, characteristics pertaining to life cycle (ranging from two years for pinks to seven years for Chinook), and migrations up rivers in order to spawn (known as runs or stocks or races). Understanding which rivers and streams "own" these runs is critical to future salmon.

From the beginning, the federal government realized that there were conflicts in the industry – between commercial fishing interests, industries such as mining and forestry that damage fish habitats, and First Nations that depend on salmon – and that regulations would be necessary. For example, the Department of Fisheries' 1875 report specifically pointed out the "decline of salmon runs" on the Fraser River owing to the destruction of spawning grounds by gold mining and the overharvesting of spawning salmon for their roe and fry by First Nations

(Commissioner of Fisheries 1875, 219). One "solution" it advocated was the artificial raising of salmon in hatcheries similar to those on the Sacramento River in the United States. Another recommendation was prohibiting commercial fishing on weekends so salmon could reach their spawning grounds.

Over the next three years, the department put regulations in place, adopting the **gatekeeper model** of management, a model based on collective ownership (that is, with salmon as a common property resource), not private ownership. Fishers were allowed to catch an unlimited number of salmon during openings (during the week) but none during closings (on the weekend). This schedule enabled salmon to swim upstream to their spawning beds. It is a simple model, but the effectiveness of closings (the key to conservation) depends on considerable research and knowledge as to when salmon are going upstream to spawn. It requires knowledge of the various races of salmon, a sense of how many fish are returning to spawn, and an idea of how many fishers will be setting their nets. Officials at the time had no evidence that a twenty-four-hour closure on weekends would allow a sufficient number of salmon to escape, guaranteeing salmon for the future. Moreover, unless fisheries officers were present, there was nothing to prevent fishers from ignoring the closures.

As the commercial salmon fishery increased, so too did the number of canneries. There were twelve by 1881 and seventeen by 1886, and although most were at the mouth of the Fraser River, rivers up the coast such as the Skeena were beginning to get their share. The federal government encouraged canning, but it meant even more salmon were being caught, and First Nations relied on it. Salmon was the basis of their culture, a fact reflected in their songs, dances, stories, and carvings. Salmon were part of their religion and critical to their diet. First Nations also employed a different management model for salmon. The whole community constructed weir and traps, and Houses that stood at weir sites regulated access (Swenerton 1993). In this sense, First Nations perceived the resource not as common property but as a communal one.

Cannery owners and non-Indigenous fishers believed First Nations were taking a large part of the commercial share of salmon and pressured the federal government to restrict Indigenous fishing and to quantify the value of fish caught by First Nations. They complained about how First Nations harvested salmon, using weirs and dams in close proximity to the spawning grounds. These harvesting techniques were banned in 1877. In 1880, the First Nations harvest was valued at $4,885,000 while the commercial share was $713,300. In 1885, the numbers were $3,257,500 for the First Nations harvest and $1,078,000 for the commercial harvest. These numbers caused cannery owners and fishers to argue that the Indigenous fishery should be restricted and the numbers reversed. For the federal government, conservation became confused with the distribution of salmon. As commercial fishers harvested more and more salmon at the mouth of rivers, fewer salmon arrived at spawning grounds, where the majority of First Nations took their share. Conflicts over entitlement to salmon increased.

Unlike salmon fishing, farming and ranching, which were confined mainly to Vancouver Island and the Lower Fraser Valley and small areas in the interior, felt the effects of the declining gold rush and recession. Access to external markets was limited by the lack of transportation routes, reducing many to a subsistence level of farming. The gold rush, from its good times until its decline, created an enduring pattern of economic boom and bust in a resource-dependent province. From the late 1860s until the end of the 1870s, industry in British Columbia was mainly a bust.

RACE-BASED POLICES AND EXCLUSION
As conflict in the salmon fishery revealed, as far as policymakers were concerned, First Nations did not fit in well with the provincial government's vision of a modern, industrial society based on resource extraction. Many First Nations continued to fish, hunt, trap, and gather to sustain their way of life; others became labourers in industries that were often short of labour but paid First Nations substantially lower wages than white workers. Because provincial politicians did not believe in Indigenous Title or that First Nations "used" the land, they, along with federal administrators, embarked on a new policy direction for First Nations – assimilation into Canadian society.

The **Indian Act** was passed in 1876 by the federal government, which had jurisdiction over First Nations. It

became the main instrument of assimilation and aliena-
tion from the land. Under the act, the federal government defined who was an "Indian" and controlled each
First Nation and its reserve through the appointment of
non-Indigenous Indian agents, and it imposed a system
of bands with an elected Chief and council. Enfranchisement was another divisive tool, because it essentially
offered individuals the choice of being a Canadian citizen
(that is, a Non-Status Indian) and offered incentives,
including small amounts of money. Non-Status Indians
were entitled to vote and to enter liquor establishments
but had no rights to live on a reserve or qualify for tax
exemptions.

In 1880, the government added a new component to
the assimilation package: "Prime Minister Sir John A.
Macdonald commissioned journalist and politician
Nicholas Flood Davin to study industrial schools for
Aboriginal children in the United States. Davin's recommendation to follow the U.S. example of 'aggressive
civilization' led to public funding for the residential
school system" (Indigenous Foundations 2009). Children were separated from families and placed in church-
run boarding schools, where Anglican, Catholic, and
Methodist missionaries tried to undermine all aspects
of Indigenous culture (Bowering 1996, 169). As a result of
this cultural genocide, a number of First Nations dialects
became extinct. Amendments to the Indian Act in 1885
resulted in a ban on the Potlatch and other traditional
ceremonies.

A similar blend of fear of competition and racism influenced the treatment of Chinese people in the province.
The gold rush had attracted a significant number of
Chinese to the province, so much so that they may have
made up one-quarter of the non-Indigenous population
by the 1860s. Some came up from California, others directly from China. To say that they were lured to the province by the prospect of getting rich is too simplistic. The
story is more complicated and helps explain anti-Chinese
and anti-immigrant sentiments today.

The Chinese who came to California and British Columbia were Cantonese rather than Mandarin-speaking
(the main language of China), which explains why Chinatowns are Cantonese-speaking to this day. They were from
southern China, which was in complete chaos during the
1850s and 1860s. There were too many people for too little

agricultural land in a primarily agrarian region, which
was suffering from famine and rebellion. It is estimated
that the Taiping Rebellion (1850–64) caused the death
of some 30 million people in the region (Hucker 1975,
154). Most of the Chinese who came to North America
were young, landless sons sent by poor families who had
pooled what little money they could to pay for passage
to this new world. It was intended that the young men
would remit any income they earned to their families in
China.

These immigrants were certainly not wanted in California, where many were badly treated, robbed, beaten,
or killed and where a host of legislation was passed that
discriminated against them on the basis of race. Conditions differed in British Columbia, but it did not take
long for California's anti-Chinese sentiment to appear.
Chinese men lived in segregated enclaves in gold-mining
communities and in communities such as Victoria and
New Westminster. They tended to prospect for gold either
in regions that white miners had abandoned or in totally
new areas where few whites were present. However, as
less and less gold was found, they relocated and found
employment in other resource industries, such as the
canneries, farming, and coal mining, or in urban-based
service industries.

White politicians and workers feared that the Chinese
would be willing to work for less money and that they
would contribute little to the economy, and they were
concerned that the Chinese were not white in what many
believed was an emerging white nation. In depressed
economic times – in an era when there was a labour
surplus, rules prohibiting unions, and no social safety
net – members of the white working class feared that
the Chinese would take their jobs. Policies that allowed
employers to hire Chinese, First Nations, and other minority ethnic groups at considerably lower wages only
compounded the problem. Although the business community enjoyed the advantages of these policies, the
media, politicians, and workers blamed the Chinese.

Policies of exclusion and discrimination produced
anti-Asian sentiment. And this sentiment increased as
the provincial government attempted to pass – in 1871,
1872, 1874, and 1878 – an act to apply some form of **head
tax** on all Chinese immigrants. In 1878, when the Supreme
Court ruled that such an act fell outside provincial

jurisdiction, the provincial government passed new legislation that banned Chinese from voting in municipal or provincial elections and from engaging in provincial contracts (Perry 2014, 124). The measures were justified on the grounds that Chinese would not assimilate. Once again, the Chinese, not the policies, were blamed – they were not entitled to assimilate.

THE CPR, THE END OF RECESSION, AND HEIGHTENED RACIAL CONFLICT

The recession finally ended in Canada when Macdonald's Conservative government came back into power in 1878 and delivered on the promised transcontinental railway, also allaying the threat of British Columbia withdrawing from Confederation. The province's population remained small but grew steadily: 49,459 people in 1881, according to the Census of Canada, and 74,000 in 1886, according to BC Stats. The railway construction boom not only drew more labourers to the province; it increased demand for railway ties, coal, food, and a host of other goods and services.

The CPR was part of the Conservatives' National Policy, which centred on government spending on westward expansion and tariffs and duties that forced Canadians to buy Canadian-made goods that were often more expensive than imports. The policies were successful and buffered the effects of the global recession. The idea of lines of steel joining British Columbia to the rest of Canada created confidence that the Prairies would remain in Canada rather than being annexed by the United States. Starting in 1871, the Prairies were subdivided by the Dominion Land Survey into homesteads of 160 acres (65 hectares) to promote agricultural settlement and growth. Railways, land surveys, and the arrival of more non-Indigenous people to the region triggered the realization that the federal government needed to "deal with" First Nations on the Prairies (most of which was called the North-West Territories at this time). The Royal Proclamation of 1763 had established the formula for extinguishing (ceding) Indigenous Title through the treaty process. As a consequence, beginning in 1871, the Numbered Treaties were negotiated throughout western Canada, clearing the way for the railway.

The CPR's last spike was driven in at Craigellachie, British Columbia, near Revelstoke, in 1885. By connecting Montreal to Port Moody, it provided the means to take advantage of the rich resources of the province. The CPR demonstrated that a rail line could be constructed across a largely vertical landscape, but it also showed that provincial and federal policies could greatly assist private rail companies. Both levels of government had awarded the CPR grants to land adjacent to the rail lines that included mineral and forestry rights; they also offered guarantees on bonds for raising money, direct cash subsidies, and other incentives. For example, the federal government initially gave the CPR 25 million acres (10,000 hectares) of farmland. Thirteen million acres (5.3 million hectares) were in BC and were free from taxation for twenty years or until sold. The CPR also received existing rail lines and stations that were worth $37.8 million and free from taxation forever. All equipment required for building the railway was duty-free, and the government guaranteed the CPR a monopoly for the first twenty years plus $25 million in cash (Chodos 1973, 22–23). The provincial government gave the CPR the most valuable land of all – 6,000 acres (over 2,400 hectares), which would be used to shift the terminus from Port Moody to Coal Harbour, where the CPR would create and build the new city of Vancouver. And this was just the beginning of the land grants and subsidies bestowed on the CPR.

By 1886, the time distance between Montreal and Vancouver had shrunk from weeks to just five days. The CPR brought British Columbia out of its recession, but it established other, longer-lasting patterns such as increased racial conflict. The rail line had been constructed from the Prairies west through the Rockies via the Kicking Horse Pass and then Rogers Pass. At the same time, construction began at Port Moody and headed east. The construction was not contiguous, though. Contracts for individual sections (such as the hazardous portions of the Fraser and Thompson Rivers) were sublet, greatly increasing the demand for labour. The subcontractor for the CPR brought in more than fifteen thousand Chinese labourers with the promise of return passage once the railway was completed. The lure was cheap labour: Chinese workers were paid $1.00 a day, with which they had to purchase their own food and camping and cooking gear. White workers, in contrast, were paid $1.50 to $2.50 a day and did not have to pay for their food or gear (Library and Archives Canada 2005).

Unfortunately for the Chinese, the promise of return passage was not fulfilled, and once the contracts were completed, surplus labour and unemployment began to increase – as did racial tensions. The Americans had passed a Chinese Exclusion Act in 1882, which prohibited most Chinese, including those from Canada, from entering the United States. Chinese men who could not afford to return to China were trapped in British Columbia.

The provincial government revived the concept of a head tax and persuaded the federal government to hold a Royal Commission on Chinese Immigration. It is notable that some of the first evidence the commission sought was from California, where it recorded statements such as "the Caucasian differs from all other races; he is humane; he is civilized and progresses" (Government of Canada 1884, 350). What this quote exposes is the fundamental belief, which is still alive today, that the white race was superior to other races. Prime Minister Macdonald espoused a similar attitude in a comment in the House of Commons in 1883: "At any moment when the Legislature of Canada chooses, it can shut down the gate and say, no more immigrants shall come here from China; and then no more immigrants will come, and those in the country at the time will rapidly disappear ... and therefore there is no fear of a permanent degradation of the country by a mongrel race" (FCCRWC 2010). The federal government imposed a $50 head tax to curb further migration, but the tax did not include Chinese already residing in British Columbia. The Chinese had little say because they were disenfranchised or, as Prime Minister Macdonald stated, they should not have the vote because they had "no British instincts or British feelings or aspirations" (Roy 1989, 152).

REGIONAL RAILS AND RESOURCE DEVELOPMENT IN THE 1880S AND '90S

The CPR provided a ribbon of connectivity between western and eastern Canada, but many more lines of steel, especially at the regional scale, were required to tap into the province's bountiful resource base. Throughout the 1880s and 1890s, regional rails were in demand, and the federal and provincial government were more than willing to offer subsidies. Vancouver Island politicians, upset that the CPR's terminus was on the mainland, got

a contract from the federal government to build a line from Esquimalt to Nanaimo (E&N). Unbelievably, the province ceded a strip of territory known as the Vancouver Island Railway Belt, which covered about a third of the island (Seager 1996, 208). This 1884 land grant, which included mineral and timber rights, was for a mere 112 kilometres (70 miles) of track; it included First Nations lands, which was not a consideration at that time.

By the late 1880s, new discoveries of coal had occurred on Vancouver Island south of Nanaimo. Because these seams were better quality than the ones north of Nanaimo, the original Wellington mines were abandoned in favour of the new South Wellington and Extension mines, which were serviced by Ladysmith, a company town. Other shafts were sunk, such as the ones on Newcastle Island and Protection Island, and a host of coal mines in the Comox Valley also became a focal point. Robert Dunsmuir, who owned many of the coal mines on Vancouver Island, was also awarded the contract to build the E&N Railroad and secured a land grant. The railway extended to the Wellington mines in 1887, facilitating the growth and development of coal mining, sawmilling, agriculture, and metal mining, which in turn attracted people.

On the mainland, Vancouver emerged as an international port connected to a national railway. Its population grew rapidly to over 13,000 in 1891 and 27,000 in 1901, making it the largest city in the province. Warehouses were built to service international trade, new hotels accommodated visitors, and sawmills located on False Creek provided lumber for buildings that spread farther and farther from the centre. Land clearing, surveying, road and streetcar building, and construction of residential, commercial, and industrial facilities required massive amounts of labour. Chinatown, which emerged to house and service labourers, soon became overcrowded.

Mining continued to be one of the greatest enticements, attracting people to the mainland. Along with coal, many outcroppings of minerals such as silver, gold, copper, lead, and zinc were staked by the 1880s, mostly in the Kootenays. Unlike during the gold rushes, these metals were processed by **lode mining**, sometimes referred to as vein mining or hard rock mining. The metals were

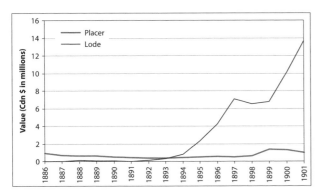

Figure 6.3 Value of placer and lode mining, 1886–1901
Source: Province of British Columbia (1902).

discovered in the bound form, which initially required miners to use a pick and shovel to follow the veins of ore into mountainsides. The crude ore then needed to be crushed, concentrated, smelted, and refined into its pure form. Smelting required coking or metallurgical coal, which was plentiful in the far southeastern Elk Valley. Fernie became a centre for this activity.

Plenty of American and British investors, along with a few Canadians, were willing to risk their capital on new ventures, and risk capital they did. Between 1886 and 1900, 1,316 mining companies were launched (Seager 1996, 212). Most were in the Kootenay region, and relatively few went into production, but all promised shareholders that they would become rich if they invested – few did. Still, unlike placer mining, lode mining boomed, particularly after 1893 (see Figure 6.3).

One of the interesting developments in the Kootenays was the intense competition that developed between Canada and the United States, not only to develop mines and build smelters but also to transport ore, refined metals, or coal either south or north. The Kootenays are rugged, framed by the Rocky Mountains to the east and, to the west, the Purcell and Selkirk Mountains and the east side of the Monashee Range. These are geologically complex mountain systems that mainly run in a north-south direction, as do the rivers and lakes in their valleys. The international boundary made no difference. While the CPR mainline went through Golden and Revelstoke to the north of the Kootenays, the Northern Pacific

Railway in the United States went through Spokane to the south. However, the Northern Pacific had the jump on the CPR because it was completed in 1883, two years before the CPR.

Transportation to mining sites was a challenge. From the beginning, rail lines were used in conjunction with paddlewheelers, and American and Canadian branch lines competed to get to mine sites. From this competition, a number of locations, including the city of Nelson, became early railway hubs. Then, in 1892, the Columbia and Kootenay Railway linked its mineral claims north to the CPR mainline. However, the following year, a Nelson and Fort Sheppard Line was built south to the United States. In 1895, Sandon became a railway hub when the Nakusp and Slocan Railway linked rich silver ores from Sandon to the CPR mainline via steamer to Revelstoke. That same year, the Kaslo and Slocan Railway also came to Sandon to transport silver ore south via Kootenay Lake to the American Great Northern Railway. Many more spur lines facilitated the transport of ore and coal both north and south.

By 1901, the metal- and coal-mining booms in the Kootenays were largely responsible for producing six of the ten largest centres in the province: Rossland, Nelson, Fernie, Revelstoke, Trail, and Greenwood (see Table 6.1).

Table 6.1

Communities with populations over 1,000, 1871–1901

Community	1871	1881	1891	1901
Vancouver	–	–	13,685	27,010
Victoria	3,270	5,925	16,841	20,919
New Westminster	1,356	1,500	6,641	6,499
Rossland	–	–	–	6,156
Nanaimo	–	1,645	4,595	6,130
Nelson	–	–	–	5,273
Fernie	–	–	–	1,640
Revelstoke	–	–	–	1,600
Kamloops	–	–	–	1,594
Trail	–	–	–	1,360
Greenwood	–	–	–	1,359
Cranbrook	–	–	–	1,196
Grand Forks	–	–	–	1,012

Source: Farley (1979).

The rail lines, mines, and smelters gave a boost to forestry, agriculture, and a host of service industries, but metal mining was a volatile industry, with mines quickly becoming exhausted and world market prices plummeting.

Mining and its associated economic spheres may have attracted many labourers to British Columbia, but many more came to farm in the 1880s. In 1881, the province had some 2,700 farms and a population of 51,000; ten years later, it had some 6,500 farms and a population of 98,000 – with approximately 22,000 of them located on farms (BCACF 2014). Of course, there was a strong relationship between the mining boom and farming. Mines meant more people to feed but, more importantly, many more rail lines to move agricultural products cheaply. Ranches in the southern interior expanded with the demand for beef to feed CPR construction workers. Farming expanded on Vancouver Island, in the Lower Fraser Valley, and throughout the river valleys of the Kootenays, and the Okanagan was beginning to gain an international reputation for its fruit farming and ranching. To facilitate these agricultural endeavours, the CPR built the Shuswap and Okanagan Railway from Sicamous on the CPR mainline to Okanagan Landing, near Vernon. Steamers then served the communities and farms along Okanagan Lake as far south as Penticton.

Unfortunately, the 1890s were not as kind to the farming industry. Although North America continued to be in the midst of a depression, British Columbia and the Prairies were largely immune because of the beneficial effects of the National Policy. However, prices for agricultural produce were depressed and made worse by overproduction in western states such as California, which were in a position to invade Canadian markets via the new transportation infrastructure.

Boom times also meant land speculation, and investors had also bought up some of the best farmland in the 1880s, not to farm but to hold onto for resale, thus driving up the price of farmland. Price increases discouraged many from coming into the province to farm. Between 1891 and 1901, the number of farms only increased by eleven, even though the population jumped from 98,000 to 178,000. As well, the Fraser River flood of 1894 (the highest ever recorded on the Fraser) had disastrous consequences for existing farms in the Fraser Valley and for the Fraser's tributary communities (e.g., Kamloops on the Thompson River). The flood waters were equally high in the Kootenay region and much of the southern half of the province, where they washed out roads, rails, and bridges, disrupting resource harvesting. The flooding of the Columbia River was particularly devastating to American communities downstream.

Although the farming industry suffered in the 1890s, the completion of the CPR gave a boost to the commercial salmon-fishing industry. By 1889, the dynamics of the industry had changed. It was now a five- to six-month industry that harvested an increasing amount of salmon to an expanding number of canneries. The canning industry had expanded to include not just sockeye salmon but also larger quantities of Chinook, coho, pink, and chum, and the railway provided access to eastern markets for both canned and fresh salmon.

With more fishers and more canneries, the federal government tried to enforce greater conservation measures. It introduced licences, but capped the number at five hundred fishers, and it introduced a host of rules regarding net size, where one could and could not fish, and when. Closures were now thirty-six hours on weekends. Complaints by fishers and cannery owners resulted in some relaxation of the regulations (e.g., reversion back to a twenty-four-hour weekend closure), but the biggest change in policy came in 1892 when licence limitations were dropped. It became an open, overcrowded fishery where anyone, not just canneries, could apply for a private, commercial fishing licence (Stacey 1982, 13). Even though the Department of Fisheries' 1890 annual report commented on reckless overfishing and the disastrous consequences of declining stocks in the United States on the Sacramento and Columbia Rivers and prophesized a similar fate for Alaskan Rivers, it instituted a similar management model for British Columbia.

While the department's policies on sustainable management were not a strong suit, its promotion of harvesting new ocean resources was. For example, Inspector Mowat reported in the annual report of 1887:

The people of this Province have not yet engaged in the whale industry, and I can scarcely understand the reason therefore, unless it be due to a scarcity of vessels or for want of properly realizing its importance. The fleet which sailed from San Francisco to the Arctic Sea last season is reported to have captured 257 whales valued at $1,285,000, or an average of over $107,000 to each vessel. Considering that our people are situated 750 miles nearer the Arctic fishing grounds than our San Francisco neighbors, I am at a loss to understand why an effort is not made to participate in this remunerative business. (Department of Fisheries and Oceans 1887, 247)

The forest industry, like the salmon industry, continued to expand in this period. Expansion was assisted by the CPR and regional rails, but the industry largely remained a coastal activity, focused on large old-growth fir, cedar, and hemlock forests that could be cut and towed to the tide line and on to large mills. With the expansion of regional rail lines through the southern half of the province, lumber mills were opened in the interior, but they served the local market for building materials. In 1885, there were only twenty-five sawmills in the province; by 1896, there were eighty-five (Gosnell 1901).

Even in these early days of logging, concerns over sustainability were expressed. In his 1901 *Yearbook of British Columbia*, R.E. Gosnell warned that "the conservation of forests becomes one of the most important subjects that can engage the attention of the legislators; but forest fires, the clearing of land, and the reckless deforesting for lumbering purposes, are having appreciable effects in reducing the supply" (Gosnell 1901). Maria Lawson and Rosalind Watson Young also commented on the rate of exploitation in *History and Geography of British Columbia*, which was published in 1906: "The sides of the coast mountains, the islands, and the uncleared valleys are covered with a magnificent growth of timber, which one might suppose would last forever, did he not know that regions which half a century ago were clothed with forests almost as vast, are now timberless" (Lawson and Young 1906, 11).

As their comment reminded readers, resource development had a dark side, which became even more evident near the close of the century, when another unforeseen event occurred – discovery of gold in the Yukon in 1896. Similar to the Fraser River–Cariboo Gold Rush of 1858, the Klondike Gold Rush attracted some thirty thousand individuals. There were only a few ways to get to the goldfields, where Dawson City, on the Yukon River, was the main centre. The most common, and cheapest, route was via the Chilkoot Pass, which involved travelling by boat to Skagway, Alaska, and then hiking over the pass. At the top, which was Canadian territory, the North-West Mounted Police made sure each miner had 1,100 pounds (550 kilograms) of provisions so as not to starve. The large tent town of Bennett arose overnight on the shores of frozen Lake Bennett. When the ice melted, boats and rafts made their way to Whitehorse and then on to Dawson City via the Yukon River.

The Klondike Gold Rush was an economic boom to Vancouver and Victoria as miners stocked up on food, especially tinned salmon, and equipment before shipping out. But as with the 1858 gold rush, it had permanent consequences. Yukon became a separate territory in 1898. Urban centres sprang up out of the wilderness, along with roads and trails and even, by 1900, a narrow-gauge railway. The White Pass and Yukon Railway was built and ran from Whitehorse to Skagway. Typically, few got rich, but some did, and some of that gold was invested in Vancouver and Victoria real estate.

The negative side of the gold rush became clear when several hundred miners made their way overland from Fort Edmonton, seriously disrupting First Nations way of life, especially in the Peace River region. Miners lived off the land and killed scarce game, leaving a fear of starvation for Indigenous Peoples. War nearly erupted in 1898. Hundreds of First Nations descended on Fort St. John and demanded that the conflict be resolved. From the federal government's perspective, the "solution" was a treaty, but because this was British Columbia territory, the province had to be consulted. Treaty 8 was signed in 1899 (see Figure 6.4), but the provincial government wanted no part of the treaty or the precedent-setting decision (Ray 1999, 38). This treaty turned out to be the best deal that any First Nation in British Columbia got: it included gifts (cash, farm implements, farm stock, and ammunition) as well as reserves based on 640 acres (259 hectares) per a family of five. Most First Nations territories were based on ten acres per family.

SUMMARY

British Columbia entered Confederation on the brink of a global recession. At the time, its resource "frontier" included Vancouver Island, which produced coal and

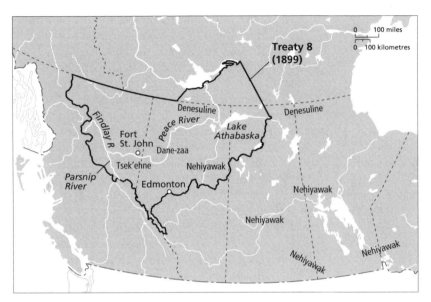

Figure 6.4 Treaty 8 in western Canada
Source: Modified from Friesen (1987).

including racial segregation. First Nations, confined to reserves, were subjected to assimilative policies by the federal government under the Indian Act and came into conflict with miners. Near war resulted in a vast northeastern British Columbia region, Treaty 8, being recognized. Racism and discrimination against Chinese and other Asians only increased as thousands of labourers moved to BC to work in mines and on railways.

Even in these early days, warnings about resource management could be heard, especially with respect to forestry and the commercial salmon-fishing industry. Politicians responsible for resource management at both the federal and provincial levels continued (as those had before them) to believe in limitless fish and forests and that the region's resource wealth was never-ending.

lumber, and the mainland, which had salmon canneries at the mouth of the Fraser River and two large sawmills on Burrard Inlet. Gold rush communities were shrinking, as less and less gold was discovered, but miners continued to search for gold in northern BC and the Kootenays. But they were few in number. Rail transportation was desperately needed to overcome the friction of distance in this vertically challenged landscape, but the transcontinental railway was delayed through scandal.

The CPR was completed in 1885, ending the recession and boosting the economy. A host of regional rails followed. The importance of these railways (which were often connected to steamers) to resource development in the province cannot be overstressed. The Kootenays, which had an abundance of metals and coal, became the dominant region, its landscape transformed by lode mining and dotted with smelters. But much of the southern half of the province also experienced growth linked to the expansion of farming and ranching, fishing and forestry. And Vancouver, a national railway terminus connected to an international port, soon became the leading urban centre in the province.

Along with new transportation patterns on the landscape came new rural, urban, and industrial patterns,

REFERENCES

BCACF (British Columbia Agriculture in the Classroom Foundation). 2014. *"Grow BC": A Guide to BC's Agricultural Resources.* bcaitc.ca/sites/default/files/Grow%20BC/Grow BC_2014.pdf.

Bowering, G. 1996. *Bowering's BC: A Swashbuckling History.* Toronto: Penguin.

Chodos, R. 1973. *The CPR: A Century of Corporate Welfare.* Toronto: James Lorimer.

Commissioner of Fisheries. 1875. *Report.* Ottawa: Minister of Marine and Fisheries. publications.gc.ca/collections/collection_2015/mpo-dfo/MA1-1875-eng.pdf.

Department of Fisheries and Oceans. 1887. *4th Annual Report.* Ottawa: Department of Fisheries. publications.gc.ca/collections/collection_2015/mpo-dfo/Fs1-1887-eng.pdf.

Dunn, William, and Linda West. 2011. "British Columbia Joins Confederation: Introduction." *Canada: A Country by Consent.* canadahistoryproject.ca/1871/.

Farley, A.L. 1979. *Atlas of British Columbia: People, Environment, and Resource Use.* Vancouver: UBC Press.

FCCRWC (Foundation to Commemorate the Chinese Railroad Workers in Canada). 2010. *The Ties That Bind.* mhso.ca/tiesthatbind/HeadTaxExclusion.php.

Friesen, G. 1987. *The Canadian Prairies.* Toronto: University of Toronto Press.

Gosnell, R.E. 1901. "Forest Wealth." *The Yearbook of British Columbia.* hp.bccna.bc.ca/Library/Yearbook/home.html.

Government of Canada. 1884. "Minutes of Evidence." *Report of the Royal Commission on Chinese Immigration.* Ottawa: Queen's Printer.

Hucker, C.O. 1975. *China to 1850: A Short History.* Stanford: Stanford University Press.

Indigenous Foundations. 2009. "The Residential School System." indigenousfoundations.arts.ubc.ca/the_residential_school_system/.

Lawson, M., and R.W. Young. 1906. *History and Geography of British Columbia.* Toronto: W.J. Gage.

Leonard, F. 2002. "'A Closed Book': The Canadian Pacific Railway Survey and North-Central British Columbia." *Western Geography* 12: 163–84.

Library and Archives Canada. 2005. "The Kids' Site of Canadian Settlement: Chinese." collectionscanada.gc.ca/settlement/kids/021013-2031.3-e.html.

Perry, J.M. 2014. "The Chinese Question: California, British Columbia, and the Making of Transnational Immigration Policy, 1847–1885." PhD diss., College of Bowling Green State University. etd.ohiolink.edu/!etd.send_file?accession=bgsu1394761542&disposition=inline.

Province of British Columbia. 1902. Sessional Papers. *Annual Report of the Minister of Mines, 1901.* Tables V and VI. cms content.nrs.gov.bc.ca/geoscience/PublicationCatalogue/AnnualReport/BCGS_AR1901.pdf.

Ray, A. 1999. "Treaty 8: A British Columbia Anomaly." *BC Studies* 123: 5–58.

Roy, P. 1989. *A White Man's Province: British Columbia Politicians and Chinese and Japanese Immigrants, 1858–1914.* Vancouver: UBC Press.

Seager, A. 1996. "The Resource Economy, 1871–1921." In *The Pacific Province: A History of British Columbia,* ed. H.J.M. Johnston, 205–52. Vancouver: Douglas and McIntyre.

Stacey, D. 1982. *Sockeye and Tinplate.* Victoria: Royal BC Museum.

Statistics Canada. 1888. *Canada Year Book.* statcan.gc.ca/eng/1888/188803460328_p.%20328.pdf.

–. 2014. "Canadian Production of Coal, 1867 to 1976." Table Q1-5. statcan.gc.ca/pub/11-516-x/sectionq/4057756-eng.htm.

Swenerton, D.M. 1993. *A History of Pacific Fisheries Policy.* Ottawa: Department of Fisheries and Oceans. dfo-mpo.gc.ca/Library/165966.pdf.

Resource Dependency and Racism in an Era of Global Chaos

7

In the first half of the twentieth century, British Columbia became increasingly dependent on external demand for resources. Population growth is one indicator of economic development, and the population increased from fewer than 200,000 people in 1901 to nearly 1.2 million by 1951. Growth was not even, however, in terms of both time and space. It reflected both global and local conditions. Few in the province predicted the profound impact that global events would have on the province during this period. In this chapter, you'll learn how local and international political developments during Richard McBride's Conservative government (1903–15) influenced the social and physical development of BC and how the Depression and two world wars influenced the evolution of the province. In particular, you'll learn how the technological innovations that were tied to these world-changing events influenced the rate of resource harvesting in the province, where people lived and worked, and how they laboured.

RAILWAYS AND THE ALIENATION OF RESOURCES

Although premiers came and went during these decades, some stayed in power considerably longer than others. Richard McBride, who is considered the father of BC's Conservative Party, lasted twelve years, from 1903 to 1915, and left behind an enormous environmental footprint and sizable provincial debt. His government also alienated plenty of Crown lands along with their resources. His election marked the beginning of party politics in the province. He understood and utilized racism and railway politics, as did premiers before and after him. Racism was popular with the voting public at this juncture in BC's history, particularly since provincial politicians could blame the federal government, which was responsible for immigration, for the arrival and presence of "cheap" Asian labour.

McBride realized that railroads were part of the economic and political formula for getting elected, but he also understood that facilitating railway expansion, along with all the other costs of government, was expensive. Therefore, some of his first initiatives to raise funds led to the sale of Crown lands for agriculture and timber. Much of the agricultural land was sold to syndicates for a dollar an acre, and the syndicates then resold the land at a large profit to individual farmers (Robin 1972, 92).

Between the outright sale of Crown land and railways selling their Crown grants, the number of farms increased dramatically from 6,500 in 1901 to nearly 17,000 by 1911. Even more Crown lands – approximately 9.6 million acres (3.9 million hectares) of prime timber land – were alienated through the granting of forest licences between 1904 and 1907 (Robin 1972, 91–92). These Crown land schemes raised revenues that were spent on infrastructure, increased bureaucracy, and resulted in "good" economic times. They also got McBride re-elected. Flush with revenues, McBride went on a railway-spending spree that triggered a demand for labour, rail ties, food to feed railway workers, and all the other inputs necessary to produce rail lines.

McBride's subsidies and guaranteed bonds influenced two transcontinental railways. In 1904, McBride discussed the possibility of encouraging the Grand Trunk Pacific Railway (GTP) to use the Yellowhead Pass to complete its terminus at Prince Rupert. The GTP was a subsidiary of the Grand Trunk Railway and a major competitor of the CPR. It had extensive rail lines in central and eastern Canada as well as in the United States. Bonds were guaranteed by the provincial government, and McBride insisted that the new transcontinental line begin construction in Prince Rupert and head east. By doing so, he ensured that the province would supply materials and labour. Only white labourers could be hired (Roy 1980, 8), and building began in 1908.

A second transcontinental railway – the Canadian Northern (CNoR) – had a network of rail lines across the Prairies that followed a more northerly route than either the GTP or CPR. It was persuaded by McBride (again, by means of a bond guarantee) to extend the line from Edmonton through the Yellowhead Pass down the North Thompson River to Kamloops and on to Vancouver. Tragically, in 1913 and 1914, the construction resulted in major landslides into the Fraser River at Hell's Gate, which had a devastating impact on sockeye salmon runs. The 1913 slide, which occurred in August, coincided with the four-year return of millions of sockeye salmon, who could not reach their spawning grounds. The Fraser River sockeye runs have never recovered. This catastrophe brought severe hardship to First Nations upriver from Hell's Gate.

While transcontinental railways were important, regional railways were perhaps higher on McBride's agenda because railway boosterism affected many more political ridings. Existing regional railways were extended, while new ones were built or promised. Some of these extensions required little provincial government assistance – for instance, the CPR takeover of the E&N railway on Vancouver Island in 1905 resulted in its gaining control of the enormous land grant, which included some of the best forest land in BC. The CPR then extended lines north to Parksville in 1910, east to Port Alberni in 1911, north to Lake Cowichan in 1913, and farther north to Courtenay in 1914.

A more ambitious project was to link the Kootenay rail systems along a southern route to Hope and the Lower Mainland. The Kettle Valley Railway (KVR) had an agreement with the CPR to continue a rail line from Midway (where the CPR ended through the Crowsnest Pass Agreement) to Penticton, thus opening up the south Okanagan. McBride persuaded the KVR with a subsidy of $5,000 per mile ($3,125/kilometre) to continue this railway from Penticton to Merritt. He then offered the KVR a subsidy of $10,000 per mile ($6,250/kilometre) to build the line through the narrow Coquihalla Valley from Merritt to Hope and $200,000 per mile to build a bridge at Hope across the Fraser River (Roy 1980, 22).

Even more grandiose was McBride's vision of railways linking Vancouver to Prince George, where they would tie into the Grand Trunk Pacific and then continue even farther north into the Peace River region. To get the line started, in 1912 he guaranteed bonds at $35,000 per mile ($22,000/kilometre) for a total of $16 million. Railways were the cornerstone of McBride's formula to grow the economy: he issued charters for fifty-two railways, but not all of them were built (Belshaw 1996, 149). By 1912, over six thousand men were engaged in railway construction, which increased prosperity and stimulated real estate speculation (Roy 1980, 14). It also ensured McBride's re-election.

INCREASING ANTI-ASIAN SENTIMENT

McBride's initiatives led to a greater demand for labour, but not Asian labour. In fact, anti-Asian sentiment only intensified during the McBride era, in part because thousands of Japanese and South Asians (mainly Sikhs) began to immigrate to British Columbia. All three groups were perceived as a threat to white workers.

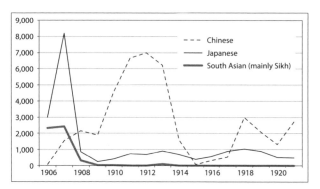

Figure 7.1 Chinese, Japanese, and South Asian immigration, 1906–21
Source: Statistics Canada (1947).

By the turn of the century, neither economic nor political conditions had improved in China, so Chinese immigrants continued to arrive in the province, paying the $50 head tax than leaving. Once again, the anti-Chinese attitudes of white workers increased, leading to renewed demands for restrictions on their immigration. The federal government complied and amended the Chinese Immigration Act to increase the head tax to $100 in 1901 and then to $500 (a huge sum of money) in 1904. In the short run, Chinese immigration reduced to a trickle, but after 1908 it resumed to thousands per year until the outbreak of the First World War in 1914 led to restrictions on travel (see Figure 7.1).

Like the Chinese, Japanese immigrants were also viewed as inferior and paid the price in policies and attitudes of racial discrimination. Even though there were few Japanese in the province before the 1890s, they were denied the vote in 1895. In 1901, there were 4,738 Japanese in the province, and many were involved in commercial salmon fishing and the canning industry centred at Steveston on the Fraser River. Others worked in the service industry in Vancouver's "Little Tokyo," in coal mines on Vancouver Island, on farms in the Fraser Valley, and in the logging industry along the coast. Unlike the Chinese, unpredictable global events gave the Japanese special status in Canada. Britain and Japan shared a common military threat – Russia – and that threat resulted in the Anglo-Japanese Alliance of 1902. Canada, as part of the British Commonwealth, could not impose a head tax or other policies of discrimination to curb Japanese immigration.

In 1907, an ugly race riot occurred in downtown Vancouver. The American-based Asiatic Exclusion League gave a speech on Labour Day that fanned the flames of racism. A white mob descended on Chinatown, doing considerable damage to buildings and businesses on Pender Street. But the rioters did not stop there. They went on to smash up Little Tokyo on Powell Street and caused significant damage to shops before the riot ended. Through the medium of the telegraph, "Race Riots in Vancouver, British Columbia" became a headline around the world. The incident embarrassed the British government.

Clearly, the white citizens of British Columbia made no distinction between different Asian groups. With the exception of those who employed them, no one wanted them in the province. Canada sent its labour minister to Japan in 1908 to make amends and structure an immigration policy that would be satisfactory for both countries. Japanese immigrants were classified into four categories: (1) prior Canadian residents, (2) domestic servants for Japanese residents, (3) contract labourers, and (4) agricultural labourers. It was agreed that categories two and four would be restricted to four individuals per year. The agreement also included a continuous-passage clause, meaning that Japanese immigrants had to come directly from Japan. The clause was intended to stop immigration from Hawaii. In 1907, for instance, over eight thousand Japanese came from Hawaii. The policy was designed to "save face" for the British. It had little impact on most Japanese immigrating to British Columbia.

By 1907, around five thousand Sikhs from northern India had immigrated to the province, and they too were subjected to policies of discrimination. They were attracted by employment in the forest industry, an industry in which many of them had experience. The McBride government did not want to allow Sikhs to enter, but halting their immigration presented a problem because India was a Commonwealth country, and Sikhs were British subjects. Their immigration should have been straightforward, but racism prevailed. The provincial government persuaded the federal government to develop immigration rules to bar their entry. The two most restrictive rules were (1) that Sikhs must come by continuous passage and (2) that they must have $200 in their possession. In the early 1900s, continuous passage

was impossible because ships could not travel between India and British Columbia without stopping for fuel and water. As well, few Sikhs had $200, a large sum of money for most in India in those days. Immigration ended, and many Sikhs in the province went back to India.

In 1914, a major incident showed how deeply rooted racism was in the province. That year, a Japanese ship, the *Komagata Maru*, arrived in Vancouver with approximately four hundred Sikhs onboard. Of course, it failed to meet the continuous-passage requirement, and none of the would-be immigrants had $200. The boat remained in the harbour for eight weeks while negotiations took place and riots raged. Few of the passengers were allowed to stay, and a navy gunboat escorted the *Komagata Maru* from Canadian waters.

FIRST NATIONS RESISTANCE TO PRESSURES ON THEIR LANDS

During the McBride years, First Nations populations continued to decline as waves of diseases – smallpox, measles, influenza, and tuberculosis – took their toll. A good deal of their Traditional Territories had been altered and reorganized into an industrial, resource-harvesting landscape. River valleys and plains had been surveyed and divided into agricultural homesteads; forest **tenures** had been let; salmon and other fish were being commercially harvested, canned, and exported; and miners had staked claims throughout the province. Paddlewheelers, roads, telegraph lines, and railways produced their own linear patterns, linking communities and resource regions to ports, further dividing the landscape.

In the more remote regions of the province, where none of these development had yet to occur, Indigenous Peoples were still labelled "pagan" and judged as uncivilized. In 1906, Maria Lawson and Rosalind Watson Young remarked: "With the exception of a few missionaries, gold-seekers, fur-hunters and employees of the salmon canneries, the great districts of Cassiar and Omineca have no civilized inhabitants" (Lawson and Young 1906, 13).

Because policy-makers believed First Nations did not fit well into this new, non-Indigenous, moulded landscape, they were segregated onto reserves, where they found it difficult to follow their seminomadic lifestyle and increasingly became dependent on Western goods, including food. Since few First Nations farmed the land, they were viewed as not "using" the land and, therefore, as an obstruction to development. They participated in the new economy as labourers (although at a lesser wage than white workers) or continued to live off the land by fishing, hunting, gathering, and trapping.

First Nations societies were often in a state of chaos because of diseases, the introduction of alcohol, the creation of Status and Non-Status Indians, the abolishment of traditional ceremonies, and the removal of children to residential schools, where they were Christianized and forced to speak only English. Frustrated by these developments, individual bands in the early 1900s began to protest, and First Nations organizations emerged to formally request that the land issue be resolved. When appeals to the provincial government fell on deaf ears, they sent petitions to Ottawa, to England, and even to the Pope in Rome – all to no avail.

Even though federally owned reserve lands were extremely small, there was relentless pressure to reduce them even further. For example, reserve lands "not occupied" were often regarded as vacant on the grounds that First Nations were a dying race. These lands, particularly those adjacent to expanding communities or proposed rail rights-of-way, were desired by the rapidly expanding non-Indigenous population. In 1906, the federal government amended the Indian Act to allow and encourage bands to relinquish their "unused" lands for a sum of money. Insufficiently used lands were surrendered through this policy. In 1911, further amendments allowed municipalities or corporations such as railways to take First Nations lands without the band surrendering them and allowed the government, through the Exchequer Court, to remove First Nations from any reserve located in a town with at least eight thousand residents (Roy 2011–12, 47).

First Nations did not give up on Indigenous Title. Delegations went all the way to Britain in an attempt to get King Edward to intercede on their behalf. When the petitions were returned to the federal government, they lobbied Prime Minister Wilfrid Laurier, with little success. Finally, in 1913, the McKenna-McBride Royal Commission was struck to "finalize" the question of Indigenous Lands. James McKenna represented the federal government, and McBride, the premier of the province, engineered it so the commission only examined reserve size, not the concept of Indigenous Title. From the outset, the terms for the McKenna-McBride Commission stated that any change to reserve size must have the consent of First Nations. When the commission was completed in 1916, its recommendations made some reserves larger, but thirty-five reserves were reduced (i.e., cut-off lands) without the consent of bands. Later, laws were passed stating that their consent had not been necessary after all (Gunn 1976, 5). Although First Nations rejected the commission's findings, the reductions were made to reserves.

RESOURCE DEVELOPMENT AND COMPETING CLAIMS

While First Nations saw the boundaries of their territories recede and their requests for treaties ignored, in 1892 Canada entered into the first commercial fishing treaty with the United States. The treaty occurred after the Americans seized British Columbia ships in 1886 and Canada asked Britain to intervene on its behalf. The conflict reached back into the 1880s, however. At issue was who was entitled to valuable fur seals, but the negotiations revealed larger issues such as the rules governing harvesting in international waters, labelling a resource as private property, and the tendency to paper over conflicts with policies of conservation.

The fur seal is somewhat unique in that its breeding grounds, or rookeries, in the Pacific are confined to two main island groups: the Commander Islands (near Kamchatka, Russia) and the Pribilof Islands (in the Bering Sea off Alaska). In the rookeries, both adult seals and pups were vulnerable and could be clubbed and their valuable skins harvested. The discovery of these rookeries in the late 1700s led to their near extinction by the early 1800s (another near tragedy). With so few fur seals to harvest, the industry abated and became commercially extinct. In the 1840s, the Russians, who owned Alaska, outlawed the hunting of females. The population rebounded by 1870 (Olesiuk 2009, 4). The hunt for the fur seal was on again, but this time around there was only one regulation: a ban on harvesting seals within sixty nautical miles of the Pribilof Islands. In reaction, a pelagic fishery emerged, meaning that seals were harvested on the open sea as they migrated from Alaska down to California and back to Alaska.

After the Americans purchased Alaska from the Russians in 1867, a new rule transformed the international conflict. Until then, the ocean's resources such as salmon and other fish had been considered common property, but the United States in 1869 turned the fur seal into a private-property resource when it gave the Alaska Commercial Company an exclusive twenty-year lease to harvest fur seals in the Pribilof Islands. Conflict occurred when the United States insisted (with considerable influence from the Alaska Commercial Company) that fur seals born in Alaskan territory belonged to the United States, even when they were in international waters (it would argue the opposite when Canadian salmon were concerned).

The seizure of three British Columbia vessels in 1886 in international waters triggered the so-called fur seal war, and more ships were seized in the years that followed (Shepard and Argue 1998). After the British navy interceded on Canada's behalf, a treaty of arbitration was established to resolve the issues. In the years that followed, it was discovered that the Americans were branding fur seals. The Department of Fisheries' annual report for 1900 recorded that BC sealers had harvested six branded seals in 1898, sixteen in 1899, and forty-five in 1900 (Department of Marine and Fisheries 1900, xxvii). There was no mention in later reports of catching branded seals, but the implication was that other nations, including Canada, were poaching or rustling American fur seals. The North Pacific Fur Seal Convention, signed in 1911 by Canada, the United States, Great Britain, Russia, and Japan, ended the pelagic hunt for all nations and likely saved the fur seal from extinction.

The commercial salmon fishery, the most lucrative commercial fishery at that time, had its own quirks when it came to entitlement. Although the British North America Act made oceans and their resources the responsibility of the federal government, the land and its resources fell under provincial jurisdiction. But it was determined that the right to fish was a right associated with the ownership of property, thus giving the provinces a role in fisheries. BC was therefore entitled to benefit from fishing revenues. In 1901, the provincial government passed the BC Fisheries Act, which created an appointed Board of Fishery Commissioners to regulate the management and conservation of fisheries (Millerd 2000).

The arrangement resulted in disjointed management, which became apparent in 1908, when cannery owners in the northern fishery (for example, from the Nass and Skeena Rivers down to Rivers Inlet) persuaded the federal and provincial governments that no new canneries should be constructed and that there should be a limit on the number of fishing vessels. Motorboats were also banned from the northern fishery, which was part of the cannery owner's plan, since they owned most of the fishing vessels. These rules emphasized the conservation of salmon, a respectable goal. However, the annual report of the Department of Marine and Fisheries (1912, 290) tells another story: "It was the desire of both the Federal and Provincial Governments to encourage White fishermen to become permanent settlers in the north; consequently, whilst the number of licenses to be issued in each area remains the same, a certain percentage was reserved for White fishermen who were British subjects owning their own boats and nets." The social engineering worked: over 450 white fishers were added to the northern fishery by 1914.

The forest industry was also managed poorly in the McBride era, even though it became the number one industry in the province, surpassing mining in terms of employment and revenues. Some of the best forest land in the province had been alienated through railway charters and forest licences, and most of the wood came from private forest lands. Forestry remained primarily a coastal activity because of the area's large old-growth fir, cedar, and hemlock forests, but lumber mills followed the expansion of rail lines throughout the southern half of the province. In addition, new technologies such as the steam engine and forestry-dedicated, narrow-gauge rail lines replaced oxen and skid roads for hauling enormous logs, producing linear patterns up river valleys. The steam donkey or donkey engine (a steam-driven winch) made it possible to clear-cut entire areas and then haul the logs to a central location, where they'd be loaded onto rail cars and then transported to the tide line. By employing diesel and electric motors, sawmills became efficient assembly lines, and railways connected them to markets on the Prairies and, increasingly, the United States.

The forest industry was dominated by lumber sales, but there was also a market for cedar shakes and shingles and for pulp and paper and the newly developed plywood.

Coastal pulp-and-paper mills opened at Port Mellon on the Sunshine Coast and at Powell River in 1908, at Woodfibre (Squamish) and Ocean Falls near Bella Coola in 1912, and in Port Alice in 1918. Although a mill opened at Swanson Bay on the north coast in 1909, poor wood supplies led to its closure in 1918, and Swanson Bay eventually became a ghost town. The first plywood plant went into production in 1913 at Fraser Mills (Coquitlam), only eight years after a plant had opened in Portland, Oregon (Griffin 2000).

The Royal Commission of Inquiry on Timber and Forestry (1910) warned that the industry would not survive if the best stands of Douglas fir continued to be cut at current rates. Natural regeneration could not keep pace with the harvesting. The commissioners recommended more tree planting, improved fire protection, and waste reduction. The commission led to the passage of BC's first Forest Act, in 1912. The new forestry regime established a uniform log scale, whereby the provincial government extracted more revenues from forest tenures and increased the number of fire wardens, but little was done about reforestation. Another tragedy of the commons was set in motion.

Industries such as farming were likewise given a boost by rail line expansion. For example, some five thousand Doukhobors, who were mainly farmers escaping religious persecution in Europe, relocated to the Kootenays between 1908 and 1913. They were persuaded to come when they were granted exemption from military duty on the grounds of their religious beliefs (contentious objector entitlement). Other farmers came from the United States seeking cheap land after homesteads there were fully claimed. Still others came from the Maritimes, Quebec, and Ontario. The proliferation of railways overcame the friction of distance and gave birth to new farming regions with access to distant markets. For example, the building of the Grand Trunk Pacific railway between 1908 and 1914 opened new farming areas such as the upper reaches of the Fraser River.

By 1911, farming occurred mainly on Vancouver Island, in the Lower Mainland, and in the Okanagan Valley and the Kootenays, and it was beginning to take hold in the Bulkley Valley and Nechako region. Ranching continued in the central interior as well as in the Okanagan, and the number of farms essentially tripled during the McBride

era. Expansion of the agricultural sector was also fuelled by a nearly 50 percent increase in the number of urban dwellers in the province and Canada, a development that increased demand for agricultural products. Specialized crops such as hops appeared in the vicinity of Agassiz and in the Okanagan district, and tobacco was cultivated in the Okanagan and manufactured into cigars in Kelowna (Gosnell 1911). However, not all farming communities chose specialty crops wisely. An attempt to turn the areas around the Thompson River at Walhachin into orchards similar to that of the Okanagan failed spectacularly. A lack of understanding of soils, climate, and irrigation were major factors in Walhachin eventually becoming a ghost town.

Coal mining also boomed, largely because of the vast increase in steam technologies that were driving other industries such as forestry. The homegrown demand for coal shifted trading patterns. Prior to 1890, the majority of Vancouver Island coal was exported to the United States. By 1914, 70 percent of it was consumed in British Columbia, 20 percent in California, and 10 percent elsewhere, mostly in Mexico (McIntosh 2000, 47). East Kootenay coal was largely consumed in coke ovens for smelting operations on both sides of the border. This was coal mining's heyday: it accounted for at least one-quarter of mineral production in the province and close to half of all mining employment (Seager 1996, 216).

Unlike coal mining, metal mining faced a number of hurdles largely because lode mining required considerable fixed capital costs at all stages – exploration and discovery; tunnelling into mountains; extracting, crushing, and concentrating ores; and smelting. Other costs included transportation and energy, equipment, and labour. Adding to the risk were two concerns: minerals are a nonrenewable resource that will eventually run out, and the value of metals, particularly copper, is highly volatile in terms of world market price.

Unlike most farming communities, many of the mining communities that emerged in this period were destined to become ghost towns. This was particularly true after the major financial crisis of 1907, which led the world market price for resources such as copper, lead, zinc, and silver to collapse. Following this downturn, a number of smelters in British Columbia shut down by 1913, including at Crofton and Ladysmith on Vancouver

Island and in Nelson in the West Kootenays. The smelter at Boundary Falls in the West Kootenays operated intermittently from 1903 to 1908 and closed permanently in 1913. On the other hand, a new copper smelter was built in 1911–12 at Anyox on Observatory Inlet near Stewart on British Columbia's remote north coast.

Premier McBride attempted to reverse the downward trend in metal mining by encouraging iron and steel production. British Columbia had plenty of iron ore, and some mining of this metal had taken place on Texada Island and near Kamloops, but deposits were also known to exist on Vancouver Island and Haida Gwaii. The importance of iron and steel to this era cannot be over-emphasized: it was in great demand to build steam engines, rails, trains, ships, and, most importantly, weapons. Iron and steel were viewed as essential commodities for the economy and for military defence. What distressed McBride and premiers that followed him was that iron and steel goods were being imported from Britain while British Columbia was exporting its ore to Washington State refineries for production. If an iron and steel industry were developed in BC, manufacturing a wide range of staple products, the industry would be able to compete in markets that were consuming about $125 million worth of British manufactures (Gosnell 1911, 198).

However, no investment was forthcoming. By 1912, the boom had begun to fade (Roy 1980, 23). Too many miles of track had been built or were in the process of construction to justify more railways, and the cost of building in the rugged landscape was extremely high. As well, McBride had relied on outside investment capital for railway construction, but global events, particularly the Crimean War (1912–13), which led to the First World War, dried up his foreign investment. As prices for many resources declined, so did the government's revenue base (Robin 1972, 137–38). McBride resigned in 1915, and the Conservatives lost the election of 1916.

THE FIRST WORLD WAR AND THE ROARING TWENTIES

The First World War produced vast economic, social, and political changes in Europe, and its reverberations rippled through the rest of the world, all the way to British Columbia. The war stymied speculative railway investment and curtailed immigration, but it set off many other unexpected changes. Like the gold rushes, the war had mythical allure for young, single men, but not in the sense of getting rich but rather because of the promise of travel to foreign, exotic countries and the respect that would come from wearing a uniform and becoming a patriotic hero for defending one's nation. But unlike the gold rushes, the war resulted in over forty thousand men and some women *leaving* the province.

Just as few gold seekers got rich, instead leaving the province disillusioned and often poorer, only a small number of soldiers returned from the war as heroes: "Few anticipated the grisly consequences for massed troop formations of industrialized warfare with the machine gun, rapid-firing long-range artillery, high explosives, chemical weapons, tanks, submarines and aircraft, all of which would be used on an unprecedented scale. Almost 14 per cent of the able-bodied men in the province in 1914 were dead or wounded by 1918" (Hume 2014). Adding to the tragedy was the fact that when the soldiers returned, many more died from the Spanish influenza of 1918–19.

The First World War brought many other changes to the province's social and physical landscapes. Although Canadian women had been struggling for the vote for nearly a century, the war proved to be the final impetus. While some argued that they gained the vote primarily in recognition that they were serving their country overseas and on the home front in their homes, factories, offices, and voluntary organizations (Dunn and West 2011), this was not the main reason for their enfranchisement. Prime Minister Robert Borden had another goal: he wanted to win the 1917 election, and he wanted to enact conscription, which was not popular, especially among those who had chosen not to enlist.

His government modified the fundamental democratic entitlement to vote through the Military Voters Act, which defined the "military voter as any British subject, male or female, who was an active or retired member of the Canadian Armed Forces, including Indian persons and persons under 21 years of age, independent of any residency requirement, as well as any British subject ordinarily resident in Canada who was on active duty in Europe in the Canadian, British or any other allied army" (Elections Canada 2007). A second act, the Wartime Elections Act, went even further and allowed any female related to anyone who served in the armed forces – alive or dead – to vote. But it disallowed conscientious objectors such as Doukhobors and Mennonites from voting, as well as some naturalized British subjects who had been born in enemy lands such as Germany or Austria (Elections Canada 2007). Borden won the election in 1917, and in 1918 all women over the age of twenty-one, but not Status Indians, were entitled to vote, but it was not until 1919 that they were entitled to run for federal office. Women gained the right to vote in BC's provincial elections in 1917.

Chinese, Japanese, and South Asian Canadians who served in the war were allowed to vote federally as a result of Borden's changes to the Military Voters Act, but they were not included in the Wartimes Election Act and could not vote provincially. The provincial government in BC remained adamant that they remain disenfranchised. Although Status Indians who served overseas could vote so long as they were in active service, Status Indians in general, as wards of the state, could note vote in federal or provincial elections.

Although the war led to some progress on the democratic front, by the time it ended, railways in British Columbia were overbuilt. With 3,000 miles (4,828 kilometres) of track (excluding logging railways), the province had far more railways than its economy could sustain (Harris 1983, 14). The main lure for railway companies had been Crown grants and provincial subsidies. There had certainly been no sound economic rationale for building two more transcontinental railways. It came as no surprise when the Canadian Northern declared bankruptcy in 1918, and the Grand Trunk Pacific in 1920. Both were taken over by the federal government to become part of the Canadian National Railway (CNR).

One of the poorest investments by the McBride government was building the Pacific Great Eastern Railway (PGE). The railway contractor went bankrupt, and the line was short. It ran only from Squamish to Clinton but had been built through some of the most difficult terrain in the province at great cost. But there were few people to serve and few resources to move. Although McBride should have known that the PGE would be a failure, he continued on with the mismanagement of the project (Stephenson 2012, iii). The next government in power compounded the problem by taking over the rail line and

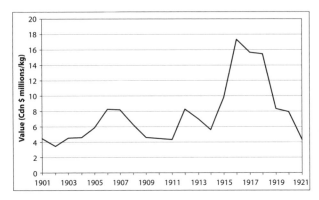

Figure 7.2 Copper prices, 1901–21
Source: BC Ministry of Energy, Mines and Mineral Resources (1989, Table 7-A).

extending it to Quesnel. Many referred to the PGE as the line that "ran from nowhere to nowhere."

Unlike railways, metal mining boomed during the war. There was a much greater demand for metals such as copper, lead, and zinc, and prices escalated (see Figure 7.2). Although gold had been important in earlier eras, gold mining remained unremarkable following the First World War and only increased slightly into the 1920s. The gold standard was in flux – for example, Canada was not on the gold standard from 1914 to 1926. And although gold remained a cornerstone in a nation's credit and a desired commodity, its rate was fixed at $21 an ounce. Once the war ended, inflated metal prices could not be sustained, and more mines and smelters closed in the West Kootenays, including those in Greenwood in 1918 and in Grand Forks in 1919.

Substantial increases in population often correlate to positive economic growth. As Table 7.1 reveals, although the province saw its population increase by over 200,000 people between 1901 and 1911, the McBride boom years, the following two decades saw a significant decline in the rate of population growth, indicated by the ten-year change. The flow of immigrants and investment from Europe was curtailed, initially because of the Balkan War and then by the First World War. There was modest growth in most sectors of the economy during the 1920s, but it was hardly the "Roaring Twenties," particularly if you lived in a community where a mine or smelter had closed. Far fewer people came to the province in the 1920s.

Table 7.1

BC population, 1901–31

	Population	Ten-year change
1901	178,657	
1911	392,480	213,823
1921	524,582	132,102
1931	694,263	169,681

Source: BC Stats (2017).

Many of the families that came to the province in this era became engaged in agriculture. Between 1911 and 1921, the number of farms increased by approximately 5,000 (from 16,958 to 21,973) (Statistics Canada 1999). A number of these farms were taken up by veterans, who were entitled to free homesteads of 160 acres (65 hectares) in the federal government's Peace River Block. Another 4,000 (from 21,973 to 26,079) farms were added in the 1920s, a decade when tractors and plows began to replace horse-drawn equipment, resulting in increased productivity and an overall increase in the value of produce.

Not only were returning soldiers offered homesteads, but they were also encouraged to become commercial salmon fishers, especially on the north coast. To accommodate them, in 1919 the federal government increased the number of salmon gill net licences issued (Department of Naval Service 1920, 43). The following year, it removed all restrictions on the number of licences – in other words, it instituted an open-door policy (Department of Naval Service 1921, 9). The northern fishery was now similar to the southern fishery, and sockeye runs collapsed throughout BC.

In response, the government took stock to assess blame. In its assessment it employed a simplistic, "smoking gun," view of causality. The Department of Naval Service's report (1923, 46–47) stated: "It has been suggested that during the period of the war when the cry was for food, more food, still more food, a much larger proportion of the salmon runs was taken than would have been the case under normal conditions and the industry is possibly now feeling the result of that intensive fishing in the brood years of the four-year cycles." There certainly was a significant increase in the amount of salmon harvested between 1913 to 1920, but if this was a lesson to reduce the harvest in years to come, it failed.

The causes of overfishing were complex. Fishers were responsible for the depletion of salmon, even though most followed the rules while some did not. The commercial fishing fleet was becoming one of the most sophisticated in the world. Similar to many other industries, wartime innovations such as ASDIC and sonar were adapted and utilized as fish finders, nylon nets replaced hemp, hydraulics lifted nets, and radio communication made the hunt easier. All these technologies increased productivity and decreased salmon stocks. There was also some concern that Americans were taking Canadian salmon.

In response, the government added more guardians and better patrol vessels, and by 1924 it was employing sea planes to enforce closings. The added bureaucracy gave the illusion that the gatekeeper model could work; at the same time, policies regarding fur seals and sea lions shifted the public's attention away from the number of fishers and the effectiveness of openings and closings to natural predators as the cause of salmon declines. Fur seals eat salmon and other commercial fish species, and they had been saved from becoming an endangered species by the North Pacific Fur Seal Convention. To stop seals from preying on salmon, the government engaged in increasingly drastic measures. In 1913, a bounty was placed on hair seals and sea lions; in 1916, bombs were placed on the sand heads where seals rested; in 1922, machine guns were employed against both seals and sea lions.

With federal policies distracting people from the real causes of overfishing, salmon were on a path to becoming the next tragedy of the commons. And another fish war was brewing with the Americans, this time over halibut, and the conflict had roots that preceded the war. The halibut fishery was primarily a winter fishery, and the main market was in central and eastern Canada and the United States because the Atlantic halibut fishery closed during the winter months and was on the brink of commercial extinction by 1900 (Thistle 2002, 10). As the Pacific halibut fishery's market share increased, so did the number of halibut fishers, including American vessels in Canadian waters. In its 1901 annual report, the Department of Marine and Fisheries (1902, 104) commented: "Considering the steady increase in this fishery, and its importance, it is very desirable that there should be no further delay in defining exactly how far Canada's exclusive rights, in the waters in which these fish are taken, extend, and in providing the necessary means to protect these rights against United States poachers. It is to be hoped that the new cruiser now being built in Vancouver may be of effective service in this direction."

Of course, as the war continued, technologies became more sophisticated as fishers from the two nations attempted to catch a greater share. Steam-powered vessels replaced sailing schooners, and these vessels were redesigned to withstand Pacific storms and stay at sea longer to harvest halibut. By 1910, fourteen Canadian or American steamers, which accounted for about 10 percent of the total fleet, was responsible for almost half the total catch. Fishers engaged in high-grading, throwing back as many halibut as they kept because they were discoloured or either too small or too large for freezing (Thistle 2002, 15). As the fishers caught more fish, it was even easier to get the fish to market. Major refrigeration complexes were built at Prince Rupert, adjacent to one of the most lucrative halibut fishing grounds in the Hecate Strait, and the Grand Trunk Pacific was competed in 1914. By the outbreak of the war, halibut stocks were going in the direction of fur seals, as the Department of Marine and Fisheries (1914, 252) acknowledged: "It has been found absolutely impossible to keep up the tonnage of the catch, although more boats and gear are being employed each year." More boats and more gear was not the right management direction to take.

The American-Canadian Fisheries Conference of 1918 attempted to draft a treaty between the two countries to impose a closed season to manage the resource. The treaty wasn't signed until 1923, and it instituted a winter closure of three months. The closure did not achieve the desired outcome (Desharnais 2001, 10). It wasn't until 1931 that a more sustainable management tool was employed – a total allowable catch. Even quotas, though, rely on substantial scientific information and enforcement.

The soldiers who returned from the war to take up farming jobs in the salmon or halibut fisheries viewed Asians, the Chinese in particular, as a threat. Chinese immigration to British Columbia increased following the war, and so did racism. Returning soldiers needed jobs, and they viewed Chinese as surplus labour. When the federal government passed the Chinese Immigration Act in 1923, also known as the Chinese Exclusion Act, no provincial or federal political party stood up for Chinese immigrants. The act essentially closed the door to Canada. As well, it intensified the hardships suffered by Chinese who remained in the province because it prohibited their wives and children from joining them.

A combination of local forces and international events also made the Japanese targets of racism and discrimination. Many were involved in the commercial fishing industry and lived in the Japanese enclave of Steveston on the Fraser River, south of Vancouver. Their dominance in the industry was resented by whites who believed that the Japanese should be entitled to far fewer fishing licences than returning soldiers who wanted to enter the fishing industry. By coincidence, the alliance between Great Britain and Japan also cooled during the war and its aftermath, when each nation grappled for control of China. When the alliance ended in 1922, so did the need to treat the Japanese as "equal" in British Columbia. That same year, a royal commission (later known as the Duff Commission) was struck to examine the domination of the Japanese in the fishing industry. It recommended the revocation of 40 percent of their gill net licences and 25 percent of their troll licences. These licences were then handed over to white and First Nations fishers. It did not end there. In 1925, Japanese seining and herring licences were reduced by 50 percent. Forced out of fishing, many Japanese took up farming in the Fraser Valley. In 1928, immigration laws were tightened further to allow only 150 Japanese in per year. The door to Canada was all but shut.

First Nations also faced a setback in their pursuit of Indigenous Title in the 1920s. They expressed their discontent over the recommendations of the McKenna-McBride Commission, which had excluded the issue of Indigenous Title, and petitioned for a royal commission to examine the question of their land not being surrendered or ceded. Unfortunately, in 1921, the provincial government persuaded the federal government to amend the Indian Act to prohibit First Nations from hiring lawyers to fight for Indigenous Title.

Coal mining also faced a setback after the war. The price of coal, similar to metals, escalated during the war even as the amount of coal mined did not (see Figure 7.3). Coal was an essential energy resource for the steam engine,

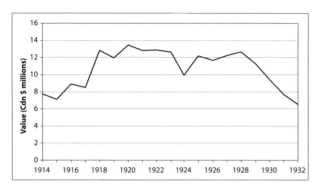

Figure 7. 3 Value of coal, 1914–32
Source: BC Ministry of Energy, Mines and Mineral Resources (1989, Table 8-A).

cooking, heating, smelting, and lighting and creating electricity. However, technological changes were beginning to make coal obsolete. The combustion engine, which ran on petroleum, replaced the steam engine, and greater amounts of electricity were being created through dams and hydro power. Coal mining was never easy and never prosperous for workers. There were no mines without casualties and some with a great many due to explosions, fires, cave-ins, and poisonous methane gas (afterdamp). Bitter strikes lasted months or, in the case of one strike in 1912, two years and involved mining families being evicted from company housing and being forced to live in tents over the winter. What made conditions even more difficult was that politicians often sided with company owners, resulting in the militia being called in, mass evictions, mass arrests, riots, gunfights, and deportations (Bowering 1996, 259). The decline in this industry following the war may have been a good thing as it was responsible for 1,147 of the 1,600 recorded mining fatalities between 1878 and 1926 (Seager 1996, 217).

In other industries, technological innovations and mechanization during the war resulted in the assembly line being used to mass produce uniforms, guns, ammunition, tanks, planes, and ships. These innovations increased the productivity of workers, and, following the war, this system of production, also known as Fordism, was applied to the production of many other consumer goods, including automobiles, fishing vessels, and bulldozers, and in the production of lumber, pulp and paper, tins of salmon, and concentrates of metals. Energy systems were also in transition. For example, electricity, used

mainly for lighting, became integral to electric motors used on the assembly line, the smelting of ores, and the production of pulp and paper. Petroleum use, essential to the automobile and diesel engines in ships, including fishing vessels, also became more common.

Forestry was one of the few industries that expanded in the 1920s. It was fuelled largely by increased domestic demand from local communities, from the Prairies and eastern Canada, particularly after the opening of the Panama Canal, and from the United States, which was experiencing a housing boom. Increased production was facilitated by the bulldozer, which made road-based rather than rail-based logging feasible. The chainsaw, which was fuelled by a combustion engine, was also incorporated into the harvesting process. Many sawmills were allowed to use river systems to transport their logs, which resulted in the construction of splash dams that impounded water and allowed log booms to race downstream, causing salmon habitats to be disrupted. There were no provincial policies protecting streams from these logging practices, which were harmful to commercial and First Nations fishing interests.

The Panama Canal, which was completed in 1914, allowed for the movement of British Columbia lumber and Prairie grain to eastern Canada and beyond. But surpassing conventional, eastward rail transportation was not a simple task, especially when the war interrupted shipping. There were no grain elevators on Vancouver's waterfront until the federal government built one in 1916, and there was some question as to whether sending grain through a subtropical region would affect its quality. On the other hand, grain growers in Alberta and Saskatchewan saw the advantage of sending their produce west, particularly when the Great Lakes froze over, thus halting the eastward movement of wheat. They were also interested in reaching a growing Asian market.

By the 1920s, grain shipments via the Panama Canal began to take off, from 0.5 million bushels in 1921 to 95.4 million bushels in 1928–29 and 96.9 million bushels in 1932–33 (Everitt and Gill 2005–06, 40). By 1929, coastal British Columbia had eight grain elevators, and subsidized rail rates (the Crow Rate) were helping prairie farmers to transport their wheat to these new terminals.

The assembly line process also made mass-produced vehicles possible, and the demand for them increased

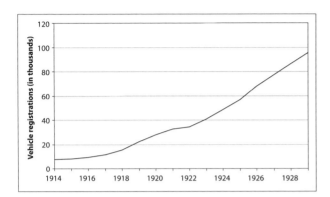

Figure 7.4 Vehicle registrations, 1914–29
Source: Statistics Canada (1932).

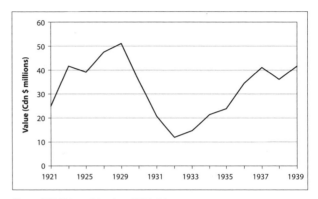

Figure 7.5 Value of lumber, 1921–39
Source: Statistics Canada, *Canada Year Book,* 1923–42, www66.statcan.gc.ca/
acyb_000-eng.htm.

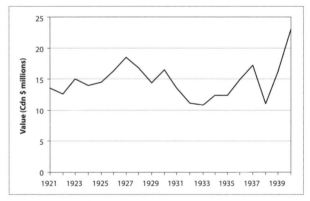

Figure 7.6 Value of pulp and paper, 1921–40
Source: BC Ministry of Forests (1940, Table 2).

during the 1920s from 18,000 to nearly 100,000 vehicles (see Figure 7.4) (Bowering 1996, 259). Increased vehicle registrations forced British Columbia to switch from left- to right-hand drive in 1922 to be consistent with the United States and the rest of Canada, although some provinces did not change until 1924. Of course, with an increase in vehicles came a demand for more roads, of which there were some 26,000 kilometres by 1922. Most were gravel or dirt roads, and most were regional in nature. Vehicles in those days were mainly used for touring, although there was some discussion of linking Vancouver to Mexico and eastern Canada, but these were just discussions. At that point, there was no provincial network, largely because of the cost of building bridges. A stop-gap measure for some regions was to use wooden planks on existing railway bridges to accommodate both vehicles and trains.

THE DEPRESSION

A number of factors combined to result in the Great Depression. There was considerable speculation in stocks that were worthless, which resulted in a major financial crisis, especially in the United States, which experienced bank failures that were felt across the border by Canada, a major trading partner. Overproduction of goods was another concern at the time, and nations such as the United States reacted with trade barriers.

The combination of the stock market crash in 1929 and tariffs resulted in a drastic reduction in the world market price for most commodities, and British Columbia's resource-dependent economy was hard hit, as were most provinces in Canada and countries around the world. The timber, fish, and mineral markets fell by two-thirds by 1923 (Bowering 1996, 264). Resource industries closed, and unemployment soared. The unemployment rate among trade union members rose from 2.6 percent in 1929 to 26 percent in 1932 (McMartin 2008). For those who were gainfully employed, wages plummeted to nearly half of what a worker had received in 1929.

The province's leading resource industry – forestry – suffered a drastic decline. House construction came to a halt throughout Canada, and the United States placed a tariff on BC lumber coming across the border (see Figure 7.5). Unfortunately, other wood products from BC such as pulp and paper (see Figure 7.6), shingles and shakes, and plywood followed a similar trajectory that led to price collapses.

Although the demand for most of BC's metals – copper, silver, lead, and zinc – increased during the 1920s as the economy picked up, prices plummeted during the Depression, and the Anyox copper smelter closed in 1935 (see Figure 7.7). The price and production of copper and other metals did not recover until the late 1930s, when another war loomed. The Depression also ended any hopes of a steel mill, even though the province had passed the Iron and Steel Bounties Act in 1929, which promised a subsidy of $3.00 per tonne for smelted BC iron ore and a further $1.50 per tonne for steel. The Mineral Survey and Development Act, also passed in 1929, provided geological information to prospectors and facilitated the sampling of ores.

By the end of the Depression, when war seemed inevitable, the search for metals was once again taken up. In 1940, the *Canada Year Book* reported: "Iron still holds its long established position as the chief war metal, but it is no more essential than the manganese required in steel, and the nickel, chromium, cobalt, and molybdenum that, used in small proportions, give strength, toughness, hardness, resistance to shock, endurance, or other properties to the many steels used in war machines" (Statistics Canada 1940, 299). In the late 1930s, the Department of Mines offered geology courses and assistance building mining roads and trails. These policies led to relatively rare metals such as molybdenum being discovered.

Gold was the only metal to increase in production during the Depression. By the early 1930s, the fixed price of gold had been adjusted upward from $21 to $35 per ounce (see Figure 7.8). A gold rush of sorts followed, but the boom had more to do with high unemployment rates and the return of desperate individuals to old placer-mining sites in the Cariboo, Kootenays, Kamloops, Cassiar, Omineca, and Atlin regions. The desire for gold also brought more corporate investment in lode mining, mainly at Bridge River near Lillooet and at Zaballos on the west coast of Vancouver Island, where a new, rich gold mine opened in the 1930s.

Farming also suffered: only a couple hundred new farms were established during the Depression, and even mechanization reversed as horse-drawn equipment

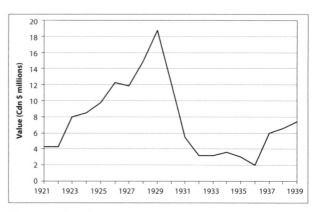

Figure 7.7 Value of copper, 1921–39
Source: BC Ministry of Energy, Mines and Mineral Resources (1989, Table 6).

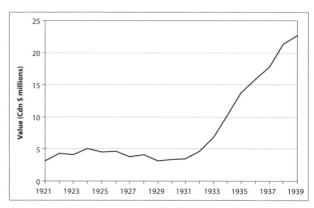

Figure 7.8 Value of gold, 1921–39
Source: BC Ministry of Energy, Mines and Mineral Resources (1989, Table 6).

began to replace tractors that cost too much to operate. Prairie farmers, who transported their grain to west coast elevators, took some of the hardest hits because of trade barriers. The price of wheat crashed from around a dollar a bushel in 1928 to around thirty cents in the 1930s (Siamandas 2012). Poor economic conditions combined with a severe drought on the Prairies forced many farmers into bankruptcy and to abandon their farms.

The commercial salmon fishery was already in serious trouble before the Depression, and developments in the 1930s only made things worse. Although Canada contended that salmon born in BC streams belonged to Canada, many of the returning sockeye and pink salmon swam through the San Juan Islands, which were owned by the United States, before reaching the Fraser River. Another fish war was brewing. Americans argued that salmon could be fished once they swam into international waters. Although the federal government imposed a number of conservation-based regulations, American fishers did not have to adhere to the rules – and they did not. They harvested a far greater quantity of Fraser River sockeye than did BC fishers.

These fish wars were not kind to fish, nor were they sustainable. They resulted in more and more fish being harvested, fewer salmon escaping to their spawning grounds, and less salmon for First Nations. In 1937, Canada threatened to send its seine boat fleet into international waters, which was only three miles offshore, to harvest all sockeye salmon before they reached the American San Juan Islands. The threat was real, and the United States agreed to sign a treaty that split the harvest 60–40 in favour of Canada.

The fact that a single fleet of seine boats could capture the entire inventory of salmon was in itself a major management problem – one of **overcapacity**. The gatekeeper model was to blame: a lack of licence restrictions, aside from those that were race-based, had resulted in intense competition and further investment in fishing technologies, technologies that allowed individual fishers to capture as many fish as possible within an opening and sometimes during a closing. Unfortunately, the Depression only put the overall sustainability of salmon in greater jeopardy as fishers caught more salmon to make up for the poor prices being offered.

With so many resource industries shut down and people unemployed, British Columbians demanded that the government do something to get people back to work. The federal government argued that providing jobs was the responsibility of the private sector, not the government, and that governments should only facilitate their endeavours. However, one "solution" posed by the provincial and federal governments was the creation of work camps for single men in the interior. Nearly 250 camps were set up, housing close to fifteen thousand men, and the costs were shared between the federal and provincial governments (Siamandas 2012). The men built roads, parks, airfields, and other public infrastructure in return for room and board and twenty cents per day. The camps created spatial patterns of hardship across the landscape

and were little better than prison camps with little pay for necessary work. Discontent led to unionization and protest. One of the largest campaigns occurred in 1935, when the unemployed jumped onto boxcars and headed east to protest government policies in the On to Ottawa Trek. The protest was halted by the RCMP in Regina, a riot broke out, and the protest contributed to the fall of R.B. Bennett's Conservative government that same year.

At the provincial level, the Liberal government of Duff Pattullo (1933–41), was somewhat of an anomaly in Canada in that Pattullo believed in stimulating the economy through government spending. In 1936, workers began construction on two major bridges – both for the automobile – in the Lower Mainland. The Pattullo Bridge, which was paid for through public spending, was completed in 1937, linking New Westminster to Surrey across the Fraser River. The second – the Lion's Gate Bridge – was sponsored by the Guinness Corporation (of beer fame). The road pushed through Vancouver's Stanley Park, and the bridge, which crossed Burrard Inlet's First Narrows, gave the Guinness family access to their residential development, the British Properties, in West Vancouver, which would house white families only. The Lion's Gate Bridge was completed in 1938, and both bridges were toll bridges.

Pattullo had other, more grand, schemes for provincial development, including developing the north, incorporating Yukon into British Columbia, constructing the Alaska Highway, and the fostering of oil development in the Peace River district (Wedley 1990–91, 64). With insufficient provincial revenues, poor prices for resources, and lack of federal government support, his vision lay dormant. Although the economy slowly rebounded in the last half of the 1930s, specifically the demand for resources, the economic and social damage of the Depression remained, and most British Columbians blamed federal government policies for the lack of employment.

THE SECOND WORLD WAR

The Depression impoverished British Columbians, and Canadians and made it clear that there were many flaws in the political system. The Second World War (1939–45) highlighted contradictions in Canadian policy-making. During the Depression, the federal government's proposed solution for unemployment had been relief camps and restraint in public spending. However, the Rowell-Sirois Commission, which was struck towards the end of the Depression, in 1937, took three years to examine federal and provincial responsibilities and concluded that although the federal government gained the most in tax revenues, provincial governments increasingly had greater costs for health, education, and social assistance (especially because of the Depression). In other words, there was no social safety net.

The war reversed the federal government's position: it created Crown corporations and engaged in massive government spending. In 1940, C.D. Howe became the minister of the Department of Munitions and Supply and "could do virtually anything he wanted in the interest of the war, including seizing private property, diverting necessary materials from civilian use to war production, and altering contracts" (Bothwell 2009). His department-controlled markets, allocated natural resources, set production volumes, and hired specialized manpower. He created twenty-eight Crown corporations to mine and refine natural resources, build tanks, produce aircraft and naval ships, and supply weapons and ammunition (Chase 2013). Unemployment disappeared, and the Canadian landscape was transformed by the federal government, the sponsor and catalyst of massive industrialization.

British Columbia did not gain factories to manufacture tanks, aircraft, or munitions, but vessels were built in the province, employing twenty-five thousand people (Bellett 2013). Equally important, there was increased demand for resources to convert into manufactured wartime items. The demand for minerals and coal increased, as did the need for forest products, energy supplies, and food, including tins of salmon. The province also benefitted as the federal government engaged in creating a social safety net. It introduced unemployment insurance in 1940 and family allowances in 1944.

Although the war resulted in over forty-five thousand Canadian deaths and even more wounded, the economy boomed largely because of government spending – the social evil of war led to a social good of full employment. But not everyone benefitted. The war was disastrous for the Japanese in BC. With the bombing of Pearl Harbour by the Japanese on 7 December 1941, the vulnerability of Canada's Pacific coast became apparent, and over twenty-two thousand Japanese were labelled "enemy aliens." Most were suspected of being spies and had their radios, cars, and fishing vessels impounded, Japanese newspapers and schools were shut down, and a curfew was imposed in communities such as Steveston. "Highly suspect" males were sent to prison-of-war camps in Ontario while others were sent to work camps in the interior. Families were gathered at the Pacific Exhibition grounds and then relocated to detention camps beyond the 100-mile demarcation line (i.e., 100 miles, or 160 kilometres, inland from the coast). Many of these camps were old mining towns in the Kootenays such as Sandon, New Denver, and Greenwood, although the largest camp, Tashme, was new and constructed just east of the Hope slide.

The threat of a Japanese invasion on the West Coast was real and only heightened with the Japanese capture in 1942 of two Alaskan Islands – Attu and Kiska. Arms and equipment needed to be transported to Alaska, which motivated the Americans to build the Alaska Highway. Mile "0" – at Dawson Creek – was sufficiently inland that bombing would be difficult, if not impossible. Similar thinking went into Canadian navy depots for Pacific armaments, which were located in bunkers at Kamloops (near the present location of Thompson Rivers University).

As the war progressed, Japanese people's houses, commercial establishments, and farms were confiscated and auctioned off – they lost everything. Because there was a steady need for labour throughout Canada, a number of Japanese families voluntarily left camps in British Columbia to work on sugar beet farms in Alberta and Manitoba. When the war ended, racism against Japanese in BC did not. Provincial politicians persuaded the federal government to close the camps so that they did not become "permanent" Japanese communities. Japanese were given a choice: locate east of the Rockies or be repatriated back to Japan. Although some four thousand did go back to Japan, most went east, to Ontario in particular. Part of the tragedy of this story was that over 60 percent of those incarcerated had been born in Canada.

The Second World War also contributed to the continued (mis)management of the commercial salmon-fishing industry. Overfishing was justified on the grounds that there was an increased demand to feed armed forces and the nation, and canned salmon was an ideal source of protein. Concerns about overfishing or underlying structural problems in the industry gave way to a single-minded desire to harvest and preserve every available fish (Meggs 1991, 153). Although the problem of over-capacity – too many technologically sophisticated fishing vessels – had been recognized before the war, the Fishing Vessel Assistance Program, introduced in 1942, simply added to it.

The war triggered numerous outside forces that would have an influence on British Columbia. The saying "necessity is the mother of invention" certainly applied to technologies spawned by the war, including the computer, the atomic bomb, plastics, synthetic rubber and fibre, trans-Atlantic aviation, radar, and the jeep (Globe and Mail 1945). Politically, the United Nations was established by 1945 as a means to prevent future wars. On the economic front, American-based institutions rose out of the Bretton-Woods Conference of 1944 – the World Bank, the International Monetary Fund (IMF), and the General Agreements on Tariffs and Trade (GATT). These institutions and agreements reinforced the economies and ideologies of Western Bloc nations and led to the industrialization of a number of Asian states: Japan, South Korea, Taiwan, Hong Kong, and Singapore (the so-called Asian Tigers). These developments would shift trading patterns for Canada from the Atlantic to the Pacific. British Columbia, long the back door to Atlantic-dominant trade in Canada, was well positioned to become Canada's front door to Pacific trade and investment. They also led to fear of nuclear annihilation and the Cold War. The North American Treaty Organization (NATO) was born in 1949 as a military pact against Soviet Bloc nations.

SUMMARY

The arrogance that motivated colonialism continued throughout the first half of the twentieth century. Racism and discrimination were tied to deep-rooted belief in a dominant, superior white society. This belief led to the exclusion of Asian minorities through manipulative institutional policies. Even though they had been in British Columbia first, First Nations were not exempt. The provincial government, convinced that they would vanish, mainly through diseases, denied Indigenous Title to their lands and supported a host of policies enacted by the federal government to assimilate Indigenous Peoples. Perhaps the most culturally destructive initiative was the Indian residential school system, which removed children from families and beat their language out of them.

Renewable resources were treated with the same colonial disregard. The belief that these resources – fish, fur seals, and forests – were endless continued, although there was often discussion about (but little action on) conservation. Elected politicians such as Richard McBride similarly relied on selling off valuable resource lands to fund transportation infrastructure to push back the so-called frontier and get re-elected; they rarely considered or practised sustainable resource development.

The First World War had unanticipated impacts in British Columbia, including an increased demand for resources and troops but a decline in British investment and immigration. The so-called Roaring Twenties saw only modest growth for some resource industries such as forestry and agriculture but more demand for petroleum, automobiles, and roads. Nonrenewable resources were not immune to the economic and political rollercoaster of global conditions. Coal mining peaked in 1912 and continued to decline as it became the casualty of new energy sources such as petroleum and electricity. When the stock market crashed in 1929, plunging the world into a Depression, the value of most commodities fell, and the province experienced shutdowns in most industries and high unemployment. The disastrous economic conditions of the Depression made coal mining all but obsolete. Although the wars stimulated metal production, lode mining, from tunnelling through to smelting, required vast capital expenditures. Depleted and abandoned mines left a legacy of ghost towns throughout British Columbia. At Trail, only the smelter remains.

During the Depression, government "solutions" such as work camps satisfied no one. But the Second World War brought a powerful lesson: government spending could fuel the economy and overcome depressions. As demand for resources escalated and Canada's economy shifted to wartime production, a more industrialized economy emerged, one that now had social safety to protect people from future recessions.

REFERENCES

BC Ministry of Energy, Mines and Mineral Resources. 1989. *British Columbia Mineral Statistics, Annual Summary Tables, Historical Mineral Production to 1990.* cmscontent. nrs.gov.bc.ca/geoscience/PublicationCatalogue/ Miscellaneous/BCGS_MP-80.pdf.

BC Ministry of Forests. 1940. "Appendix to Forest Branch Annual Report, 1940." for.gov.bc.ca/hfd/pubs/docs/mr/ annual/arstats_1912-40.pdf.

BC Stats. 2017. "Population Estimates." gov.bc.ca/gov/ content/data/statistics/people-population-community/ population/population-estimates.

Bellett, G. 2013. "Contracts Mark Boom Time Again for British Columbia's Boom-and-Bust Shipbuilding Industry." *Vancouver Sun,* 7 October.

Belshaw, J.D. 1996. "Provincial Politics, 1871–1916." In *The Pacific Province: A History of British Columbia,* ed. H.J.M. Johnston, 134–64. Vancouver: Douglas and McIntyre.

Bothwell, R. 2009. "Howe, Clarence Decatur." *Dictionary of Canadian Biography,* Vol. 18, *1951–1960.* Toronto: University of Toronto Press. biographi.ca/en/bio/howe_clarence _decatur_18E.html.

Bowering, G. 1996. *Bowering's BC: A Swashbuckling History.* Toronto: Penguin.

Chase, S. 2013. "C.D. Howe: 'The Minister of Everything.'" *Daily Observer,* 14 March.

Department of Marine and Fisheries. 1900. *Thirty-Second Annual Report of the Department of Marine and Fisheries, 1899.* publications.gc.ca/collections/collection_2015/ mpo-dfo/MA1-1-1899-eng.pdf.

–. 1902. *Thirty-Fourth Annual Report of the Department of Marine and Fisheries, 1901.* publications.gc.ca/collections/ collection_2015/mpo-dfo/MA1-1-1901-eng.pdf.

–. 1912. *Forty-Fourth Annual Report of the Department of Marine and Fisheries, 1910–11.* publications.gc.ca/ collections/collection_2015/mpo-dfo/MA1-1-1911-eng. pdf.

–. 1914. *Forty-Seventh Annual Report of the Department of Marine and Fisheries, 1913–14.* publications.gc.ca/ collections/collection_2015/mpo-dfo/MA1-1-1914-eng. pdf.

Department of Naval Service. 1920. *53rd Annual Report of the Fisheries Branch for 1919.* publications.gc.ca/ collections/collection_2015/mpo-dfo/MA1-1-1919-eng. pdf.

–. 1921. *54th Annual Report of the Fisheries Branch for 1920.* publications.gc.ca/collections/collection_2015/mpo -dfo/MA1-1-1920-eng.pdf.

–. 1923. *55th Annual Report of the Fisheries Branch for 1921–22.* publications.gc.ca/collections/collection_2015/ mpo-dfo/MA1-1-1921-eng.pdf.

Desharnais, C. 2001. "The Pacific Halibut Fishery: Success and Failure under Regulation, 1930–1960: The Canadian Experience." Master's thesis, Simon Fraser University.

Dunn, W., and L. West. 2011. "Women Get the Vote, 1916– 1919." *Canada: A Country by Consent.* canadahistory project.ca/1914/1914-08-women-vote.html.

Elections Canada. 2007. "From a Privilege to a Right, 1867– 1919." Chapter 2, *A History of the Vote in Canada.* elections. ca/content.aspx?section=res&dir=his&document=chap2 &lang=e.

Everitt, J., and W. Gill. 2005–06. "The Early Development of Terminal Grain Elevators on Canada's Pacific Coast." *Western Geography* 15–16: 28–52.

Globe and Mail. 1945. "The Second World War," 7 May.

Gosnell, R.E. 1911. *The Year Book of British Columbia.* archive.org/stream/TheYearBookOfBritishColumbia/ YearbookofBC1911_djvu.txt.

Griffin, B. 2000. "The Plywood Industry in BC: Part 1, The Rise and Boom." *Royal BC Museum.* arch.mcgill.ca/prof/ sijpkes/abc-structures-2005/Lectures-2005/lecture-6/ plywoodpart1.html.

Gunn, A. 1976. "The Lost Lands." *The Province* (Vancouver), 20 January.

Harris, C. 1983. "Moving amid the Mountains." *BC Studies* 58: 3–39.

Hume, S. 2014. "B.C. Streets Filled with Patriotism, Then Flooded Like Rivers of Grief as First World War Began." *Vancouver Sun,* 2 August.

Lawson, M., and R.W. Young. 1906. *History and Geography of British Columbia.* Toronto: W.J. Gage.

McIntosh, R. 2000. *Boys in the Pits: Child Labour in Coal Mines.* Montreal and Kingston: McGill-Queen's University Press.

McMartin, W. 2008. "The Great Depression in BC." *The Tyee,* 26 November.

Meggs, G. 1991. *Salmon: The Decline of the British Columbia Fishery.* Vancouver: Douglas and McIntyre.

Millerd, F. 2000. "The Evolution of Management Institutions for the British Columbia Salmon Fishery, 1900 to 1930." researchgate.net/publication/239571925_The_ Evolution_of_Management_Institutions_for_the_British_ Columbia_Salmon_Fishery_1900_to_1930.

Olesiuk, P.F. 2009. "Preliminary Assessment of the Recovery Potential of Northern Fur Seals (Xallorhinus Ursinus) in British Columbia." Fisheries and Oceans Canada, Canadian Science Advisory Secretariat, Research Doc. 2007/076. waves-vagues.dfo-mpo.gc.ca/Library/337174.pdf.

Robin, M. 1972. *The Rush for Spoils: The Company Province, 1871–1933.* Toronto: McClelland and Stewart.

Roy, P.E. 1980. "Progress, Prosperity and Politics: The Railway Policies of Richard McBride." *BC Studies* 47: 3–28.

–. 2011–12. "McBride of McKenna-McBride: Premier Richard McBride and the Indian Question in British Columbia." *BC Studies* 172: 35–47.

Seager, A. 1996. "The Resource Economy, 1871–1921." In *The Pacific Province: A History of British Columbia,* ed. H.J.M. Johnston, 205–52. Vancouver: Douglas and McIntyre.

Shepard, M.P., and A.W. Argue. 1998. *Ocean Pasturage in the Pacific Salmon Treaty: Fact or Fiction?* Canadian Industry Report of Fisheries and Aquatic Science, 242. publications.gc.ca/collections/collection_2007/dfo-mpo/Fs97-14-242-1998E.pdf.

Siamandas, G. 2012. "Prairie Farmers and the Great Depression: Farmers Feel the Tripple Whammy." timemachine.siamandas.com/PAGES/institutions/Prairie_farmers_great%20depression.htm.

Statistics Canada. 1932. *Canada Year Book.* statcan.gc.ca/eng/1932/193206060568_p.%20568.pdf.

–. 1940. *Canada Year Book.* statcan.gc.ca/eng/acyb_c1940-eng.aspx.

–. 1947. *Canada Year Book.* statcan.gc.ca/eng/acyb_c1947-eng.aspx.

–. 1999. "Table M12-22, Farm Holdings, Census Data, Canada and by Province, 1871 to 1971." statcan.gc.ca/n1/pub/11-516-x/sectionm/4057754-eng.htm.

Stephenson, P.J. 2012. "The Pacific Great Eastern Railway and British Columbia." PhD diss., UBC Okanagan.

Thistle, J.R. 2002. "Sea Change: An Environmental History of the Pacific Halibut Fishery, 1878–1960." Master's thesis, UBC.

Wedley, J.R. 1990–91. "Laying the Golden Egg: The Coalition Government's Role in Post-war Northern Development." *BC Studies* 88: 58–92.

Changing Values during the Postwar Boom

8

The Second World War ushered in an era shaped by a number of developments: unprecedented economic expansion, massive government spending, the baby boom and a new youth culture, increased foreign immigration, the birth of car culture and consumerism, suburbanization, the Cold War, high employment levels, and the transformation of Canada's resource-based economy into an industrial one. In this chapter, you'll learn how all of these changes influenced the development of British Columbia – from new, more positive attitudes towards racial and ethnic minorities to unprecedented exploitation of renewable and nonrenewable resources.

The province's contribution to postwar expansion was resources, but it was also during this period that BC transitioned from a rural, resource-dependent province to an urban, service-oriented one. Premier W.A.C. Bennett was in power from 1952 to 1972, the longest running premier in history, and his Social Credit government had to deal with enormous social, technological, environmental, demographic, and global changes. It experimented with old and new ways of thinking and doing things – from new rail- and road-building initiatives to Crown corporations and megaprojects – that helped win elections but altered the landscape beyond recognition, ultimately fuelling the environmental and **Indigenous Rights** movements, which emerged to protest the loss of land and resources.

POLICIES OF PROGRESS IN THE BENNETT ERA

Following the war, the demand for resources and manufactured goods throughout North America only increased. The baby boom, greater disposable income, and substantial immigration meant an increased need for single-family dwellings, which were being constructed farther and farther from urban centres in new suburbs. As the building of new housing accelerated, so too did the demand for consumer goods such as telephones, furniture, and appliances, and new suburbs required automobiles, which increased demand for roads, bridges, and petroleum. Manufacturing boomed within Canada, although much of the production of consumer goods was carried out in the East. But provinces such as British Columbia experienced an extraordinary demand for their resources.

However, the economic "good times" were not experienced equally throughout the province. Postwar BC was profoundly rural; there was no provincial grid for electricity and no highway network. Many communities relied on diesel generators for their electricity, and wood was common for heating and cooking. In some locations, coal was substituted for wood. Communities in the interior, including First Nations, were still pretty much in poverty in comparison to modern, urban Vancouver and Victoria, and they resented the disparities.

Other values that coincided with the postwar economic boom included a new awareness of the many acts and policies that had institutionalized racism and discrimination in the province and Canada as a whole. Good economic times brought new and more tolerant views towards those who were subjected to racism and discrimination – and new policies were forged. Following the war, the Chinese Immigration Act was repealed, and Chinese were entitled to vote in 1947. The continuous-passage rule that prevented South Asians from immigrating was also repealed, and South Asians also gained the right to vote. Japanese immigrants were finally accepted as citizens in 1949, and they were entitled to vote and allowed to return to the coastal region of BC.

For First Nations, things were considerably more complex because of the province's nonrecognition of Indigenous Title and the federal Indian Act. On the positive side, Status Indians were allowed to vote in the provincial election of 1949, even though they could not vote in federal elections until 1960. The ban on the Potlatch and other ceremonies was lifted in 1951, along with prohibitions against hiring lawyers to pursue Indigenous Title.

RAILWAYS, ROADS, HIGHWAYS, AND FERRIES

Provincial policies are often driven by political leaders who have a "vision" or a dream for the future. Premier W.A.C. Bennett's vision was to invest in BC's interior and outlying regions rather than rapidly growing metropolitan centres such as Vancouver and Victoria. Bennett came from Kelowna in the interior of the province and understood the importance of overcoming the friction of distance in a vertically challenged province. In the elections of 1952 and 1953, he also understood that rural ridings outnumbered Greater Vancouver and Greater Victoria ridings by a margin of twenty-nine to nineteen. Providing access to outlying areas would connect the interior and rural British Columbia to national and international markets. Vancouver and Victoria citizens generally and the entrepreneurial and administrative elite in particular recognized that Vancouver's business sector and Victoria's administration would only be enhanced by infrastructure investment, no matter where it occurred in the province.

Bennett's vision took him in a number of directions, including new railways and road initiatives. The Pacific Great Eastern (PGE) was a provincial asset but a public liability: few people lived adjacent to it, and there was little cargo to transport between Squamish and Quesnel. The line was extended to Prince George in 1952 (where it intercepted the CNR, which ran between Prince Rupert and Edmonton and then east to central Canada) and to North Vancouver in time for the 1956 election, thereby linking Vancouver directly north to the CNR. By 1958, the rails were pushed north of Prince George through the Pine Pass into the Peace River region to Fort St. John and Dawson Creek.

In the 1960s, Bennett's railway plans included a spur line to the new town of Mackenzie in 1966, to Fort St. James in 1968, and to Takla Landing in 1969. The intention was to push the line to Dease Lake and then north to the Yukon, where Bennett had a vision of annexing the mineral-rich territory. The right-of-way was cleared, but the rail line was never built. Finally, the PGE was extended from Fort St. John north to Fort Nelson in 1971. All of these additions conformed to Bennett's dream of opening up the north, but they were expensive because they ran hundreds of kilometres of rail through rugged terrain where there were few people and few resources.

By contrast, roads were needed everywhere in the province. Vehicle registrations increased from approximately 135,000 in 1945, to over 400,000 in 1955, to more than 0.5 million in 1960, and to over 1 million in 1970 (see Figure 8.1). Moreover, this change reflected a significant increase in the number of commercial vehicles such as taxis, trucks, and buses. Although there were some 22,000 miles (35,000 kilometres) of roads by the end of the Second World War, they consisted mostly of dirt and

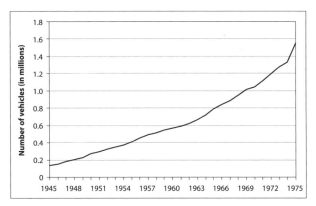

Figure 8.1 Vehicle registrations, 1945–75
Source: Statistics Canada (2014).

gravel and were used mainly for short-haul trips to railway loading stations. There was no provincial highway network, and vehicles and roads were mainly considered a mechanism for touring. By 1965, roads had only increased by 6,500 miles (10,500 kilometres), but rough roads had been transformed into a paved highway network that collapsed time and space in a considerably more flexible manner than rails ever could.

In overcoming the friction of distance, bridges and tunnels represented some of the greatest road transportation costs, but they were obvious monuments to progress. As a consequence, a pattern emerged. The government would announce major infrastructure projects prior to an election, tolls would be initiated with the completion of a project, and then the tolls would be removed before the next election. For example, the four-lane Oak Street Bridge, which facilitated a freeway to the United States and suburban development to Richmond, was announced prior to the 1956 election, opened with tolls in 1957, and then had the tolls removed in 1959, just in time for the 1960 election.

One of the most ambitious highway projects was launched in 1956 and completed in 1962 – the combined federal-provincial building of the Rogers Pass from Revelstoke to Golden. This section of the Trans-Canada Highway replaced the road that followed the Columbia River north of Revelstoke and then made the "big bend" south to Golden. It cost $50 million and cut about 160 kilometres and seven hours off the travelling time (BC Ministry of Forests, Lands and Natural Resources 1963, 20).

The politics and policies of pavement led to improved roads and new highways connected by wide bridges, tunnels, and ferries. Until the 1950s, the coast and islands of BC were accessible mainly by steam vessel. When two privately owned ferry companies – Black Ball Ferries Ltd. and Canadian Pacific Steamships – went on strike in 1958, they stranded ferry-dependent people on the coast, as well as on the many islands off the mainland, including Vancouver Island. In response, the Social Credit government expropriated the ferry service and declared it part of the highway system. Terminals were built at Tsawwassen and Swartz Bay, then the Upper Levels Highway was built through North and West Vancouver to Horseshoe Bay, where crossings to Nanaimo (Departure Bay), Bowen Island, and the Sunshine Coast (Langdale) occurred. Over time, more vessels and more terminals were constructed, but ferry service was no longer controlled by private interests.

Bennett's pro-business government did not hesitate to "nationalize" the private sector, and although these developments were tangible evidence of government spending in the name of progress, new transportation networks posed a challenge to the economic viability of railway passenger service and to rail-based cargo shipping. Public money spent on roads undermined public money spent on rails.

HYDROELECTRIC POWER AND MEGAPROJECTS

Bennett's dream of opening up the interior and the north included much more than extending the PGE and building roads. Alcan's Kemano hydroelectric project became an icon of industrial development on the so-called frontier. In 1950, Alcan Aluminum had agreements with the federal and provincial governments to build the Kenny Dam on the Nechako River and reverse its flow via a tunnel through the Coast Mountains. A hydroelectric plant was established at Kemano, and a transmission line was built to the newly planned town of Kitimat, which had an aluminum smelter. In this instance, hydroelectricity was the catalyst for remote northern development.

Until the 1960s, electricity was supplied in the province mainly through private corporations (BC Electric being the largest supplier), but the provincial government, through its Water Rights Branch, had been involved with electricity production for some time. Because the use of water required a licence, private hydroelectric power generation contributed to provincial revenues. Moreover, the Water Rights Branch promoted hydroelectric development by providing assessments on various rivers throughout the province. For example, by 1925, it had completed some 120 reports on potential power sites and, by 1950, a further 75 reports that included one on the potential of the Nechako River (Northwest Power and Conservation Council 2008). The Water Rights Branch had another mandate: to predict flood events and implement policies to reduce risk. The branch was well aware of major flooding in the Fraser and Columbia River watersheds in 1894 and 1948 and that dams had a two-fold purpose: to hold back the river discharge at critical high-water periods and to produce electricity.

In response to the 1948 flood, one of Bennett's energy policies in the early 1950s included granting permission for an American company, Moran Development Corporation, to revisit its 1934 Moran Dam plan and build one of the largest dams in the world on the Fraser River, just north of Lillooet. But the excessive cost of construction, the purchase of properties upstream from the reservoir that would be flooded, and potential sediment issues put the proposal on hold. In 1954, Bennett supported a project to build a second aluminum plant for the Kootenay region. The Kaiser Aluminum and Chemical Company would construct a $30 million dam on the Columbia River at Castlegar. Unfortunately, the agreement stalled when Parliament passed a law requiring a federal licence (Robin 1973, 208–9).

Essentially, the federal government trumped the provincial government in terms of gaining export revenues. The province viewed the federal government as another hurdle in its plans for energy production. While the Fraser River watershed is almost entirely contained within provincial boundaries, the Columbia and Peace Rivers are not and therefore fall under federal jurisdiction. The export of electricity (or any resources) beyond provincial borders requires federal government approval, and the provincial and federal governments had separate agendas where the export of energy resources was concerned. The second aluminum plant was shelved.

There were other plans for and interest in hydroelectric megaprojects in the mid-1950s. A Washington State electric company proposed building the Mica Dam on the Columbia River to generate power for both the province and Washington. A more contentious and somewhat extreme scheme was the McNaughton Plan, which recommended diverting Columbia River flows to the Fraser River, where dams would then be constructed.

Bennett's ambitions to open up and modernize the interior did not stop with these two watersheds. In 1956, the Social Credit government signed a memorandum of intention with the Wenner-Gren British Columbia Development Company. Like railway grants in the past, this land grant included forest, mineral, and water rights – to nearly one-tenth of the province. The 40,000 square mile (104,000 square kilometres) region stretched from north of Prince George to the Yukon border. Wenner-Gren conjured up a vision of mining and pulp mill towns and hydroelectric development, all connected by a high-speed monorail system (Tomblin 1990, 52). This image of progress and modernity hinged on converting an isolated, rural landscape into the modern-day world. Developing hydroelectric power on the Findlay, Parsnip, and Peace Rivers would mean new industries, new towns, and more jobs. This concept appealed to Bennett and to British Columbians.

By the early 1960s, the Two Rivers Policy became Bennett's grand plan for electrical production. He envisaged dams on the Peace and Columbia Rivers. Power generation from the Peace would be used by British Columbians, while the downstream benefits of damming the Columbia River would be sold to the Americans, who wanted both electrical energy and flood protection. Again, the federal government was viewed as obstructionist when it argued that there would be a glut in electrical production if both rivers were developed at the same time. This was not the only obstacle. The Wenner-Gren surveys identified the electrical potential of the Peace River but few other resources to warrant investment.

Bennett, however, was committed to damming the Peace River. The scheme depended on a private corporation – BC Electric – purchasing the power, which it initially did. But it refused to do so in 1960 after adopting "the position that other sources of energy, including Hat Creek thermal and Columbia hydro power, were more cost-effective" (Ostergaard 2002, 219). What was unusual and bold at that time was Bennett's decision to take over BC Electric, to buy out Wenner-Gren, and to create a new Crown corporation – BC Hydro – yet another "nationalization" of the private sector.

One of the advantages of the province controlling electrical production was that, unlike a private utility, a Crown corporation did not have to pay taxes to the federal government. Through this move, in 1961, Bennett persuaded the federal government to sign the Columbia River Treaty, by which British Columbia received downstream benefits from the Americans for flood protection and energy production. The construction of the W.A.C. Bennett Dam began in 1961 and was completed in 1968. The Columbia River Treaty was ratified in 1964, and four more dams were built – Duncan (1967), Arrow (1968), and Mica (1973) in BC and Libby (1977) in Washington State.

Providing electricity entails delivery, and, in this case, transmission lines and the establishment of a provincial grid were required. In 1964, BC Hydro built a transmission line to Prince George from the Bridge River plants near Lillooet. The line's arrival replaced gas-fired generation in Quesnel and Prince George and integrated the Peace River projects in an emerging provincial transmission grid (Ostergaard 2002, 220). The government, through BC Hydro, also had other grand proposals for electricity megaprojects on the Fraser, Liard, Homathko, Skeena, Nass, and Stikine Rivers (Ostergaard 2002, 218–19). These were grand schemes, indeed, schemes built on the assumption that there would be a steady demand for cheap electricity that, in turn, would industrialize the province and the interior. The realization that these speculative megaprojects would put the provincial government billions of dollars into debt led to them being put on hold.

OIL AND GAS AND PIPELINES

Building dams and providing electricity were not the only energy projects promoted by the Bennett government. The province, along with much of the world, was shifting to petroleum and the combustion engine. There was a proposal, initiated in the late 1940s, to run an oil pipeline from Edmonton, Alberta, to Vancouver's Burrard Inlet, but there was one complication – it would have to be built through Jasper National Park. In response, the federal government implemented the Pipelines Act in 1949. It made interprovincial pipelines a federal responsibility.

It was American pressure that provided the push to construct the Trans Mountain Pipeline. The United States needed a west coast refinery to aid its involvement in the Korean War. The pipeline to Burrard Inlet was completed in 1953, and a spur line ran south of the border to Puget Sound in Washington State. Refineries were then constructed in both locations. In 1961, when more oil was discovered in the Fort St. John region, a second oil pipeline was built from Taylor, on the Peace River, to Kamloops, where it intercepted the Trans Mountain line. British Columbia no longer needed to rely on California oil transported by tankers, and refineries soon appeared in Kamloops, Prince George, and Taylor.

The quest for oil led to exploration of the sedimentary basins off the coast. Coastal communities – including First Nations, commercial fishing interests, recreationalists, and environmentalists – all raised concerns about offshore drilling in this seismically active region. In 1959, the provincial government placed a moratorium on offshore drilling, but the federal government intervened, claiming that the ocean, including the ocean floor, was federal jurisdiction. Shell began staking offshore claims from the early to the mid-1960s. The provincial government's response was to go to the Supreme Court, claiming it had jurisdiction to the seabed and its resources. The provincial government did not want to prevent exploration; it wanted to collect revenue from exploration permits. The Supreme Court ruled in favour of the federal government in 1967, and exploratory drilling of the continental shelf began.

Drilling for oil in the far northeast region of the province carried on, but it was natural gas that was more frequently discovered. The Alaska Highway facilitated the movement of equipment into the region. The West Coast Transmission System was built between 1955 and 1957 from Taylor to deliver natural gas to Lower Mainland customers. It also exported natural gas to the American northwest, thus stimulating further exploration in the region. The supply of natural gas to interior communities such as Prince George and Quesnel gave its residents and businesses a new option for energy outside of diesel. It led to the construction of gas-fired electrical generation plants, which were connected by transmission lines to communities such as Vanderhoof and Fort St. James (Tomblin 1990, 60). By the 1960s, a pipeline linked Kamloops to the Okanagan and Kootenays, and a line was later extended to Prince Rupert and Kitimat. For much of the province, natural gas came from the Peace River region, but one of the largest single users was Burrard Thermal in Port Moody. The gas-fired generators were built with a capacity to provide nearly 1,000 megawatts of electricity to the Lower Mainland.

FORESTRY AND MINING

Despite these developments, forestry provided the provincial government with the greatest revenues. Wood products continued to be produced in coastal regions in the 1950s – mainly from private, Crown-granted forest lands – but the shift to road-based logging resulted in a dramatic increase in the harvesting of interior forests.

Large multinational corporations were interested in investing in BC's resource frontier, and the provincial government encouraged regional economic development. A host of new pulp mills were constructed, mainly in the interior at Castlegar, Skookumchuck (near Cranbrook), Kamloops, Quesnel, Prince George, and Mackenzie. Two new mills were built on the coast – Gold River on Vancouver Island and Kitimat on the north coast. These large corporate entities had a strategy to diversify and integrate their forest interests. In other words, they were not simply single-pulp mill companies; they owned paper mills, sawmills, plywood mills, and, later, oriented-strand-board mills. Moreover, the waste material from a sawmill, such as slabs, could be chipped and utilized as pulp. The forest industry was becoming more efficient in wood utilization as it increased the size and number of clear-cut operations. Assembly lines mass produced products such as two-by-fours and two-by-sixes, and bales of pulp were produced mainly for an American market.

Metal and nonmetal resources such as asbestos were also high on the provincial government's promotion agenda, and the expanded road, rail, and energy systems provided the means to open up these mines. Again, multinational corporations employed enormous earth-moving equipment and state-of-the-art concentrators to process lower and lower grade metals (e.g., copper went

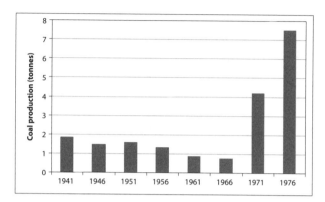

Figure 8.2 Coal production, 1941–76
Source: BC Stats (2009).

from 10 percent down to 0.5 percent by the 1960s). Many of these megaprojects resulted in open-pit mines – mountains were removed, tailings filled up valleys, and concentrates were exported out of the province and, in most instances, out of the country.

By the mid-1960s, the coal industry had also made a huge recovery, but demand was no longer for the steam engine but for metallurgical (or coking) coal, which was required to produce steel goods (see Figure 8.2). Thermal coal, or coal used to produce electricity, was also in demand, but not for British Columbia. These types of coal were in the Crowsnest Pass region of the east Kootenays, and demand came from Japan. However, Crowsnest Industries of Fernie Ltd. planned to transport the coal on American rail lines to a port in Washington State. Bennett formulated a new policy: coal would be transported by rail through BC to a new coal "super" port at Roberts Bank just north of the Tsawwassen Ferry Terminal. The problem with the plan was that the federal government had jurisdiction over ports and did not want another port facility competing with the federal-run coal facilities on Burrard Inlet. Bennett threatened to build the port facility regardless, but eventually backed away. The provincial government did manage to influence the timing and planning of the megaproject, and the port was completed in 1970 (Tomblin 1990, 60).

When new mines and pulp mills were built in unpopulated areas, new communities were required to service them, but the provincial government did not want company towns to reappear. As past events had demonstrated,

they were not particularly healthy communities, either in terms of the job site or social conditions, and they existed outside the Municipal Act. Kitimat, built as a planned town in the wilderness during the 1950s, represented another option. It conformed to the Municipal Act in that there were elected councils, workers owned their own homes (although apartments and mobile homes were options), and retail franchises were independent of the major employer, Alcan, as were schools. In many ways, it was a "normal" community.

In 1965, the Social Credit government passed the Instant Town Act, which required new towns to follow the Kitimat model. Gold River, which centred on a pulp mill on Vancouver Island's west side, was the first **instant town**, and it was followed in quick succession by more towns: Mackenzie north of Prince George (lumber plus pulp and paper), Sparwood in the East Kootenays (coal), Fraser Lake west of Prince George (molybdenum), Logan Lake near Kamloops (copper), Tahsis on the west coast of Vancouver Island (lumber), Elkford in the east Kootenays (coal), and Granisle north of Houston (copper). All of these towns had a common element. They were single-industry towns dependent to a great degree on the continued operation of the mine or mill. Their promotion by the government gave the illusion that they were permanent communities. Employees invested in housing with the expectation of long-term stability.

FISHERIES AND WHALING

Although commercial fishing came under federal jurisdiction, it remained a provincial interest, particularly when it came to regulating the overcapacity of technologically sophisticated fishing vessels such as seine boats. In the mid-1950s, the commercial salmon-fishing industry experienced yet another crisis, and Sol Sinclair was commissioned in 1958 to recommend policies to remedy the situation. Sinclair understood the problem: "In a fishery with unrestricted entry, fishing will always be intensified beyond the optimum economic level and most likely to the point where the net yield is wholly dissipated" (Sinclair 1960, 12). He proposed restricting vessel licences and putting levies on catches (Ross 1987, 189). But making a policy recommendation through a commission is one thing; having politicians implement it is another.

The whaling industry had even fewer rules than the salmon industry. Although it was a relatively new industry in the province, catch records from 1908 to 1967 reveal that nearly twenty-five thousand whales were caught (Nichol et al. 2002, vi). This relentless pursuit of whales, in BC and around the world, resulted in most whales being relegated to the endangered species list. Coal Harbour on Vancouver Island, the last whaling station in the province, closed in 1967.

Trawl fishers, also referred to as draggers, likewise transformed the region and its resources. Trawlers were used by BC and US fishers, but foreign vessels also became significant players in the 1960s in international waters off the coast. These vessels were designed to harvest groundfish such as cod, hake, and halibut, but they actually harvested anything close to the ocean floor and didn't distinguish between types or ages of fish. As a consequence, immature halibut and species not deemed economically viable become bycatch and were thrown back – dead. Halibut stocks declined rapidly (Pearse 1982, 15). Trawlers are not sustainable technologies and only increase the complexity of managing a fishery such as halibut, which requires specific quotas.

CULTURAL AND POLITICAL SHIFTS

BC's population nearly doubled during this era as a result of the baby boom and immigration from Europe and from across Canada (see Table 8.1). People were attracted to the region because of economic opportunities, and the demographic transition led to considerably younger population as well as a shift from a rural to an urban lifestyle and employment. By 1971, the majority of the population was under forty, and the baby boom generation had different values than their parents' generation, which had been shaped by the impoverishment of the Depression and the distress of the Second World War.

With the rise of the feminist movement, the invention of the birth control pill, and urbanization, the "traditional" role of women changed greatly. Education and careers became a priority for both women and men, and families had fewer children. The Department of Industrial Development, Trade and Commerce (1968, 18) commented, "A phenomenon of labour-force growth in British Columbia has been a rapid increase in female employment during the past decade. The number of females

Table 8.1

BC population, number and percentage, rural and urban, 1951–71

	Population	Ten-year change	% rural	% urban
1951	1,165,210		47.2	52.8
1961	1,629,082	463,872	27.4	72.6
1971	2,184,621	555,539	24.3	75.7

Source: Department of Industrial Development, Trade and Commerce (1968).

employed increased by 81.6 per cent between 1957 and 1967." Fuelled by these changes, Victoria College became a university in 1963, Simon Fraser University opened in 1965, and community colleges appeared throughout the province to facilitate higher education.

The baby boom generation grew up watching television and being bombarded with media images of what was happening nationally and globally. Flying and international flights also became more common during the 1960s, allowing young people to experience other parts of the world and other ideas. One of the overriding global concerns was the Cold War, which posed a threat to Canada because the shortest distance between the Soviet Union and the United States was over the North Pole. In the early 1950s, a series of radar stations, named the Pine Tree Line, was established across the country between 50° and 53° north latitude. Six of the thirty-nine Canadian stations were in British Columbia, and many were operated by American personnel, a reminder of the imminent danger that Canada faced as the middle ground between two nuclear powers.

The threat of attack became more serious with the emergence of new technologies for bombing. The intercontinental ballistic missile (ICBM) was much more efficient than jet fighters. ICBMs made the Pine Tree Line largely obsolete because the warning would be too late for antiballistic missiles to be launched. As a consequence, two new radar lines were completed farther north by 1957. The Distance Early Warning Line was a series of radar stations in the Far North (roughly at the sixty-ninth parallel). They were built and operated by the United States, which raised concerns about Canadian sovereignty. A third line – the Mid-Canada Line (about 55° north latitude) – was a series of radar stations built by Canada at the same time as the DEW line, in part because of sovereignty issues.

Partly fuelled by the Cold War, many colonies throughout the world demanded to become independent nation-states. The original 51 members of the United Nations in 1945 grew to 132 members in 1971; some decolonized smoothly, but others were influenced by the Cold War partition between Communist and capitalist bloc nations. Cuba was aligned with the Communist bloc, and the missile crisis of 1962, when Soviet missiles were to be installed essentially next door to the United States, only heightened anxiety of another war until Russia backed off. But the assassination of President Kennedy the following year renewed fear of war.

The decolonization of Vietnam proved to be particularly contentious. France did not want to give up this colony, and SEATO (Southeast Asia Treaty Organization) was formed in 1954 to restrain the spread of Communism to locations such as Vietnam. The Vietnam War began in late 1955. It was initially fought by French soldiers, but with the French losing a number of battles and Americans fearing the spear of Communism, Americans became involved in the mid-1960s until the war ended in 1975. One of the most contentious issues for the American public was the introduction of a lottery-based conscription system. The so-called draft was heavily criticized and protested in the United States and led to "draft dodgers" coming to provinces such as British Columbia. Once they arrived, many of these young, well-educated people held antiwar views and were devoted conservationists who exerted an enormous influenced on the province.

The music of the 1960s and 1970s also reflected antiwar sentiments and concerns about the environment, and the fact that more people had more leisure time meant they placed more importance on recreation, where their food came from, and preserving nature and wilderness. They demanded sustainable resource practices and more transparency in policy-making.

THE IMPACT OF RESOURCE EXPLOITATION

Historically, there had been little interest in preserving wilderness or protecting the environment in BC. The only exception had been when one industry such as forestry or mining had a negative effect on another industry such as commercial fishing. When Strathcona Park, the first

provincial park, was created Vancouver Island in 1911, the goal was to protect the land from industrial development. However, pre-existing mining claims and timber holdings were exempted from the legislation (BC Parks 2007). By the 1950s, the "park" had been seriously altered from its original configuration by mining, forestry, and hydroelectric dams, which caused a few to protest. It should have been an embarrassment to the province, but neither the provincial government nor the majority of the public were overly concerned at that time.

More provincial parks were added after 1911, but it wasn't until 1957 that parks were separated from the Ministry of Forests – an obvious conflict of interest. Clearly, parks were not a priority during the Bennett era. They were not created to conserve wilderness but to accommodate automobile touring, where one could pull off the road to enjoy a rest. Parks increased in number from 59 in 1949 to over 300 in 1972, giving the appearance of increased environmental protection (BC Parks 2007). But the opposite was true. There were over 3,600 hectares of park in 1949 but only 2,900 in 1972. In 1940, parks had been classified into three categories: A (more protection), B (mining and forestry may be allowed), and C (small recreation areas). When the system was reassessed in the 1960s, it was revealed that abuse of the classification system included altering the status of parks and eliminating them altogether. For example, the Kenny Dam, which created the Nechako Reservoir, removed land from Tweedsmuir Park, and the Mica Dam, which flooded the Columbia Valley, reduced the size of Hamber Park. In 1965, the Park Act was modified to allow mining in Class A parks if they were over 2,000 hectares in size. Preserving wilderness was not a priority for the provincial government.

As these examples suggest, one of the biggest differences between resource exploitation before and after the Second World War was not the policies but the development of technologies that ramped up the scale of resource harvesting, leaving a bigger footprint on the land. Economically, the province was booming, but on another level the cumulative impact of projects by big business and big government had led to massive environmental degradation of the air, water, and land.

However, in its desire to grow the economy, the Bennett government worried little about the environment. For example, in its haste to dam the Nechako River for Alcan's Kemano aluminum project, it showed little concern for the environment or local people. At the proposal stage, there was opposition from commercial fishing interests regarding the damming of a salmon-bearing river. The dam destroyed salmon runs on the Nechako and had other negative impacts: it resulted in a slower Fraser River, which in turn increased the mortality of salmon fry. What others found more upsetting was that the forests had not been harvested before they were flooded to form Ootsa Lake or Nechako Reservoir. Another issue, although not a great concern at the time, was the relocation of over seventy-five Cheslatta Dakelh (Carrier) First Nations. Their gravesites were flooded, and their hunting and trapping lands were radically altered. Non-Indigenous settlers were also displaced.

Again, the environment was not much of a concern for the provincial government when it built the W.A.C. Bennett Dam in the 1960s. The dam created Williston Lake, the largest lake or reservoir in the province – rendering productive valleys into single-purpose basins to produce hydroelectricity. More than 1,700 square kilometres of forest were flooded, creating an unstable aquatic environment in which fish are still laden with mercury. An Elder of the Kwadacha Nation reported that nine people drowned as the community tried to navigate the reservoir on flat-bottomed boats designed for river travel (Wakefield 2016).

Later, it was acknowledged that the downstream impact of the W.A.C. Bennett Dam included a drastic reduction in flows of the Peace River, which caused wetlands in the Athabasca Delta in Alberta to dry up, which seriously reduced the area's biodiversity and caused hardship to those who depended on its flora and fauna. Both non-Indigenous people and the Tsek'ehne First Nations were displaced. Living off the land by hunting, fishing, and trapping was considerably more difficult as animal habitats were permanently altered. It is ironic that one of the province's bragging rights at that time was that the dam would be "the province's biggest – so big, in fact, that it would change the climate of the north" (Sherman 1966, 220).

Although the Columbia River Treaty dictated that logging should occur before dams were built on the river, scarce agricultural land and forests were lost, biodiversity diminished, and some 2,300 non-Indigenous settlers and First Nations were relocated, creating the need for ferries to be installed to facilitate transportation across new reservoirs. People in the Peace River region and the Kootenays resented that their regions were paying a heavy environmental and social price for hydroelectricity. The W.A.C. Bennett Dam created electricity for the south of the province, not the north; similarly, the Columbia River Treaty dams benefitted the United States, not the Kootenays. Moreover, there was province-wide apprehension because BC Hydro and the provincial government had published plans to alter many more British Columbia watersheds in the pursuit of even more electricity.

There was also growing unease about the rate at which forests were being clear-cut. Economically, the forest industry brought in the greatest revenues for the province, but there had been warnings from the early 1900s that forestry practices were not sustainable. These warning led to the first Forest Act in 1912, but little was done with respect to reforestation. Dissatisfaction with the amount of timber cut, the allocation of rights to Crown timber supplies, and the sophistication of harvesting technologies led to the 1945 Royal Commission on the Forest Resources of British Columbia (Sloan Commission), followed by a second Sloan Commission in 1955. These commissions led a change in the Forest Act and formed the basis of our present tenure system, which was intended to encourage large corporations to invest through long-term access to fibre supply. The assumption was that these companies would reinvest in plants and equipment to make themselves even more competitive and not funnel profits and wealth out of the country. By viewing forests as fibre, the Sloan commission gave a scientific stamp to the view that old-growth forests should be harvested before they fell down, and it set the stage for foreign control of the forest sector.

Other developments contributed to overharvesting and environmental damage. Road-based logging presented an attack on old-growth forests. Logging roads, no longer confined to the linear patterns of rail-based logging, became switchback roads that climbed ever steeper slopes. These roads often resulted in slides and debris flows that deposited major accumulations of sediment in fish-bearing streams. By the 1960s, combustion-based technologies, including the chainsaw, had also facilitated an ever-increasing allowable annual cut and a shift to the largely untapped forests of the interior. When the provincial government permitted the use of rivers for log drives, the result was yet more damage to salmon-spawning channels, which brought opposition from the commercial fishing industry, sports-fishing interests, First Nations, and conservationists.

There were plenty of sawmills in the interior by the 1960s, but no pulp mills until a mill was constructed at Castlegar in 1961. Road-based forestry facilitated an explosion of pulp mills, which spread throughout the interior to Cranbrook, Kamloops, Quesnel (two mills), Prince George (three mills), and Mackenzie. Two new mills were constructed on the coast at Gold River and Kitimat. These mills were controlled by large corporations that consolidated and integrated the various components of forest products (pulp, paper, lumber, plywood, shingles, and so on). As well, these pulp mills were responsible for introducing effluent into the air and water, even as the provincial minister of highways, Phil Gaglardi, argued that pulp mills were not the smell of pollution but "the smell of money" (Sandborn 2010).

Forest products were only one of many postwar resources in demand. New copper mines opened, but this time around, large, earth-moving technologies resulted in open-pit mining wherever lower grades of minerals became economically viable. Although much of the prospecting had occurred in the 1950s and earlier, production increased considerably in the 1960s in places such as Endako, Afton, Craigmont, Boss Mountain, and Granduc. It left giant scars on the landscape as mountains were torn down and valleys filled with tailings, adversely affecting water quality. Fuelled by Asian demand for coking or metallurgical coal, mining companies in the Kootenays employed similar technologies.

The search for hydrocarbons left its own footprint. New seismic technologies produced linear corridors through the forests of northeastern BC in an attempt to detect oil and natural gas deposits. Pipelines created other patterns that interfered with water systems and animal habitats. Moreover, increased dependence on oil in the 1960s fuelled public concern about offshore oil drilling and oil tankers. A number of catastrophes put pressure on both the provincial and federal governments to enact policies to halt these activities. An oil barge sank off Bowen Island and contaminated local beaches in 1964; an offshore oil well exploded near Santa Barbara, California, in 1969, causing major environmental damage; and an oil tanker, *Arrow,* ran aground in Chedabucto Bay, Nova Scotia, in 1970 – causing the worst oil spill in Canada. These disasters and considerable political pressure persuaded the provincial government in 1971 to establish a policy banning tanker traffic on the West Coast. In 1972, the federal government enacted a similar policy, particularly in the treacherous northern waters of British Columbia, and it followed it up with a moratorium on offshore drilling.

THE ENVIRONMENTAL AND INDIGENOUS RIGHTS MOVEMENTS

The wilderness was receding, and the environment was being degraded with the rapid development of hydro-electric megaprojects, the exploration and discovery of oil and natural gas, open-pit mining, clear-cut logging, new pulp-and-paper mills, and new roads, rails, transmission lines, and pipelines. The environmental concerns did not end there: agricultural land was rapidly being transformed for urban, commercial, recreational, and industrial uses, and many were questioning the increased use of herbicides and pesticides. More and more, there was a growing awareness that "solutions" were not necessarily going to come from science and technology; rather, a change in attitude was required, one that questioned how we harvest and process resources, including the food we eat. Rachael Carson put much of this in perspective with her 1962 book, *Silent Spring,* which revealed the consequences to human health of bioaccumulation.

Environmental concerns had been raised prior to the 1950s and 1960s. Some of the most vocal groups were the United Fisherman and Allied Workers Union (UFAWU, formed in 1945) and First Nations, who led the charge over degraded salmon-spawning channels and the fishing industry in general. The BC Wildlife Federation, which traced its roots back to the 1890s, and local fish and game clubs were also concerned about fish habitats, as well as habitats for all wildlife.

Environmental organizations in the United States were also beginning to have an impact on industrial policies. For example, the US Sierra Club, which also began in the 1890s, was active by the 1960s in stopping dams that flooded river valleys and served as a model for political involvement to thwart policies that harmed the environment. BC had its own provincial environmental organizations by the late 1960s and 1970s, including the Society Promoting Environmental Conservation (SPEC) and Greenpeace. Moreover, the media was instrumental in spreading the views of Beatniks, Hippies, and environmentalists who expressed antiwar sentiment and questioned the status quo in terms of resource development.

It was not simply environmental organizations that expressed concern over the deteriorating condition of BC's landscape. First Nations were greatly affected, which only increased their resolve to settle the issue of Indigenous Title. The organization of First Nations, and their objection to the taking of their land, has a long history, including First Nations delegations going to Victoria, Ottawa, and England to demand the recognition of Indigenous Title, all to no avail.

However, in the late 1960s, the federal government attempted to resolve the issue, not just for First Nations in British Columbia, but nationally. The **1969 White Paper,** a piece of proposed legislation, projected fairly radical policies with respect to First Nations and the responsibilities of the federal government. In effect, the federal government proposed the elimination of the Indian Act and, therefore, the concept of Status Indians. All reserves would be divided up as private property among band members, and all existing treaties would come to an end. First Nations reacted rapidly and powerfully. New provincial and federal Indigenous organizations formed and came together to protest the fact that the 1969 White Paper ignored Indigenous Rights and Title, failed to address discrepancies in reserve size, and did nothing to reverse the historical marginalization of First Nations across Canada. The federal government listened to the objections and withdrew the White Paper.

When it came to Indigenous Title, the *Calder* case had long-term implications both provincially and nationally. Frank Calder, Chief of the Nisga'a, whose Traditional Territories lay in the northwestern region of the province in the Nass River watershed, argued that Indigenous Title (called "aboriginal title" in the court cases) had never been given up or ceded through a treaty and that the Nisga'a continued to have title to their lands. The BC Supreme Court and the BC Court of Appeal disagreed. The Supreme Court of Canada's decision, which came down in 1973, recognized that Indigenous Title did exist historically, but it could not decide whether it still existed. Three judges agreed that Indigenous Title continued to exist, and three stated that it had been extinguished. A seventh judge disqualified the case on a technicality. Even though no decision had been made as to present-day Indigenous Title, the ruling persuaded the federal government to adopt policies on comprehensive and specific "claims." **Comprehensive claims** constituted recognition that First Nations had title to Crown lands where no treaties had been signed; therefore, the modern treaty process evolved to surrender these lands through negotiation. The concept of **specific claims** applied to issues dealing with existing treaty lands or to specific reserves where lands had been taken away unjustly (e.g., cut-off lands, see Chapter 7).

Conflicts in BC only became more intense in the late 1960s and '70s when the provincial government rejected the notion of comprehensive and specific claims, insisting that Indigenous Title had been ceded to the Crown at Confederation. The provincial government's concern was that giving up Crown lands meant giving up its resource base. Many confrontations evolved over resource development on Indigenous Title lands and, inevitably, First Nations joined forces with the environmental movement.

Bennett's Social Credit government treated First Nations, nature, and resources in a manner similar to the premiers before him. It argued that First Nations did not have rights to the land, that resources were never-ending, and that the endless "frontier" was there for exploitation. However, Bennett did recognize that society was becoming aware and concerned about environmental deterioration and land-use conflicts stemming from resource development and urban sprawl. His government passed the Pollution Control Act in 1967 and created the Land Use Committee in 1969. The committee consisted of five cabinet ministers who were to resolve environmental and land-use conflicts. Two years later, the Environment and Land Use Act came into effect, "but the early performance of the Committee indicated a rather conservative and restricted interpretation of the scope of the Act" (Crook 1975).

URBANIZATION AND THE SERVICE ECONOMY

By any measure, the Bennett era witnessed profound changes to the landscape. The government created "modernization" through a host of infrastructure and resource-based developments such as a provincial electric grid, a paved road network, more rail lines, natural gas pipelines, and massive resource developments in forestry and mining. However, the downside to these many developments was their impact on the environment and the people. Bennett played to the underdeveloped interior of the province and engaged in a politics of pavement that unlocked isolated interior communities but, at the same time, fuelled urbanization. By 1971, over 75 percent of the population lived in urban areas.

In 1951, only three municipalities outside of the Greater Vancouver and Greater Victoria core regions exceeded 10,000 people – Kamloops, Penticton, and Trail. Twenty years later, there were nineteen. Urbanization spread up the Fraser Valley to include four centres – Chilliwack, Langley, Matsqui/Abbotsford, and Mission – while Vancouver Island saw a similar extension north of Victoria to North Cowichan and Nanaimo and the resource municipalities of Port Alberni and Campbell River. Powell River, on the Sunshine Coast, emerged as another industrial and service centre. The Okanagan, a popular destination, had three main cities – Vernon, Kelowna, and Penticton – and Kamloops became a major service and transportation centre for the South Central Interior. There were two major centres in the Kootenays – Trail and Cranbrook – and Prince George became an important location for the North Central Interior. Prince Rupert and Kitimat became ports for the north coast, and Dawson Creek and Fort St. John became service centres for the Peace River region. The Greater Vancouver Regional District (Metro Vancouver), which had over 1 million people, dominated the province in terms of population and its economy. The Capital Regional District, with Victoria as its principal municipality, had nearly two hundred thousand people and was the second largest urban centre in the province.

Cities in these decades underwent a major transformation in their form and structure. Mass production of and dependence on the automobile, along with urban policies in the form of zoning, promoted the development of residential suburbs, shopping centres, and industrial parks farther and farther from central business districts, blurring the distinction between rural and urban. As cities sprawled into the surrounding countryside, agricultural land was consumed for nonagricultural uses.

In response to urbanization and population growth, the province began to create regional districts in 1965, especially for areas outside of incorporated municipalities. The boundaries for twenty-nine regional districts were drawn over the next few years, giving the total population, whether an incorporated municipality or a rural unincorporated region (electoral area), political representation. Each regional district became an amalgam (or federation) of municipal representation and rural electoral area representation, and the federation was represented by an elected regional director. For example, the Sunshine Coast is a relatively small regional district north of the Greater Vancouver Regional District. Its board was made up of representatives from two municipalities – Gibsons and Sechelt – and six electoral Areas. Decisions here ranged from local and regional fire protection, house numbering, regional water supply, land-use planning, and many more services.

One of the most important functions of the regional districts was the financing of projects – parks, sewage systems, water systems, and other infrastructure projects. If a municipality faltered in its loan payments, the regional district would provide financial backing. The Municipal Finance Authority, created in 1971, ensured that municipal or regional district borrowing was at the most favourable interest rates.

For people, the shift to urban living also involved a fundamental change from goods-based employment to services-based employment, which only enhanced urbanization. The goods-producing sector, which was resource-related, dominated in 1951, representing 53.1 percent of the labour force (see Table 8.2). By 1970, however, it had declined to 42.5 percent (see Tables 8.3 and 8.4). Provincial revenues for 1971 followed pace: by then, natural resources amounted to only 10.6 percent (see Table 8.5). Of course, indirect revenues were generated through general taxes and income taxes, but resource revenues in general were on a downward path. What these data collectively indicate is rapid urbanization in the short space of twenty years; equally important, but not widely recognized, is that the role of resources in providing jobs and provincial revenues shrank.

Table 8.2

Labour force, by industry, 1951

Goods	%
Agriculture	6.4
Fishing, hunting, and trapping	1.2
Logging and forestry	4.0
Manufacturing and mechanical occupations	13.3
Mining and quarrying	1.6
Transportation	8.2
Communications	1.6
Manual labour	7.2
Electric light and power and stationary engineering	1.8
Construction	6.4
Not stated	1.4
Total	*53.1*

Services	
Proprietary and managerial	8.9
Professional	7.5
Clerical	11.0
Commercial	6.6
Financial	1.0
Service	11.9
Total	*46.9*

Source: Department of Industrial Development, Trade and Commerce (1958).

SUMMARY

Between 1952 and 1972, W.A.C. Bennett's government had a profound impact on resource development. His policies were little different than those of previous eras in that his government was not above awarding Crown grants or investing public money in infrastructure projects to encourage private investment in the resource sector and win elections in the process. But several new directions and developments set the Bennett era apart. Bennett's Social Credit government went beyond traditional investment in rails and roads to engage in large megaprojects and take control of electricity production and delivery. The creation of BC Hydro was an industrial strategy to provide energy for mining and forestry ventures. The creation of BC Ferries constituted another annexation of the private sector wrapped up in the

Table 8.3

Labour force, by industry, 1961

Goods	N	%
Agriculture	23,290	4.0
Forestry	21,068	3.6
Fishing and trapping	4,478	0.8
Mines, quarries, oil wells	8,179	1.4
Manufacturing	113,019	19.6
Construction	36,338	6.3
Transportation, communications, and other utilities	62,806	10.9
Total	*269,178*	*46.6*

Services	N	%
Trade	99,278	17.2
Finance, insurance, real estate	22,642	3.9
Community, business, and personal service	123,782	21.4
Public administration and defence	46,001	8.0
Unspecified or undefined	16,767	2.9
Total	*308,470*	*53.4*
TOTAL, ALL INDUSTRIES	577,648	100.0

Source: Department of Industrial Development, Trade and Commerce (1968).

Table 8.4

Labour force, by industry, 1970

Goods	N	%
Agriculture	23,000	2.8
Other primary	41,000	5.1
Manufacturing	141,000	17.4
Construction	53,000	6.5
Transportation, communications, and other utilities	87,000	10.7
Total	*345,000*	*42.5*

Services	N	%
Trade	155,000	19.1
Finance, insurance, real estate	42,000	5.2
Community, business, and personal service	223,000	27.5
Public administration	46,000	5.7
Total	*466,000*	*57.5*
TOTAL, ALL INDUSTRIES	811,000	100.0

Source: Department of Industrial Development, Trade and Commerce (1972, 18).

Table 8.5

Provincial revenues, 1970–71

	$	%
General taxes	241,473,357	19.2
Income and succession	339,539,840	27.0
Natural resources	133,992,294	10.6
Motor vehicles	117,862,240	9.4
Federal government	251,660,029	20.0
Government enterprises	66,031,334	5.2
Other	107,658,644	8.6
Total	1,258,217,738	100.0

Source: Department of Industrial Development, Trade and Commerce (1972, 23).

promise that it was simply an extension of the highway system that would reduce the isolation of coastal and island communities.

The provincial government's encouragement of oil and natural gas pipelines ensured that much of the province had a diversified and relatively inexpensive energy base. As well, Bennett's insistence that a "super" port be built for coal at Roberts Bank resulted in greater control over the export and import of resources. Policies such as the Instant Town Act were also structured so that the province could work closely with large corporations in the creation of new resource communities. More than anything, Bennett played to the underdeveloped interior of the province but, in doing so, facilitated the urbanization of the province. The province went from 52 percent urban in 1951 to over 75 percent by 1971. These strategies increased the public debt but resulted in successive wins at the ballot box.

However, the Bennett government didn't anticipate the forces of the new global economy, the changing values of a society dominated by a younger demographic in which more women worked and families had fewer children, or the effective organization of environmental groups and First Nations to demand recognition of Indigenous Title and better care of the environment. Certainly, "good" economic times contributed to a reduction in institutional racism – fewer working people felt threatened by "others" taking their jobs. Acts were passed that allowed Asians into British Columbia and Canada,

and they gained the right to vote. First Nations were a thornier issue for the government. Although they did get the vote and had their right to perform ceremonies restored, the provincial government failed to recognize Indigenous Title.

The environmental movement grew in reaction to the unprecedented scale of exploitation in the megaproject era. Open-pit mines left giant holes in the earth, deposited tailings on the landscape, and often put the quality of water in jeopardy. The forest industry's clear-cuts pushed back the "frontier" in the interior, and pulp-and-paper mills fouled the air and water with their pollution and damaged salmon habitat. Hydroelectric projects flooded productive valleys and created single-purpose lakes or reservoirs that became some of the largest in the province. Oil and natural gas pipelines radiated throughout the province, and offshore exploration for hydrocarbons raised the risk of earthquakes in a seismically active region. And all this activity took place as renewable resources were being harvested at an unsustainable pace. Whales became an endangered species, and salmon continued to meet crisis after crisis because of insufficient salmon returning to spawn. Only one fishery – halibut – gave some direction in terms of sustainability. Fuelled by the new youth culture, environmental and conservation organizations formed, and the public demanded more accountability, setting in motion a new, more difficult political equation for governments: how to balance resource development with conservation. When the voting age was lowered from twenty-one to eighteen in 1970, the younger generation had a larger voice (Elections Canada 2007), and it voted Bennett out.

REFERENCES

BC Ministry of Forests, Lands and Natural Resources. 1963. *Water Resources Service: Annual Report.* Victoria: Queen's Printer.

BC Parks. 2007. "The History of BC Parks." env.gov.bc.ca/bcparks/aboutBCParks/history.html.

BC Stats. 2009. "British Columbia Annual Coal Production and Value from 1866 Onwards." catalogue.data.gov.bc.ca/dataset/bc-annual-coal-production-from-1866-onwards/resource/00d7f4ae-092f-4361-84ac-45bf3cf87a10.

Crook, C.S. 1975. "Environment and Land Use Policies and Practices of the Province of British Columbia." Report for the British Columbia Institute for Economic Policy Analysis, Victoria. for.gov.bc.ca/hfd/library/documents/bib50069_vol1.pdf.

Department of Industrial Development, Trade and Commerce. 1958. *British Columbia Facts and Statistics, 1958*. Victoria: Queen's Printer. llbc.leg.bc.ca/public/pubdocs/bcdocs2014_2/123586/1958.pdf.

–. 1968. *British Columbia Facts and Statistics, 1968*. Victoria: Queen's Printer. llbc.leg.bc.ca/public/pubdocs/bcdocs2014_2/123586/1968.pdf.

–. 1972. *British Columbia Facts and Statistics, 1972*. llbc.leg.bc.ca/public/pubdocs/bcdocs2014_2/123586/1972.pdf.

Elections Canada. 2007. "Modernization, 1920–1981." Chapter 3, *A History of the Vote in Canada*. elections.ca/content.aspx?section=res&dir=his&document=chap3&lang=e.

Nichol, L.M., E.J. Gregr, R. Flinn, J.K.B. Ford, R. Gurney, L. Michaluk, and A. Peacock. 2002. "British Columbia Commercial Whaling Catch Data, 1908 to 1967: A Detailed Description of the BC Historical Whaling Data Base." Report for Fisheries and Oceans Canada. dfo-mpo.gc.ca/Library/264154.pdf.

Northwest Power and Conservation Council. 2008. "British Columbia." nwcouncil.org/history/BritishColumbia.

Ostergaard, P. 2002. "Energy Resources in BC's Central Interior." *Western Geography* 12: 216–29.

Pearse, P. 1982. *Turning the Tide: A New Policy for Canada's Pacific Fisheries*. Vancouver: Commission on Pacific Fisheries Policy.

Robin, M. 1973. *Pillars of Profit: The Company Province, 1934–72*. Toronto: McClelland and Stewart.

Ross, W.M. 1987. "Fisheries." *In British Columbia: Its Resources and People*, ed. C.N. Forward, 179–96. Western Geographical Series, vol. 22. Victoria: University of Victoria.

Sandborn, T. 2010. "Making BC a Green Jobs Machine." *The Tyee*, 21 October. thetyee.ca/News/2010/10/21/BCGreenJobs/print.html.

Sherman, P. 1966. *Bennett*. Toronto: McClelland and Stewart.

Sinclair, S. 1960. *Licence Limitation – British Columbia: A Method of Economic Fisheries Management*. Ottawa: Department of Fisheries and Oceans.

Statistics Canada. 2014. "Roads and Road Transport." Series T142–194. statcan.gc.ca/n1/en/pub/11-516-x/pdf/5220021-eng.pdf?st=LX2cR7kP.

Tomblin, S.G. 1990. "W.A.C. Bennett and Province-Building in British Columbia." *BC Studies* 85: 45–61.

Wakefield, J. 2016. "BC Hydro Acknowledges W.A.C. Bennett Dam's Dark Side." *Dawson Creek Mirror*, 10 June.

Resource Uncertainty in the Late Twentieth Century

9

In 1972, W.A.C. Bennett's government was ousted from power, and British Columbians elected their first NDP government. The government instituted a number of forward-looking, some would say "socialist," initiatives that would eventually have a significant impact on the province. Even though the Social Credit Party came back to power in 1975, its victory coincided with the 1970s energy crisis and a major downturn in the global economy during the 1980s recession.

In this chapter, you'll learn how the energy crisis of the 1970s and recession of the 1980s ignited a revolution in transportation technologies that collapsed time and space. These technologies combined with digital technologies to fundamentally alter, through globalization, the production of goods and services, and it was the resource sector in British Columbia that was the most negatively affected. You will also learn about different political parties' opposing ideologies, which have resulted in policies detrimental to the environment and opposition to modern-day treaties.

When the NDP came back to power in the 1990s, the Cold War had ended, and a new era of free trade began. The NDP was fully committed to ensuring environmental protection and was the first provincial government to recognize Indigenous Title, but it was forced to deal with problems both new and old. Resource industries were increasingly volatile and subject to new technologies and restructuring, which put new pressures on the environment, increased job insecurity, and contributed to the continued urbanization of the province.

First Nations continued to protest the degradation of their lands and turned successfully to the courts to regain Indigenous Title to their lands and resources. The NDP government passed legislation to protect the environment and gained the first modern-day treaty. But like governments before it, the NDP also consciously promoted resource development as if it continued to be the lead employment base in the province, which was no longer the case.

"SOCIALIST" POLICIES IN A RESOURCE-DEPENDENT PROVINCE

In 1972, British Columbia voted W.A.C. Bennett's Social Credit Party out and, for the first time, the New Democratic Party (NDP), led by David Barrett, in. Many considered Barrett's government socialist, and even though it was only in power from 1972 to 1975, it left its mark on the landscape.

The NDP addressed many of the issues pertaining to resource exploitation and the environment that the Bennett government had ignored, including the consumption of agricultural land for nonagricultural uses. For example, the Environment and Land Use Committee (ELUC) was given much more power to resolve conflicts. In 1973, through the Park Act, it phased out mining in provincial parks to preserve wilderness (Crook 1975, 69). Nine new parks were added under the NDP. ELUC also undertook the decentralization of land-use planning and resource development through the creation of resource management regions through which the public could be more involved. The overharvesting of forests was another concern for the NDP because reports showed that the province's cutting allowances were not sustainable (Crook 1975, 46). Increased control of forest tenure (e.g., through tree farm licences) by only a few large multinational corporations led to a royal commission on forestry, which was headed by Peter Pearse. The commission commented on the intensity of multinational corporate control: the ten largest companies controlled over 53 percent of the forest tenures. Another aspect of the industry was that it produced poor economic returns, which meant that it would not be able to attract capital for either expansion or maintenance (Schwindt 1979, 13). Finally, the industry was dependent on US markets for lumber, pulp, and paper sales. Pearse recommended more forest-resources planning, "including the need to address water, fisheries, recreation, wildlife, and other concerns." These concerns were included in the 1978 Forestry Act, passed by the next government (Vyse et al. 2006, 129).

The mining industry, which had a significant environmental footprint within the province, was not exempt from scrutiny. The Mineral Royalties Act, passed in 1974, resulted in a super tax on mineral royalties. The belief was that mining companies were not paying their fair share to the province, especially when world market prices for commodities such as copper were reaching record levels.

The NDP's most important environmental initiative was saving agricultural land for agricultural purposes. The increase in roads and vehicles throughout Vancouver

Island, the Lower Mainland, the Okanagan, and the Kootenays in particular had put the most productive farmlands in jeopardy, not just for urban and industrial expansion but also for recreational uses such as golf courses, theme parks, motels, and private camps. These activities occurred in other regions of the province but at a slower rate. The 1973 Land Commission Act employed the federal government's system of classifying agricultural land on a scale from Class 1, the best, to Class 7, no agriculture capabilities. It "froze" all Class 1 to 4 lands, creating the Agricultural Land Reserve for future generations. The act had its critics, including the Social Credit Party.

Despite this progress, the NDP, like the Social Credit government before it, was unwilling to recognize Indigenous Title, which continued to frustrate First Nations. However, building on earlier policy changes, the federal government in 1974 committed to negotiating the Nisga'a comprehensive claim, and it opened the door to specific claims by passing the BC Indian Cut-Off Lands Settlement Act to resolve the issue of lost reserve lands following the McKenna-McBride Commission, back in 1916. First Nations were slowly being conceded rights and title to their lands.

Although the NDP lost the 1975 election for a number of reasons, including being labelled as opposed to free enterprise – one thing was certain: environmental conservation would be a significant factor in future elections and policy-making. Politicians and opponents of environmentalism feared the environmental movement would either derail future development projects or add huge costs to them, particularly when the public was starting to demand that Crown corporations and municipalities be environmentally responsible. In this new political era, you were either for the environment or for prosperity, economic opportunity, and jobs. A looming global economic crisis brought on by fear that the world was running out of oil helped tip the scale towards environmentalism.

THE 1970S ENERGY CRISIS AND THE SHIFT TO CONSERVATION

The Social Credit Party's victory in 1975 coincided with a major global event over which it had little control – the 1970s energy crisis. The crisis began in 1973 when the Organization of the Petroleum Exporting Countries

(OPEC) proclaimed an oil embargo. The price of oil quadrupled, shifting from $3 per barrel in 1973 to $12 a barrel in 1974. Prices levelled off but then jumped to nearly $40 per barrel by 1979. For oil-rich Alberta, the embargo was a windfall; for British Columbia, where approximately three-quarters of oil was imported, it was a disaster. In response, the federal government introduced the National Energy Policy (NEP) in 1980. For Alberta, one of the most contentious aspects of the policy was that the price of oil would be subsidized – that is, Canadians would pay less than the world market price. Further, the federal government would be entitled to royalties and taxes on oil, which meant that Alberta would gain less in revenues. The NEP also entailed promoting more exploration for oil and gas in the Arctic so that Canada would become energy (read "oil") self-sufficient.

The crisis made it clear that much of the world, including Canada, was heavily dependent on oil, not just for transportation but also because of the petrochemical industry, which was producing a myriad of new products, from plastics and paints to polyesters. As oil prices rose, so did the price of petroleum products. Of greater concern was the fear that the world would run out of oil. A number of nations instituted gas rationing. Inflation was rampant. For example, goods and services worth $1.00 in 1970 increased to $2.17 in 1980 (Crompton and Vickers 2000). As prices increased, so did the demand for wages, sparking a spiral of higher wages, higher prices, and outrageous interest rates.

As individuals began to recognize that energy was lifestyle and that petroleum energy was expensive, a host of unpredictable energy-related changes occurred throughout the province. Conservation of energy became part of the environmental movement, and in people's everyday lives it translated into turning off lights, turning down thermostats, improving insulation, installing air-tight wood-burning stoves, installing solar panels (mainly for hot water), building homes with passive solar principles, and demanding vehicles that consumed less gasoline. In other words, individuals initiated many of these energy-saving projects.

Governments were slow to develop policies to conserve energy or develop alternative energy sources. However, BC Hydro did promote what it saw as an alternative to oil – a massive shift to electricity. BC Hydro forecast an increase in demand of 10 to 12 percent per year. Its solution was no small plan. It would involve the damming of many salmon-bearing rivers – including the Fraser, Skeena, Nass, Stikine, and Iskut – as well as two more dams on the Peace River, others in the Kootenays, and two dams on the Liard River in the extreme north. But according to BC Hydro's projections, even this would not be enough electricity: coal-fired generators would need to be constructed on Vancouver Island, in the east Kootenays, and at Hat Creek near Cache Creek. The proposal also called for nuclear power plants for Vancouver Island and the Lower Mainland.

Of course, the environmental footprint would be enormous, and the protests started to mount. Environmentalists, commercial-fishing interests, the tourism industry, recreational organizations, First Nations, and many others reacted negatively to BC Hydro's projections and warned of the environmental consequences of reservoirs and transmission lines to fish, forests, agricultural land, and animal habitats generally. They objected to coal-fired thermal plants because of the pollution they would create (acid rain in particular) and to the prospect of nuclear plants being in proximity to inhabited areas. The disaster at Three Mile Island's nuclear plant in Pennsylvania in 1979 only heightened the fear.

BC Hydro was attacked for its exaggerated projections and was taken to task by the independent British Columbia Energy Commission (the BC Utilities Committee was not created until 1980), which stated that a 4 to 6 percent increase in demand may be more realistic. Environmental groups suggested that the future direction should be conservation and promoting renewable energy sources, which would result in a 1 to 2 percent per year increase in electrical energy use.

What BC Hydro did not take into account in its exaggerated projections was that individual ingenuity could combat high energy costs, nor did it anticipate the reaction of the largest energy-consuming industry in the province – the forest industry. The high price of oil forced the industry to seek new energy sources, and it did not view BC Hydro's "solution" as the answer. Rather, it determined that wood wastes from sawmills and black liquor (waste from the pulping process) could be burned as hog fuel to produce the energy the industry required. Pulp mills became more energy self-sufficient.

Other industries likewise began to implement energy-conservation technologies or switched to natural gas, which was abundant. In short, the price of oil forced a change in energy consumption, but not necessarily to electricity, as BC Hydro had forecast.

The high price of oil and fear of shortages also caused the government to reassess tankers and pipelines. Even though the provincial and federal governments had policies against oil tanker traffic on the West Coast, they began to explore its potential during the crisis. There was also some apprehension in British Columbia about the United States building a pipeline that would connect north shore Alaskan oil from Prudhoe Bay to the port of Valdez in the Gulf of Alaska, where it would then be shipped to the Cherry Point Refinery in Puget Sound. In 1976, one year before the pipeline was completed, American demand changed, and the focus shifted to transporting Alaskan oil to either Prince Rupert or Kitimat, where a pipeline would then be constructed to move the oil south.

The federal government, in reaction to the idea of potential tanker traffic on the West Coast, launched the West Coast Oil Ports Inquiry in 1977. It dismissed the proposed port facility at Kitimat as being too high a risk. Similar environmental pressures occurred south of the border. Congress ruled that the expansion of the Cherry Point terminal would be environmentally unacceptable (Thompson 1978). There was still concern, however, that Alaskan oil would be transported by tanker past British Columbia to US western ports. A Tanker Exclusion Zone was created in 1977 to keep these vessels 100 miles (160 kilometres) to the west of Haida Gwaii, thereby reducing the risk of an oil spill on its shoreline.

GOVERNMENT CUTBACKS AND RELAXED REGULATION DURING THE RECESSION

By the early 1980s, monetary policies imposed to curb runaway inflation during the energy crisis had sent Canada and much of the world into a major recession. The demand for and price of commodities fell, industries shut down, the need for energy plummeted, and unemployment rose. The recession forced widespread concern about BC's dependence on a limited range of natural resources. Unemployment in the province increased from 6.7 percent in 1981 to 14.7 percent in 1984. In other words,

79,000 people lost their jobs – 3,100 worked in forestry, 10,400 in mining, 23,600 in construction, and 37,700 in manufacturing (half of which worked in sawmills or pulp mills) (Belshaw and Mitchell 1996; Barnes and Hayter 1993). This was a serious recession, reminiscent for many of the Depression.

As the demand for, production of, and price of resources fell, so did provincial revenues. The newly re-elected Social Credit government reacted with policies to cut government costs and stimulate the economy by relaxing environmental regulations. Its budget included measures to reduce the role of government in natural resources management. Within one year, it reduced the staff of the Ministry of the Environment by 30 percent, abolished regional resource management committees, and reduced the number of resource management regions from eight to five (Dorgey 1987, 21). The same government that had developed policies to promote resource extraction and processing was the same government responsible for developing sustainable practices and policies to curb pollution. And those government jobs were being cut. The provincial government made it possible for corporations to harvest resources more easily.

Although the major dams proposed by BC Hydro were delayed because of the recession, protests, and fear of incurring horrendous public debt, three dams were eventually built. Two were medium-sized – one was located in the Kootenays (Seven Mile on the Pend Oreille River) and another on the Peace River below the W.A.C. Bennett Dam (Site B or the Peace Canyon Dam). The largest was the Revelstoke Dam on the Columbia River.

The provincial government also continued to pursue its interest in offshore oil drilling. However, moving forward would require the federal government to lift its moratorium, which it declined to do. The provincial government then went to the Supreme Court in 1982 to challenge federal jurisdiction to the inside waters and continental shelf between Vancouver Island and the mainland. The Supreme Court ruled in 1984 that British Columbia had jurisdiction over the Strait of Georgia; however, there was little prospect of significant deposits of oil or natural gas in that small sedimentary basin. Moreover, the energy crisis was over: the price and demand for oil, and most other energy sources, were in steady decline.

There were a number of energy-related catastrophes during the 1980s that reinforced concerns about nuclear reactors and transporting oil by tanker. The *Arco Anchorage* ran aground near Port Angeles (off Washington State and across from Victoria) in 1985, spilling over 200,000 gallons (700,000 litres) of oil, which then washed up on local beaches. Although the spill was disastrous, it paled in comparison to the grounding of the *Exxon Valdez* in Alaska, which polluted some 1,500 miles (over 2,400 kilometres) of coastline in 1989. The explosion of the nuclear plant in Chernobyl in 1986 caused much of the world to reassess the risks of using nuclear power to produce electricity. On the other hand, environmentalists were pleased when the provincial government placed a five-year ban on offshore drilling in 1989 and when BC Hydro launched its Power Smart Program. In this same year, a new environmental think tank was created – the David Suzuki Foundation.

But the Social Credit government's policies of restraint included cutting funding to many social programs established when the NDP was in power. Some programs were dismantled; others had their budgets drastically reduced. And the government instituted a zero-wage increase policy to fix incomes. The policy attacked long-standing collective labour rights and "initiated a considerable dismantling of the welfare state" (Couture and Macdonald 2013). A near–provincial-wide general strike organized by the Solidarity Coalition was narrowly averted in 1983–84 through negotiations with union organizers. In the past, Asians who earned low raises had been blamed for declining employment – now it was unions.

The reorganization of the government by the Social Credit government included more centralized decision making, deregulation, and privatization. One of its schemes was to give away or sell marginal and failing industries, such as the pulp mill at Ocean Falls, which had been propped up by the NDP. In 1979, the province gave everyone with a driver's licence five shares in British Columbia Resources Investment Corporation (BCRIC), which eventually became worthless. It was the Social Credit's way of making everyone seem like a shareholder in provincial resources. The Ocean Falls mill, one of the BCRIC stocks, was closed in 1980.

The government also partially dismantled the highly contentious Agricultural Land Reserve created by the

NDP. The majority of the population believed in the wisdom of reserving good agricultural land, which is akin to a nonrenewable resource, for agricultural endeavours. The government, however, altered the appeals policy to make it easier to extract agricultural land from the reserve for nonagricultural purposes. Some lands were excluded, but intimidating media headlines linking land extraction to the Social Credit Party curtailed a flood of exclusion applications.

The recession also coincided with the digital revolution and a revolution in transportation. Innovations such as satellites, coaxial cables, the internet, jet air cargo carriers, and supertankers allowed goods to be moved rapidly around the globe and information to move in an instant. New players entered the production world, particularly in Asia, where workers could be hired for very low rates of pay. The traditional assembly line for mass-produced goods such as two-by-fours was under attack, and so were union jobs. The term "restructuring" was used to describe how multinational corporations fragmented the production process for both goods and services into its many components and then used the technologies of movement to outsource the production of these components to least-cost locations around the globe, only to bring them back together for assembly.

As in previous eras, all of these forces – the oil crisis, recession, restructuring, and environmentalism – challenged and transformed the harvesting of natural resources in the province.

RESTRUCTURING FORESTRY
Forestry is a good example of an industry where bad things occurred because of a clash of outside forces. By the 1970s, the province's number one industry in terms of employment and provincial revenues had undergone a significant transition and expansion, made possible by road-based logging. For the first time, the largest volume of wood came from the interior rather than the coast. The energy crisis, however, had a negative effect on the industry: lumber production decreased by 25 percent and pulp production by 18 percent.

The recession resulted in even further losses: production declined by 15 percent between 1980 and 1982 alone (Carrol and Ratner 2007, 46). British Columbia was losing its market share, timber supply was declining, and international competition for lumber and pulp and paper was fierce owing to considerably lower wage rates in other lumber-producing regions of the world. The combination of these factors resulted in a major restructuring of the fairly labour-intensive assembly line process for producing wood products. Mills throughout the province were restructured, resulting in lost jobs. For example, in Alberni on Vancouver Island, the number of jobs declined from 5,358 in 1980 to 3,904 in 1986 (Hayter 2000, 300).

During the recession, the provincial government reacted with forest policies that reduced bureaucracy and allowed large multinational forest corporations to gain even greater control of the forests. In 1982, for instance, the Ministry of Forests proposed a policy change that would allow forest licences and timber sale licences of short duration (but under the government's control) to be converted to tree farm licences, which would mean even greater control of the forests by relatively few large multinational corporations (Geography Open Textbook Collective 2014). Public pressure resulted in a reversal of this policy.

MINE CLOSURES AND MISGUIDED EXPANSION IN THE COAL INDUSTRY
The mining industry was also hit by the recession. The price of metals declined even as foreign countries continued to produce metals such as copper. In response, BC metal mines cut back on production, employed more capital-intensive technologies, and, in some instances, stopped mining altogether. For instance, Kitsault, an instant town built in 1979 in the remote region north of Prince Rupert to house workers for a molybdenum mine, closed in 1982 with the collapse of the world market price for molybdenum. Other mines followed, including Granduc and Scottie near Stewart, Highmont (now part of Highland Valley Copper) near Ashcroft, Horn Silver near Penticton, Carolin near Hope, and the Goldstream mine near Revelstoke.

The coal industry responded to the recession with questionable expansion in a desperate bid to provide resource jobs, regardless of the cost. One coal mine – Wolf Mountain near Nanaimo – closed down in 1985, but both the federal and provincial governments were lured by proposed Japanese investment to develop the metallurgical coal fields in the northeast region of the province.

In other words, if the provincial and federal governments provided the infrastructure, the Japanese would purchase the coal.

Throughout the 1970s, during the energy crisis, the Japanese had given exaggerated projections of their need for metallurgical coal for steel production. Japan, however, was fully dependent on imported petroleum, which continued to escalate in price. As a consequence, the cost of steel production in Japan continued to spiral upward, decreasing demand for steel and, therefore, metallurgical coal. Unfortunately for Canada, both the federal and provincial governments had already invested heavily in the northeast coal development scheme. This scheme included Tumbler Ridge, the instant town built to house workers for the Quintette and Bullmoose mines, known as Northeast Coal, which were linked by a new electric rail line to the main BC Rail line at Chetwynd. The CNR was double-tracked to handle more coal trains running from Prince George to Prince Rupert, and a new super port was built at Prince Rupert. The provincial government invested $277 million in the construction of a grain terminal (Young 2008, 35), and a coal terminal was also completed in 1984. Policies to support northeast coal development cost both the province and the federal government hundreds of millions of dollars. Japan bought the metallurgical coal as promised, but within a few years it reneged on the price it had guaranteed ($110 per tonne, instead paying $82.40 per tonne) and reduced its annual import volume. Quintette closed in 1998, followed by Bullmoose in 2003.

REGULATION AND RESISTANCE IN THE FISHING INDUSTRY

The rising price of oil only marginally impacted the commercial fishery, which continued to be a mismanaged disaster. Federal commissions signal crisis, and the fishing industry had more commissions than all other resources combined. A major problem was the gatekeeper model adopted by the federal government. This management system, which revolves around openings and closings, is highly dependent on knowledge of salmon runs on specific rivers or streams, the technologies used to harvest salmon, and the number of fishers participating in an opening. The system also encouraged ever more sophisticated vessel technologies, especially seine boats.

Although the Sinclair Commission in 1960 had pointed out the need to reduce fleet capacity, the federal government ignored the advice for nearly ten years, and commercial salmon fishing continued down the path towards tragedy until Jack Davis, federal minister of fisheries, came up with the Davis Plan in 1968 to decrease the number of fishing vessels. The buy-back program he instituted reduced the commercial fleet from 6,104 vessels in 1969 to 4,707 in 1980. The initiative appeared to accomplish the goal of capacity reduction; however, Sinclair had recommended that licences should be nontransferable. Had the advice been followed, the number of vessels would have been reduced by attrition, removing the need for a buy-back program (Swenerton 1993, 53). Instead, the federal government allowed old vessels to be replaced and upgraded, effectively doubling or even trebling the capacity of the fleet (Ross 1987, 189). In fact, the seine fleet increased from 286 to 316 vessels when the new rules allowed old gill-net licences to be combined into a seiner licence (Meggs 1991, 95). What is astounding is not just that the Davis Plan backfired but that the federal government provided the incentives to increase capacity: "In direct contrast to the aims of the Davis Plan, the federal government provided Pacific fish harvesters with $133 million from 1974 to 1981, twice the amount that had been provided from 1955 to 1974" (Swenerton 1993, 59).

By the late 1970s and early 1980s, the fishery was in an even worse state of crisis, characterized by declining income for commercial fishers and reduced quantities of salmon for the sports fishery and Indigenous Peoples. Another commission was struck in 1981. This time, it was led by UBC economist Peter Pearse, who had also been a commissioner for the forest industry inquiry. His recommendations were wide-ranging and, in some instances, extremely controversial, particularly since, in his words, the "inquiry pointed inescapably to deficiencies of government policy" (Pearse 1982, 3).

The commission brought the destruction of salmon-spawning streams and the deterioration of estuaries by mining, forestry, and hydroelectric dams to public attention: "These estuaries and foreshore areas are also the main centres of settlement, port and shipping facilities, marinas and industrial developments; and they are the scene of active dyking, filling, dredging, log storage and other operations" (Pearse 1982, 21). He also made note of the issues of predation and increasing levels of pollution in fresh- and saltwater environments. His recommendations included more research on salmon ecosystems, which led to greater investment in the Salmon Enhancement Program, which had commenced in 1977. He recommended a full range of enhancement measures, including the restoration of spawning streams, the creation of new spawning channels, the fertilization of lakes, and an increase in hatcheries. However, he also recommended the introduction of fish farms.

Somewhat more contentious were his recommendations that First Nations become more involved in all aspects of commercial fishing, from hatchery operations to holding more fishing licences. These recommendations coincided with the federal government's establishment, in 1982, of the Northern Native Fishing Corporation. The federal government was attempting to live up to its fiduciary trust as well as its recognition of Indigenous Title. In response, BC Packers Ltd. sold 243 vessels and 252 licences to the new corporation (Cohen 2012, 85).

Other significant changes followed the Pearse Commission, including more enforcement of regulations, licences for sports fishers, a system of restrictions (and even tags) on the daily quota of salmon caught, and no more subsidies for new vessels or improvements to old ones. The "radical" end of Pearse's recommendations included a 50 percent reduction of licences within ten years to reduce capacity, the separation of licences from vessels, and individual catch quotas that required individual fishers to bid for a ten-year licence. Separating licences from vessels was the antithesis of the gatekeeper model, which rendered salmon a common-property resource that could be fished without limits during openings. By contrast, individual catch quotas would privatize salmon and result in the allocation of fixed amounts of salmon to a specific licence – as a consequence, the competition for salmon would come to an end. The fishing industry was overwhelming in its opposition to the proposed changes, and the federal government backed away from any action. Commercial fishers believed they were entitled to as many salmon as they could catch in an opening regardless of the consequences, and the federal government acquiesced.

Desperate conditions often result in desperate political policies. During this period, the province encouraged any

private-sector endeavour that would create jobs, including fish farms in the mid-1980s. Aquaculture, a technology developed mainly in Norway that involved raising salmon in net pens in protected bays and inlets, had been recommended by the Pearse Commission.

Because salmon are raised in salt water, fish farms could have been a federal responsibility; however, the net pens were anchored to the foreshore, so the federal government allowed the provincial government to become the lead agency in awarding foreshore leases for fish farms. Initially, most of these farms were located on the Sunshine Coast north of Vancouver, but considerable opposition mounted and centred on fear of contaminating local waters (although no baseline studies existed), disease transfer to wild salmon, and the privatization of the foreshore. In the late 1980s, the problem was resolved for the Sunshine Coast when the waters became too warm in the summer, leading to plankton blooms and the death of hundreds of thousands of farmed salmon, which were transferred to the local landfill. The industry transferred to the north end of Vancouver Island, as well as to the Pacific side of Vancouver Island, where the water temperatures are more consistent on a year-round basis.

The issue of the international harvesting of salmon was also ongoing. By 1977, Canada's boundaries extended to the present 200-mile (322-kilometre) limit. The establishment of borders, however, had little effect on salmon returning to the Fraser system through the San Juan Islands or other runs returning to BC's rivers via Alaska. Although a portion of the conflict with the United States had been resolved through treaties – for sockeye salmon in 1937 and pink salmon in 1957 – uncertainty about other species continued to result in fish wars, which made management of salmon difficult as both parties tended to overharvest their share. In 1985, however, a treaty was signed that included all species of salmon.

TOURISM

Another industry that was not affected by the recession was tourism. People continued to take holidays because many had more free time. As a result, the provincial government promoted this vibrant resource by bidding to host Expo 86, a project that involved considerable infrastructure spending. Expositions are major events that attract local, national, and foreign tourists, and they occur every four years. Expo 86 celebrated the one hundredth anniversary of Vancouver's incorporation, but it was not a typical World's Fair. It fell on an off year. New Orleans was awarded the 1984 World's Fair, and Brisbane, Australia, hosted it in 1988. Any region can host a World's Fair on an off year, but it must be a theme fair, and Vancouver's theme was transportation. Transportation infrastructure needed to be constructed and showcased. The Coquihalla Highway from Hope to Merritt was completed in time for the exhibition, and it was extended to Kamloops the following year. It took until 1990 to open the Kelowna Connector from Merritt to Peachland. Vancouver's SkyTrain rapid transit system was another major transportation undertaking at the time.

Within Vancouver, Expo 86 centred on False Creek. The neighbourhood was transformed from a traditionally heavy industry area to a trendy residential, recreational, and tourist-oriented destination. Expo 86 was a great success, and tourism increased significantly following the event. However, the promotion of the industry had other implications – tourism and recreation were on a collision course with the traditional extractive resource industries.

ENVIRONMENTALISM AND INDIGENOUS RIGHTS

The Social Credit government's schemes to deal with the energy crisis – building more dams, coal-fired electrical generators, nuclear power plants, fish farms, the promotion of offshore oil and gas drilling, reduction in environmental staff, and weakening the agriculture land reserve – resulted in immense public concerns for the environment. However, the provincial government did create institutions to protect the environment. The Energy Commission, which was established by the NDP in 1973, was replaced by the British Columbia Utilities Commission (BCUC) in 1980, and legislation required proposed energy projects to include environmental assessments (Haddock 2010, 13). After 1979, pesticides and wastes disposed in aquatic and terrestrial systems fell under the newly created Waste Management Act, which was amended to include air pollution in 1982 (Dorgey 1987, 16).

Some environmental issues were site-specific, such as the prospect of the City of Seattle building the High Ross Dam, which would flood the Skagit Valley south of Hope. Protests resulted in the High Ross Treaty being signed in

1984 to put the construction off for eighty years. When Alcan proposed building Kemano II, a dam on the Nanika River, a tributary to the important salmon-bearing Skeena River, to power its new pulp-and-paper mill at Kitimat, protests likewise erupted. Again, opposition from many factions caused the provincial government to cancel the project in 1987 in favour of the Kemano completion project. This undertaking did not require another dam (or an environmental review); rather, it required another tunnel to be built through the Coast Range Mountains to the existing power-generating station at Kemano. The existing Nechako Reservoir could be drawn down to turn more turbines, but even more of the Nechako River's flows would be diverted from entering the Fraser system – posing another serious impediment to Fraser River sockeye runs.

First Nations were also concerned about the environment – their environment. Even though the provincial government was in denial of Indigenous Title, First Nations continued to press for their rights by blockading resource activities, petitioning the federal government, and turning to the courts. And many of these actions dovetailed with the environmental movement. The Constitution Act, 1982 – specifically section 35, which recognized Aboriginal Rights (now often referred to as Indigenous Rights), including Indigenous Title – reinforced their determination to settle the land question. However, the act did not define what Indigenous Rights included – that was left to the courts to decide.

In the mid-1970s, for instance, a dispute occurred on South Moresby Island, Haida Gwaii, when unsustainable logging plans led to a public proposal to protect the area. In 1985, the Haida Nation designated the area a Haida Heritage Site and put up a blockade on Lyell Island (Parks Canada 2012). The conflict was not resolved until a memorandum of understanding was signed in 1987. Conflict between First Nations and the Social Credit government triggered forty blockades between 1984 and 1990, most over logging (Blomley 1996, 31–34). Another notable conflict occurred in 1984, when one of the major forest companies, MacMillan Bloedel, planned to log Meares Island, near Tofino on Vancouver Island. The island had been part of the Tla-o-qui-aht First Nation's Traditional Territory, and the nation declared it a Tribal Park. Environmental groups did not want to see old-growth forests clear-cut and began to spike trees. In the courts, the conflict went all the way to the Supreme Court of British Columbia and then to the BC Court of Appeal, where Justice Seaton ordered an injunction to stop the logging of Meares Island.

In this era, the Stein Valley near Lytton was the last watershed in the region that had not been logged, and environmental groups began lobbying to preserve this wilderness area. It was also a region with significant archaeological evidence of First Nations habitation. In December 1984, the Nlaka'pamux Nation Tribal Council filed a comprehensive claim to the entire watershed (M'Gonigle 1988, 112). Blockades, the organization of a festival in the valley, and a court injunction in 1985 halted the logging. First Nations entitlement to their land and resources was being recognized by the most conservative of institutions – the Supreme Courts.

By 1990, the Social Credit government was mired in protests and blockades that created great insecurity in the forestry sector, which was struggling to be recognized as the number one goods-producing industry. Commercial fishers were not happy with the 1990 *Sparrow* decision, which confirmed the Indigenous Right to fish for food and for social and ceremonial purposes and stated that only conservation could take priority over this right. Those concerned about agricultural land were protesting legislative amendments that allowed the land to be converted into golf courses.

The premier, Bill Vander Zalm, did not want to recognize Indigenous Title, but his government did create a Ministry of Native Affairs, through which some meaningful negotiations over land-use conflicts could be initiated. This policy was an important first step, but the more important one followed the election of the NDP in 1991. One of the NDP's election platforms was to recognize Indigenous Title and engage in **modern-day treaties**. The Social Credit era was over.

POPULATION GROWTH AND THE BIRTH OF FREE TRADE

The population of British Columbia nearly doubled between 1971 and 2001, and it became increasingly urban (see Table 9.1), which was one indicator of the increasing divide between the services-producing (mainly urban) and the goods-producing (mainly rural) sectors of

Table 9.1

BC population, urban and rural, 1971–2001

	Population	Ten-year change	Rural (%)	Urban (%)
1971	2,184,621		24.3	75.7
1981	2,744,465	559,844	22.0	78.0
1991	3,282,061	537,596	19.6	80.4
2001	4,078,447	796,386	18.0	82.0

Source: Statistics Canada (2011).

employment. A geographical divide emerged as well: the Lower Mainland made up nearly 60 percent of the population, and the greater Victoria area added another 5 percent. Although medium-sized centres (75,000 to 100,000 people) such as Kelowna, Kamloops, Prince George, and Nanaimo accounted for many service-related jobs, much of the rest of the province was made up of considerably smaller communities that often relied on resource extraction. This uneven distribution of jobs led the NDP, like governments before it, to consciously promote resource development as if it were the lead employment base in the province and paramount to provincial revenues.

The world was no less turbulent when the NDP came to power in 1991. The Cold War had ended, the Soviet Union had become Russia, former Communist countries had turned capitalist, and the global economic system had become considerably larger and somewhat safer with the reduced threat of nuclear war. The Brundtland Commission had been formed in 1987 in response to worldwide environmental concerns. It published its findings as *Our Common Future*, which advocated for a more sustainable future with respect to resource use as well as the need to protect wilderness. Technologies that collapsed space and time had only increased, facilitating a global division of labour and a shift in the production of goods and services to Asia and other low-cost regions around the world. Japan and the Asian "Tigers" (Korea, Taiwan, Hong Kong, and Singapore) had become more dominant in trade and investment, as had China. For Canada, Atlantic trade patterns had finally been eclipsed by those in the Pacific, and British Columbia had become the front door to Pacific trade.

Free trade agreements to avert protectionism had initially been signed between Canada and the United

States. In 1994, the agreement expanded to include Mexico and was renamed the North American Free Trade Agreement, or NAFTA. The European Union (EU) had signed a free trade agreement in 1992. These agreements had little direct impact on British Columbia, although they did provide an incentive for the province to engage in the competitive wine industry. The General Agreements on Tariffs and Trade (GATT) – another one of the Bretton Woods agreements signed following the Second World War – was discarded in 1993, and the World Trade Organization (WTO) emerged in its place to establish rules to facilitate global trade. Recessions occurred in the free trade era, and the one that hit Asia in the late 1990s curtailed British Columbia's export of resources to that region. The WTO was viewed by many as having excessive political power, promoting anti-union practices, encouraging exploitive working conditions in poorer countries, and lowering global environmental standards, sparking protests, many of them violent (such as in Seattle in 1999).

ENVIRONMENTAL POLICIES AND PROCESSES IN THE NDP ERA

A host of environmental issues dominated during the NDP era. One of the platforms for the newly elected provincial government was a strong stance on the environment, including the preservation of more wilderness areas. Although the NDP was sympathetic to the environmental movement, it could not satisfy its many demands. To formulate environmental policies, the government opted for a decentralized process that was often viewed as bureaucratic, time-consuming, and frustrating. But the NDP did react quickly to preserve the original intention of the Agricultural Land Reserve. Golf courses were removed from the equation, as was the appeal process, which promoted the removal of agricultural land. Legislation that entitled "good" farmland to be placed in the reserve saved agricultural land but not necessarily the farmer, as there were no guarantees that farmers could make a living from their farms. The Right to Farm Act (1996) removed some of the tension, usually over noise and smells, between farmers and adjacent residential dwellers.

The NDP created the Commission on Resources and Environment (CORE) in 1992, and one of the first disputes it dealt with was the proposed Windy Craggy mine. This large copper-rich mineral deposit was located in the remote and rugged Tatshenshini-Alsek region in the northwestern part of the province. A mining plan was put in place in the late 1980s, but it was controversial, particularly because of the proposed treatment of mine tailings or waste rock, which would be deposited on a nearby glacier. The concern was that acid rock deposition could destroy fish populations not only in BC but also in adjacent Alaska. There were also environmental concerns about wildlife populations being affected by the building of roads and pipelines and of concentrates moving out and diesel fuel moving in. CORE made a number of recommendations, and the provincial government turned the region into a Class A park in 1993. In this instance, the environment and retaining wilderness won out over resource development; however, from the perspective of the free enterprise sector, which viewed resources as the driving force of the British Columbia economy, the environment had won out over jobs.

Much of the tension in the province during the 1990s concerned the practices of the forestry industry – excessive rates of harvesting, clear-cutting techniques, the elimination of old-growth forests, erosion from logging roads, and clear-cuts that resulted in deteriorating water quality for fish and human consumption. Others wanted protection of wildlife and the elimination of conflict with First Nations over Indigenous Title. CORE formed four regional round tables – East Kootenays, West Kootenays, Cariboo-Chilcotin, and Vancouver Island – to bring the various stakeholders together to formulate sustainable resource development plans. Unlike the Social Credit government, the NDP decentralized policy-making by involving the public. The government's 1993 Protected Areas Strategy called for protected area to increase from 6 percent to 12 percent, as the Brundtland Report had recommended. CORE's land-use strategies for the four selected regions were completed in 1996, and although the process was intended to end confrontation, it led to mixed feelings as to its success (Anderson 1997, 145).

This was not the end of the planning process but the beginning. By 1995, the Land Use Coordination Office had shifted the focus to smaller regions than those tackled by the CORE process, and each region developed land and resource management plans (LRMPs). Although CORE and LRMP land-use plans evolved out of decentralized round tables, the public did not always feel that they were valued or listened to (Booth and Halseth 2011). The plans developed at these two scales evolved into more detailed, on-the-ground plans through sustainable resource management plans (SRMPs). Moreover, issues dealing specifically with forestry in a region were managed through local resource use plans (LRUPs), and conflicts over water resulted in integrated watershed management plans (IWMPs). All of these round tables took considerable time and commitment by all the stakeholders. Decisions were forged, but many expressed frustration over the process. Decentralizing the entitlement process and relying on consensus decision making created a process that was neither rapid nor easy.

The mandate of the Protected Areas Strategy, which included the preservation of 12 percent of the province's wilderness, was accomplished and (at least in some regions) brought a sense of finality to controversial conflicts over logging through the creation of parks. Meares Island Tribal Park was established in 1991 and followed by Gwaii Haanas in 1993 and Stein Valley Nlaka'pamux Heritage Park in 1995. Many other parks and wilderness areas were created, such as the Great Bear Rainforest, one of the last coastal old-growth forests. It entered the LRMP process, which halted logging plans in 1997, but the conflict wasn't resolved until 2016. Another large tract of land (6.4 million hectares) in the northeast part of the province, known as the Muskwa-Kechika, also came under the LRMP process. The 1998 Muskwa-Kechika Act in 1998 resulted in an integrated resource plan that allowed for both mining and forestry activities and wilderness preserves. But the mining industry walked out of the process, stating that it was too restrictive. By doing so, it reinforced public perception (for some) that resource extraction should not be restricted.

The CORE process did not satisfy all environmental movements. The most notable centred on Clayoquot Sound on Vancouver Island. This old-growth forest region had not been preserved as a protected area by CORE, and protests by the public along with Nuu-chah-nulth First Nations resulted in over eight hundred people being arrested in 1993. The conflict gained worldwide attention as pictures and stories circulated that stated that the province was "the Brazil of the North." Demonstrators demanded a boycott on BC lumber in Europe and the United States. First Nations and environmental organizations had some success when the region became part of modern-day treaty negotiations, which led to the formation of Iisaak, a tree farm licence, in which the majority shareholder (51 percent) was the Nuu-chah-nulth. Furthermore, an agreement in 1999 between Iisaak and five environmental groups – Greenpeace Canada, Greenpeace International, the Natural Resources Defense Council, the Sierra Club of British Columbia, and the Western Canada Wilderness Committee – resulted in sustainable harvesting of the Clayoquot as well as the establishment of protected areas. The outcome satisfied most people, but the conflict required to achieve the outcome left a residual sense of distrust in the government's CORE process.

Environmental concerns about other practices led to the Forestry Practices Code in 1994, which became part of the CORE round table negotiations on land use. The code detailed prescriptions for logging, including for **silviculture**, limitations on the size of clear-cuts, riparian buffer zones to reduce silt in streams and rivers, and a prohibition that no more than 10 percent of soil could be disturbed in a logged area (Hsu 2007, 2). The NDP also raised stumpage rates or taxes on the harvesting of the forests in an attempt to encourage value-added manufacturing (Hayter 2000, 10). The government's 1994 Environmental Assessment Act was even broader in that it consolidated a host of previous legislation to make projects that affect the environment more transparent. The Reviewable Projects Regulations set thresholds that triggered environmental assessments for projects based on their proposed size or production capacity (Haddock 2010, 13).

COMPENSATION FOR PAST PROJECTS AND NEW DILEMMAS

The NDP government had to make a number of decisions on ongoing or past projects and, in some instance, compensate parties. For instance, the federal government gave Alcan's Kemano completion project permission to proceed with no environmental review. Local communities, recreationalists, fishers, and organizations such as the Rivers Defence Coalition, with the support of West Coast Environmental Law, took the federal government to court (West Coast Environmental Law 2012, 5). They lost but pursued a review by the British Columbia Utilities Commission, which concluded that the project would have serious implications for Upper Fraser River salmon. The provincial government cancelled the project in 1995, which prompted lawsuits by Alcan. Alcan and the province came to an agreement in 1997. The aluminum company was compensated, and the agreement established a Northern Development Fund to economically assist First Nations displaced by the creation of the Nechako Reservoir back in the early 1950s. The government also had to pay compensation for past decisions in the Kootenays, where the Columbia River Treaty dams and reservoirs had caused the displacement of people and distress to communities. The Columbia Basin Trust was created in 1995 to fund projects to improve the environment in regions affected.

Past energy-related conflicts were significant to policy-making, as the creation of the Northern Development Fund and Columbia Basin Trust examples attest, but the NDP also had to cope with new developments. Global demand for a cleaner, safer environment was affecting the petroleum-refining industry, which was under pressure to remove lead from gasoline and reduce sulphur emissions. These pressures resulted in Edmonton capturing much of the refining business for western Canada while refineries in Kamloops, in Taylor, and on Burrard Inlet shut down. Only Husky Oil in Prince George and Chevron on Burrard Inlet remained by the early 2000s.

A crisis occurred on 1 August 2000, when the oil pipeline from Taylor to Kamloops ruptured near the Pine River, about 90 kilometres upstream of Chetwynd. It was probably the most expensive oil spill in Canadian history: cleanup and restoration costs exceeded $26 million, and several million dollars were paid out for remediation and to settle claims (Ostergaard 2002, 223). Adding to the tragedy was the fact that the same pipeline had been breached in 1974 near Prince George, leading to a number of recommendations, including the installation of automatic shut-off valves on both sides of the Pine River. However, the company did not install them, nor did the federal government, who was responsible for regulations, force the company to comply with the recommendations. Corporate entitlement trumped environmental safeguards, and those in the Chetwynd region paid the environmental price.

There were also concerns about older coastal pulp-and-paper mills using Bunker-C oil, which emitted a lot of carbon, as a major energy source. In response, the provincial government initiated the construction of a natural gas pipeline, which was completed in 1995, from the Vancouver area to Whistler, Squamish, the Sunshine Coast, and then across to Vancouver Island. The pipeline allowed pulp mills to replace Bunker-C oil with lower carbon-emitting natural gas and gave residential, commercial, and industrial sectors in these regions an alternative energy source.

Electrical energy was another issue that the NDP had to tackle. As the province's population increased from 2.7 million in 1981 to 4 million in 2001, the demand for electricity also increased (see Figure 9.1). The province faced a new dilemma: how to add more electrical energy to the grid without causing further harm to the environment. Conservation was encouraged, and BC Hydro's Power Smart programs provided practical guidelines to reduce energy demand. But there was a shortfall on the supply side. Building another major dam on the Peace River at Site C was proposed as a significant supply "solution," but the displacement of people and flooding of some of the best farmland in the province was untenable. As hydroelectricity production plateaued, the NDP decided to use thermal electric sources (mostly natural gas) to fulfill demand (see Figure 9.2). It planned to produce 2,600 megawatts (MW) of electricity through the burning of natural gas, a process referred to as **cogeneration** (producing electricity and useful steam for other heating functions). The rationale for cogeneration was the surplus of natural gas in the province, its relatively low price, and a belief that carbon emissions were not as environmentally damaging as flooding river valleys. New plants at Taylor and Fort Nelson and upgrades to improve emissions at the large Burrard Thermal generator in the Lower Mainland resulted in more electricity being added to the grid.

The NDP faced another issue: what to do with transmission cables under Georgia Strait that carried electricity to Vancouver Island. They were old and needed to be replaced by 2007. Following its policy to provide electricity via cogeneration, the province, through BC Hydro, contracted Georgia Strait Crossing Pipeline (also known as GSX) to build an undersea pipeline to Vancouver Island and to construct a gas-fired plant in Port Alberni.

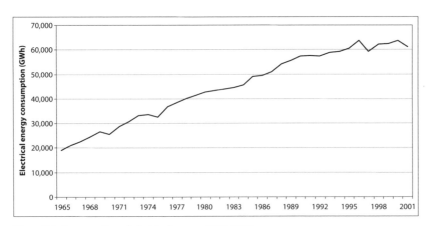

Figure 9.1 Consumption of electrical energy, 1965–2001
Source: Ministry of Economic Development (1977).

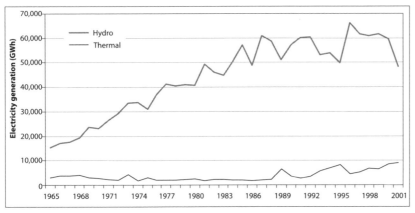

Figure 9.2 Hydroelectric versus thermal generation of electricity, 1965–2001
Source: Ministry of Economic Development (1977).

energy is a predicament for governments when energy options such as large dams or natural gas can produce "cheap" energy, reducing costs for consumers.

To produce electricity, the government also allowed **independent power producers** to develop electrical energy, which BC Hydro then purchased. Twenty-four producers were in place by 2000, but most were small (less than 5 MW) run-of-river projects. An argument in favour of this strategy was that the private corporations would bear the cost and risks, they would have to comply with environmental regulations, and they would create a considerably smaller footprint than megaprojects.

The government also had to deal with the commercial salmon-fishing industry, which had been significant in terms of providing government revenues and employment to fishers and cannery workers but was now failing badly. Although fisheries were mainly a federal responsibility, the provincial government had its share of environmental concerns. By the mid-1990s, considerably fewer salmon were returning to British Columbia rivers than had been predicted, triggering another set of policies referred to as the Pacific Salmon Revitalization Strategy or the Mifflin Plan, named after Fred Mifflin, the federal minister of fisheries at the time. The changes were substantial and resulted in a 50 percent reduction in licences by 2001 through a government buy-back program. Licences were also tied to a specific area or region: the coast was divided into two areas for seiners, three for gillnetters, and three for trollers (although with different delineations). However, the plan allowed licence stacking, which meant fishers could fish in another area if they bought a licence in that other area, thus reducing the overall fleet – a process labelled "cannibalism."

There were protests over cogeneration plants, especially since Canada had signed the Convention on Climate Change in Rio de Janeiro in 1992, which showcased the global consequences of greenhouse gases and the use of fossil fuels. The federal government had also signed the Kyoto Protocol in 1997, committing to a framework to reduce greenhouse gases to below 1990 levels. Also in that year, the provincial and federal governments had struck an agreement on the management of Pacific fisheries, the Fish Protection Act, which prohibited dam construction on the Fraser mainstem and many northwest rivers (Ostergaard 2002, 221). Producing clean

There was certainly a negative reaction to the loss of licences, vessels, and, most notable, jobs. Moreover, licence stacking resulted in more affluent fishers gaining licences, and these fishers rarely came from small coastal communities. The combination of buyouts and licence stacking had a negative impact on rural coastal areas. Moreover, the plan did not tackle the fundamental source of mismanagement – the gatekeeper model. The Mifflin Plan reduced capacity, but, tragically, the salmon did not return. There were many other variables that required attention if salmon fishing were to become sustainable. One of them was the destruction of streams where salmon spawned. A 1996 study found that habitat degradation from logging, urbanization, and hydropower development had caused 142 documented stock extinctions over the century while 624 stocks remained at high risk, 78 were at moderate risk, and 230 were of special concern. There was no doubt that overfishing had also resulted in severe stock depressions (Slaney et al. 1996, 20).

Although the commercial salmon-fishing industry was in major decline, the opposite was true of the fish-farming industry, a provincial responsibility. Fish farm licences had increased from 70 in 1986 to 135 by 1989, and raising farmed salmon took off in the 1990s. By 2000, farm-raised salmon was worth approximately three times the value of wild salmon. From an economic perspective, Atlantic salmon grow more rapidly than Pacific varieties of salmon, reducing costs, and the most valuable species of Pacific salmon – sockeye – cannot be farmed successfully.

But environmentalists raised a host of concerns about the impact of fish farms on the environment, from the use of pesticides and chemicals to fish feces and food pellets falling below nets. The farms required foreshore bays and inlets to be transformed into a private-property resource, and the shift to Atlantic salmon came with the risk of disease transfer to wild salmon and the escape of Atlantic salmon. Escaped fish could contaminate the gene pool of wild salmon, compete with wild salmon for food, and prey on juvenile wild salmon. Fish farms also led to the extermination of animals such as seals and eagles that preyed on farmed salmon and an increase in sea lice, which can have a devastating impact on wild salmon smolts. The economic law of supply and demand also reduced the value of Pacific salmon to fisherman (Beamish et al. 2004, 358). Moreover, because salmon are carnivores, feeding them required harvesting other species of fish to be pelletized for fish food. In 1995, faced with these concerns, the NDP government placed a moratorium on the number of fish farms. Only 121 tenures were allowed, but this restriction did not establish size limits of the net pens or the number of salmon for each farm. Production increased.

What occurred in the halibut fishery may be a lesson for the commercial salmon fishery. Even though the fishery was governed by a treaty with the United States, which established a quota, until 1979 it had remained an open fishery in which anyone could obtain a licence. There were over four hundred licences, and the gatekeeper model remained in effect. The fishing season lasted only ten days, but the total allowable catch often exceeded 200,000 pounds. By 1991, however, halibut was managed as a private-property resource through a system of individual vessel quotas. The season lasted 250 days, catches remained within the prescribed limits, the quality of the harvested fish improved, and market prices increased (Department of Fisheries and Oceans 2011). Clearly, this was a much more sustainable model.

RECOGNITION OF INDIGENOUS TITLE

Perhaps the NDP's most notable policy decision was to recognize Indigenous Title in 1992 and set up a six-stage BC Treaty Commission (BCTC). The policy shift slowed the number of blockades, protests, and court challenges, but it did not eliminate them.

A significant and influential court case – *Delgamuukw* – occurred just as the NDP came to power, but the BC Supreme Court's 1991 decision was viewed with disdain by First Nations. Gitxsan and Wet'suwet'en Hereditary Chiefs from the Bulkley River region, including Smithers, had gone to the Supreme Court to gain compensation for resource exploitation on Traditional Lands that had never been ceded to the Crown. In a setback for all First Nations, Justice McEachern ruled that Indigenous Title (called "aboriginal title" in the court ruling) had been extinguished when British Columbia became a province. The appeal, in 1993, reversed this decision, and a landmark decision by the Supreme Court of Canada in 1997 made it clear that Indigenous Title meant ownership of Crown lands. The decision did not award Indigenous Title to the Gitxsan and Wet'suwet'en – that was left to the treaty process and further court cases. Most importantly, *Delgamuukw* set a precedent for consultation and compensation for economic activities on unceded First Nations territories.

The BCTC's six-stage process established a means for First Nations to negotiate treaties that included entitlement to compensation for ceded lands, self-government, and other concessions to make First Nations economically viable. The process involved all three levels of government – provincial, federal, and First Nations – and it was slow and often frustrating. First Nations frequently charged that the governments were not bargaining in good faith. And not all First Nations bought into the modern treaty-making process. A rather violent protest that almost escalated to warfare occurred in 1995 during the Gustafsen Lake standoff, when some 450 RCMP and the military were deployed. At issue was a private ranch that sat on unceded land that had once been a First Nations sacred site and subject to Indigenous Title. There were other, less violent, confrontations, including Stó:lō and Gitxsan blockades of the CNR to disrupt the movement of minerals and forest products (Blomley 1996, 18).

While the NDP was in power, only one treaty was completed. The negotiation of the Nisga'a Treaty, which was signed in 2000, was fast-tracked (i.e., it did not go through the six stages), largely because it was the Nisga'a's court case, *Calder*, that had resulted in the federal government initiating comprehensive claims back in 1973. However, there was a flaw in the negotiations that has yet to be resolved. When the BCTC was formed, Premier Mike Harcourt arbitrarily stated that treaties would entitle First Nations to "less than 5% of British Columbia" (Grant 2012, 4). First Nations interpreted the statement as policy, and as consequence of the 5 percent benchmark, it was in each First Nation's interest, as each entered into the treaty process, to claim large territorial boundaries.

The boundaries of the Nisga'a Nation's territory were defined as extending from the source to the mouth of the Nass River. The treaty ceded 92 percent of the total area, leaving the remainder – the mouth of the Nass, where historical villages had been located – as the Nisga'a Homeland. This Homeland is owned in fee simple by the Nisga'a Nation and includes both surface and subsurface rights. The Indian Act no longer applies, and the tax exemptions under the act were phased out over eight to twelve years. The cash settlement included $190 million over fifteen years, although another $11.5 million was set aside to purchase commercial fishing licences. Other money was allocated to purchase a forest licence and to assist the nation in establishing its economic base. The agreement also included wildlife management and harvesting rights and justiciability, the right to establish their own court system.

Somewhat controversial was a separate own-source revenue agreement that entitled the Nisga'a to a percentage of Nass River salmon for commercial harvest. The implication was that salmon, which are a common property resource in the commercial industry, would become privatized in this instance and prioritized for the Nisga'a. This policy posed a threat to many commercial fishers, who recognized that the treaty process could allocate more salmon, already a scarce resource, to First Nations.

Even more controversial was the Nisga'a negotiators' insistence that the Indigenous Title boundaries include the entire Nass watershed (Grant 2012, 5). Of course, it was in the Nisga'a's interest that the boundaries be as expansive as possible. Both the provincial and federal governments agreed to the boundaries; the Nisga'a were compensated and gained wildlife management rights for the entire watershed. Unfortunately, the upper Nass watershed is claimed by other First Nations that have historical rights that include hunting, fishing, gathering, and trapping. It will likely take further court cases to resolve this issue.

The Nisga'a Treaty gives some notion as to what is included in a modern-day treaty, although it must be recognized that each First Nation is unique, and the Nisga'a Treaty is in no way a template for future treaties. The Nisga'a did not go through the six-stage process; had they done so, the boundary issue discussed above may have been resolved at the second stage. It is at this stage of the process that each First Nation must establish its Indigenous Title boundaries and have the consent of adjacent First Nations before moving to the next stage.

TURMOIL IN THE RESOURCE SECTOR AND THE END OF THE NDP ERA

As the NDP went into the 2001 election, the party was besieged by a number of issues, including an alleged bribery scandal, the building of new aluminum "fast ferries" that ran into technical problems, and a general concern about their environmental policies. Resource industries, which mainly affected the hinterland but had considerable media attention, were not doing well. The two coal mines at Tumbler Ridge had failed, and the community was facing ghost town status. Metal mining fared little better. During their tenure, nineteen mines closed while only four opened.

The number one industry – forestry – also faced some grave concerns. Restructuring through the introduction of capital-intensive technologies had reduced employment and contracting out. Another threat that needed to be resolved had originated in 1986, when the United States imposed a 15 percent countervail tax on lumber. American lumber producers convinced their government that British Columbia was marketing subsidized lumber. The dispute revolved around BC's tenure system, which allowed for long-term access to timber supplies, and the province's method of taxing corporations for the logs harvested. The latter system, referred to as stumpage, is a complex taxation system based on tenure, species, size, quality, end use, and even terrain difficulty. The countervail issue was not resolved until 1990, when the provincial government increased stumpage rates by 15 percent. The United States' threat of unilateral action remained and led to the Canada–United States Softwood Lumber Agreement in 1996, which limited British Columbia lumber producers to a quota. The agreement ended in 2001 with a new round of US-imposed taxes that amounted to 32 percent.

The woes for the forest industry did not end there. Forest-destroying insects such as the spruce bud worm come in cycles and have always been a concern for the industry. But this time around it was the mountain pine beetle outbreak, which began in the 1990s. By 2001, the beetles had consumed a great deal of the forests in the interior, with no end in sight. The principal target for the beetle was mature (eighty plus years) lodgepole pine, the dominant tree species in the interior. Historically, outbreaks of mountain pine beetle had occurred, but they had been held in check by forest fires and extremely cold winters; however, fire suppression, which was a policy from the first Forest Act in 1912, reduced the number of forest fires. Although lodgepole pine trees are easily killed by fire, they release seeds from serotinous cones

during the process (Taylor and Carroll 2004, 42). In other words, because their cones are fire-resistant, a young, even-aged pine forest will spring up following a forest fire. However, many interior regions have mature lodgepole pine forests. The second factor – warmer winters – has occurred since the 1990s, and this condition is directly related to climate change. Moreover, average winter temperatures are predicted to rise. To make matters worse, dead and decaying pine forests lead to considerably higher risk for forest fires and an overall increase in carbon dioxide, the main cause of climate change.

During the election, the NDP suffered a decisive loss to the Liberal Party, which was largely a reconstituted Social Credit Party. It won seventy-seven of the seventy-nine seats. One of the Liberal Party's platforms was to have the Nisga'a Treaty annulled as it believed Indigenous Title had been extinguished with Confederation. The party also promised to launch a province-wide referendum on treaty rights to see if the majority of the population opposed negotiated treaties.

LABOUR FORCE TRENDS AND URBANIZATION

From the 1970s to the turn of the twenty-first century, the volatility of world market prices, global competition, and the ability to fragment the production process and organize on a global scale resulted in economic uncertainty that had a profound impact on BC's labour force. Between 1976 and 2001, the goods-producing sector continued to decrease considerably in comparison to services-oriented employment (see Table 9.2). Primary industries made up a portion of the goods-producing sector, and it is no surprise that employment in agriculture steadily declined. However, it was not because of a decrease in the number of farms but rather because of mechanization. What is surprising is that Statistics Canada grouped together "forestry, fishing, mining, quarrying, oil and gas" and that employment in that category in 2001 was only 2.1 percent. Employment in construction and manufacturing also steadily declined.

But population and employment were far from being evenly distributed throughout the province. Table 9.3 delineates the various regional districts (see Figure 9.3) that make up the larger economic regions used by Statistics Canada. The regional imbalance is striking: the smallest economic region – Lower Mainland–Southwest – has

Table 9.2

Labour force survey estimates, in thousands, 1976–2001

	1976		1981		1986		1991		2001	
	N (thousands)	%	N (thousands)	%	N (thousands)	%	N (thousands)	%	N (thousands)	%
Total employed, all industries	1,060.9		1,319.6		1,331.8		1,577.5		1,920.9	
Goods-producing sector	323.8	30.5	390.2	29.6	322.0	24.2	377.7	23.9	383.8	20.0
Agriculture	17.6	1.7	20.5	1.6	33.3	2.5	30.3	1.9	25.6	1.3
Forestry, fishing, mining, quarrying, and oil and gas	49.2	4.6	69.2	5.2	53.1	4.0	51.4	3.3	41.2	2.1
Utilities	12.7	1.2	13.8	1.0	8.9	0.7	12.4	0.8	10.6	0.6
Construction	85.6	8.1	110.2	8.4	72.5	5.4	113.2	7.2	112.3	5.8
Manufacturing	158.7	15.0	176.4	13.3	154.3	11.6	170.5	10.8	194.1	10.1
Service sector	737.1	69.5	929.4	70.4	1,009.8	75.8	1,199.8	76.1	1,537.2	80.0
Wholesale and retail trade	178.4	16.8	212.7	16.1	225.7	16.9	259.4	16.4	303.3	15.7
Transportation and warehousing	78.4	2.4	94.1	7.1	90.3	6.8	97.3	6.2	112.7	5.9
Finance, insurance, real estate, rental, and leasing	66.1	6.2	83.7	6.3	87.9	6.6	103.6	6.6	120.2	6.3
Professional, scientific, and technical services	31.2	2.9	57.8	4.4	61.2	4.6	93.9	6.0	139.2	7.2
Business, building and other support services	16.9	1.6	22.9	1.7	33.8	2.5	38.8	2.5	70.5	3.7
Educational services	73.8	7.0	74.2	5.6	86.3	6.5	98.3	6.2	137.3	7.1
Health care and social assistance	79.8	7.5	104.7	7.9	120.1	9.0	155.7	9.9	197.2	10.3
Information, culture, and recreation	45.9	4.3	53.1	4.0	62.7	4.7	67.9	4.3	103.7	5.4
Accommodation and food services	60.7	5.7	84.4	6.4	94.2	7.1	119.1	7.5	164.0	8.5
Other services (except public administration)	41.5	3.9	60.1	4.6	68.7	5.2	74.7	4.7	98.6	5.1
Public administration	64.5	6.1	81.8	6.2	79.0	5.9	91.1	5.8	90.3	4.7

Source: Statistics Canada (2019).

nearly 60 percent of the population. Vancouver Island and Coast is second in population with slightly over 17 percent of the total; however, an examination of the regional districts involved indicates a much higher concentration of the population in the southern half of the island. The Thompson-Okanagan region was another growth region, with nearly 12 percent of the provincial population. The other four regions accounted for over three-quarters of the province's area, but their populations were relatively low: Kootenay (3.6 percent), Cariboo (4.1 percent), North Coast–Nechako (2.6 percent), and Northeast (1.5 percent). Table 9.4 shows employment by sector in each of the economic regions.

The Lower Mainland–Southwest's dominance of the province in terms of population and employment was due to the expansion of the service sector. Furthermore,

Table 9.3

Economic regions, regional districts, and their populations, 2001

Economic region	Regional district	Population	%
Vancouver Island and Coast (9.8% of land)	Capital	340,672	
	Cowichan Valley	73,925	
	Nanaimo	129,838	
	Alberni-Clayoquot	30,741	
	Central Coast	3,899	
	Comox-Strathcona	98,168	
	Powell River	19,957	
	Mount Waddington	13,667	
	Total population	710,867	17.4
Lower Mainland–Southwest (4.2% of land)	Greater Vancouver	2,093,125	
	Fraser Valley	249,202	
	Squamish-Lillooet	34,675	
	Sunshine Coast	25,947	
	Total population	2,402,949	58.9
Thompson-Okanagan (10.8% of land)	Central Okanagan	154,473	
	Columbia-Shuswap	49,016	
	North Okanagan	75,221	
	Okanagan-Similkameen	77,424	
	Thompson-Nicola	123,925	
	Total population	480,059	11.8
Kootenay (6.7% of land)	Central Kootenay	57,914	
	East Kootenay	57,022	
	Kootenay-Boundary	32,105	
	Total population	147,041	3.6
Cariboo (13.5% of land)	Cariboo	67,443	
	Fraser–Fort George	98,690	
	Total population	166,133	4.1
North Coast and Nechako (33.58% of land)	Bulkley-Nechako	41,261	
	Kitimat-Stikine	43,295	
	Skeena–Queen Charlotte	22,325	
	Stikine	851	
	Total population	107,732	2.6
Northeast (21.5% of land)	Peace River	56,119	
	Northern Rockies	5,981	
	Total population	62,100	1.5
Total BC population		4,076,881	

Source: BC Stats (2019).

most of the goods produced in this region were in construction and manufacturing, so few of the jobs in this sector were related to resource industries. Vancouver Island and Coast was a growth region in terms of population but not because of the goods-producing sector. This region had a long history of forestry activity; however, restructuring and closures reduced employment in the industry. Coal and metal mining met with similar fates. The growth component corresponds to increased services-related jobs and to an ever-increasing retirement population, tourism, and administrative functions.

Another growing region in terms of population was Thompson-Okanagan. Again, much of the growth was related to an increase in the region's retirement population and tourism; although, in this instance, tourism also had a primary resource connection – the grape and wine industries. Goods-producing was somewhat higher here than in British Columbia overall (e.g., farming at 2.6 percent and forestry at 3.8 percent). The region had several emerging urban service centres – Kamloops, Vernon, Kelowna, and Penticton.

Although the Kootenay region was known primarily for its mining, forestry, and some farming, the goods-producing sector declined rapidly as mines closed and the forest industry was restructured, but 7.2 percent of people continued to be employed in forestry, whereas the overall provincial rate was 2.1 percent. With the exception of 1997, agricultural employment was too low to be included in the data. Employment had remained steady over the six years recorded, but the goods-producing sector had declined in that time. The region's economy was in transition, and it began to relying to a greater degree on tourism and recreation, using its ski facilities and golf courses to attract visitors from Alberta in particular.

The final three regions (Cariboo, North Coast and Nechako, and Northeast) are relatively sparse in population but vast in size. Historically, they were BC's "frontier" and dependent upon resources and the goods-producing sector. This continued to be somewhat true, but resource employment subsided considerably. In terms of overall employment, the Cariboo region experienced some fluctuations, but essentially employment numbers remained the same: the goods-producing sector remained at over 30 percent, while forestry (at 9.0 percent) and

Table 9.4

Employment by economic region, 2000

Sector	N (thousands)	%	Sector	N (thousands)	%	Sector	N (thousands)	%
Vancouver Island and Coast			**Kootenay**			**North Coast and Nechako**		
Goods	67.7	20.7	Goods	17.6	24.8	Goods	15.5	32.7
Agriculture	5.3	1.6	Agriculture	×	×	Agriculture	×	×
Forestry, etc.	14.8	4.5	Forestry, etc.	5.1	7.2	Forestry, etc.	5.0	10.5
Utilities	1.6	0.5	Utilities	×	×	Utilities	×	×
Construction	22.7	6.9	Construction	5.1	7.2	Construction	×	×
Manufacturing	23.3	7.1	Manufacturing	6.6	9.3	Manufacturing	8.1	17.1
Services	259.5	79.3	Services	53.4	75.2	Services	31.8	67.3
Total employment	327.2		*Total employment*	71.1		*Total employment*	47.4	
Lower Mainland, Southwest			**Cariboo**			**Northeast**		
Goods	218.7	18.8	Goods	24.9	31.5	Goods	15.5	32.7
Agriculture	15.6	1.3	Agriculture	1.8	2.3	Agriculture	×	×
Forestry, etc.	8.2	0.7	Forestry, etc.	7.1	9.0	Forestry, etc.	5.0	10.5
Utilities	5.5	0.5	Utilities	×	×	Utilities	×	×
Construction	63.2	5.4	Construction	4.5	5.7	Construction	×	×
Manufacturing	126.2	10.8	Manufacturing	11.2	14.2	Manufacturing	8.1	17.1
Services	945.2	81.2	Services	54.1	68.5	Services	20.7	64.9
Total employment	1,163.9		*Total employment*	79.0		*Total employment*	31.9	
Thompson–Okanagan						**Total employment (British Columbia)**	1,930.8	
Goods	51.6	24.5						
Agriculture	5.4	2.6						
Forestry, etc.	8.0	3.8						
Utilities	1.5	0.7						
Construction	12.1	5.7						
Manufacturing	24.7	11.7						
Services	158.8	75.5						
Total employment	210.4							

Note: The numbers have been rounded, and an "x" indicates populations that are too small to record for privacy reasons. Therefore, the totals do not always add up exactly.
Source: Statistics Canada (2019).

manufacturing (at 14.2 percent) were considerably higher than the provincial average. Prince George was the region's major service centre.

North Coast and Nechako also had a goods-producing sector (32.8 percent) somewhat similar to that of the Cariboo, but even more people were employed in forestry (10.5 percent) and manufacturing (17.1 percent). Although these industries remained dominant, mine closures and forestry restructuring took their toll on employment. Finally, the Northeast region, home to Northeast Coal, contained 100 percent of oil and natural gas production, a diversified forest industry, and considerable agricultural production. It is north and east of the Rockies, however, which means extremely cold and long winters. The goods-producing sector, at 35.1 percent, is the highest of all the regions, and the forestry industry leads the category with 13.8 percent. Resources continued to play a significant role in all three of these regions.

These population trends corresponded to another extraordinary transition within the province – increased urbanization. Between 1971 and 2001, the percentage of the population that lived in urban centres increased from 76 percent to 82 percent (see Table 9.1), and the number of cars more than doubled from 1.1 million to 2.4 million (BC Stats 2017; Garrish n.d.-a, n.d.-b). Suburbanization caused enormous problems for high-growth regions such as the Lower Mainland, the Victoria-Saanich Peninsula on Vancouver Island, and the Okanagan. Agricultural land, never in abundance in British Columbia, was consumed for urban-industrial uses at a rapid rate. The Agricultural Land Reserve was designed in part to curb suburban expansion on to agricultural land. Servicing low-density regions with water, sewer, and solid waste facilities presented other challenges.

Between 1971 and 2001 the number of municipalities with populations over 10,000 increased from nineteen to twenty eight (see Table 9.5). More individuals were moving to cities and engaging in service-oriented occupations, but the growth of many of these centres also reflected boundary changes or the amalgamation of rural and urban areas or the uniting of incorporated urban areas. But not all urban centres grew. Resource-dependent cities such as Trail, Kitimat, Dawson Creek, Powell River, Prince Rupert, and Port Alberni lost people in this period.

Figure 9.3 Regional districts and economic regions

SUMMARY

Was British Columbia still a resource-dependent province in 2001? In terms of employment, resources certainly played less of a role, even in the more remote regions, which have the highest number of single-resource communities. However, unlike in the past, when economic downturns resulted in a large tally of ghost towns, these communities remained resilient and innovative, albeit with reduced populations. For example, communities such as Gold River, Tahsis, and Granisle saw their pulp mills, sawmills, and mines close, but tourism, recreation, the relocation of retirees, and individual entrepreneurship kept the communities thriving.

Table 9.5

Municipalities with more than 10,000 people, 1971 and 2001

Municipality	1971	2001
1. Vancouver, CMA	1,082,352	2,092,902
2. Victoria, CMA	195,800	326,753
3. Abbotsford/Mission, CMA	33,774	156,073
4. Kelowna, CMA	19,412	154,241
5. Kamloops	26,168	81,153
6. Prince George	33,101	75,206
7. Nanaimo	14,948	75,106
8. Chilliwack	23,739	65,228
9. Vernon	13,283	34,548
10. Penticton	18,146	31,642
11. Campbell River	10,000	29,181
12. North Cowichan	12,170	27,158
13. Courtney	7,152	18,865
14. Cranbrook	12,000	18,721
15. Port Alberni	20,063	17,795
16. Fort St. John	8,264	16,437
17. Salmon Arm	7,793	15,540
18. Prince Rupert	15,747	15,032
19. Squamish	6,121	14,867
20. Powell River	13,726	13,085
21. Terrace	9,991	12,376
22. Williams Lake	4,072	11,596
23. Comox	3,980	11,316
24. Summerland	5,551	10,824
25. Dawson Creek	11,885	10,788
26. Kitimat	11,803	10,775
27. Parksville	2,169	10,451
28. Quesnel	6,252	10,271

Source: BC Stats (2012).

However, resource extraction as the engine of economic growth for British Columbia was not questioned by the provincial government throughout this era, which explains why it continued to promote resource extraction in the name of creating jobs.

As in previous eras, the resource sector influenced jobs and settlement patterns, and the environmental footprint of industrial activities triggered public protest. However, in this period, the energy crisis of the 1970s and the recession in the 1980s led to environmental awareness at the local, national, and global levels, causing people to pressure the provincial government to develop policies to harvest resources more sustainably and to preserve wilderness areas. In response, parks expanded from 6 percent to 12 percent of the province. With the creation of the Agricultural Land Reserve, agricultural land became largely untouchable for endeavours other than agriculture. The economy also diversified. Tourism evolved and became widely recognized as an important employment and revenue industry. In many ways, Expo 86 highlighted and showcased "supernatural British Columbia."

The energy crisis also unhinged the assumption that government and corporations – that is, centralized decision making – could provide province-wide energy "solutions." Individuals embraced conservation measures, and individual industries made energy decisions in their own interests. For example, to reduce its dependence on oil, the forest industry created its own energy from hog fuel. The public firmly rejected BC Hydro's plan to electrify the province by means of major hydroelectric dams, coal-fired thermal plants, and nuclear power plants. However, by the end of the era two problems remained – how to supply Vancouver Island with more energy when undersea electric cables were approaching the end of their life span and how to supply the whole province with more electricity.

The environmental movement also dovetailed with First Nations protests over resource harvesting on unceded Crown lands, and First Nations made significant gains during this period, particularly after the federal government recognized Aboriginal Rights in the Constitution Act, 1982, and after the election, in 1991, of a provincial government willing to recognize Indigenous Title and to establish a process for treaty negotiations. The Supreme Court's ruling in *Delgamuukw* (1997) reinforced First Nations ownership of Crown lands. Although it appeared that the "land issue" had been settled, these developments led to protracted negotiations while the provincial governments acted as though Crown lands were theirs to grant without consulting First Nations.

REFERENCES

Anderson, J. 1997. "Consensus Based Land Use Planning: Success and Failure of British Columbia's Commission on Resources and Environment's Shared Decision Making Model." Master's thesis, University of Calgary.

Barnes, T., and R. Hayter. 1993. "British Columbia's Private Sector in Recession, 1981–86: Employment Flexibility without Trade Diversification?" *BC Studies* 98: 20–42.

BC Stats. 2012. "British Columbia Municipal Census Populations (1921–2011)." bcstats.bc.ca/StatisticsBy Subject/Census/MunicipalPopulations.

–. 2017. "Annual Population, July 1, 1867–2017." gov.bc. ca/gov/content/data/statistics/people-population-community/population/population-estimates.

–. 2019. "Municipalities, Regional Districts and Development Regions, 2001 to 2011." gov.bc.ca/gov/content/data/statistics/people-population-community/population/population-estimates.

Beamish, R.J., A.J. Benson, R.M. Sweeting, and C.M. Neville. 2004. "Regimes and the History of the Major Fisheries off Canada's West Coast." *Progress in Oceanography* 60: 355–85. pac.dfo-mpo.gc.ca/science/people-gens/beamish/PDF_files/7.pdf.

Belshaw, J.D., and D.J. Mitchell. 1996. "The Economy since the Great War." In *The Pacific Province: A History of British Columbia*, ed. H.J.M. Johnston, 313–42. Vancouver: Douglas and McIntyre.

Blomley, N. 1996. "'Shut the Province Down': First Nations Blockades in British Columbia, 1984–1995." *BC Studies* 3: 5–35.

Booth, A., and G. Halseth 2011. "Why the Public Thinks Natural Resources Public Participation Processes Fail: A Case Study of British Columbia Communities." *Land Use Policy* 28: 898–906. unbc.ca/assets/annie_booth/what_the_public_thinks_final.pdf.

Carrol, W.K.l., and R.S. Ratner. 2007. "Ambivalent Allies: Social Democratic Regimes and Social Movements." *BC Studies* 154: 41–66.

Cohen, B. 2012. *The Uncertain Future of Fraser River Sockeye.* Ottawa: Minister of Public Works and Government Services Canada. cohencommission.ca/en/FinalReport/.

Couture L., and R. Macdonald. 2013. "The Great U.S. Recession and Canadian Forest Products." Analytical paper, Economic Insights. statcan.gc.ca/pub/11-626-x/11-626-x2013028-eng.pdf.

Crompton, S., and M. Vickers. 2000. "One Hundred Years of Labour Force." *Canadian Social Trends* 57: 2–14. publications.gc.ca/collections/Collection-R/Statcan/11-008-XIE/0010011-008-XIE.pdf.

Crook, C.S. 1975. "Environment and Land Use Policies and Practices of the Province of British Columbia." Vol. 1. Report for British Columbia Institute for Economic Policy Analysis. for.gov.bc.ca/hfd/library/documents/bib50069_vo11.pdf.

Department of Fisheries and Oceans. 2011. "Pacific Commercial Halibut Fishery." Online media statement.

Dorgey, A.H.J. 1987. "The Management of Super, Natural British Columbia." *BC Studies* 73: 14–42.

Garrish, C.J. n.d.-a. "British Columbia Licensing Data, 1950–1954." *Archival Records of the British Columbia Motor Vehicle Branch.* bcpl8s.ca/BC-Licensing-Data-1950-1954.html.

–. n.d.-b. "British Columbia Licensing Data, 1970–1974." *Archival Records of the British Columbia Motor Vehicle Branch.* bcpl8s.ca/BC-Licensing-Data-1970-1974.html.

Geography Open Textbook Collective. 2014. *British Columbia in a Global Context.* Version 1.5. Victoria: BCcampus. opentextbc.ca/geography.

Grant, P.R. 2012. "The Legacy of Treaty Making: Reconciliation or a New Era of 'Divide and Conquer.'" grantnativelaw.com/wordpress/wp-content/uploads/2015/08/The_Legacy_of_Treaty_Making.pdf.

Haddock, M. 2010. *Environmental Assessment in British Columbia.* Victoria: University of Victoria Environmental Law Centre. elc.uvic.ca/documents/ELC_EA-IN-BC_Nov2010.pdf.

Hayter, R. 2000. *Flexible Crossroads: The Restructuring of British Columbia's Forest Economy.* Vancouver: UBC Press.

Hsu, S. 2007. "Forest Tenure Reform in BC." Case study, University of British Columbia Faculty of Law. allard.ubc.ca/sites/www.allard.ubc.ca/files/uploads/enlaw/pdfs/forestry04_20_09.pdf.

Meggs, G. 1991. *Salmon: The Decline of the British Columbia Fishery.* Vancouver: Douglas and McIntyre.

M'Gonigle, R.M. 1988. "Native Rights and Environmental Sustainability: Lessons from the British Columbia Wilderness." brandonu.ca/cjns/8.1/mgonigle.pdf.

Ministry of Economic Development. 1977. "Gross Energy Generation and Net Requirements in British Columbia, 1965–76." *British Columbia Facts and Statistics* 25: 54.

llbc.leg.bc.ca/public/pubdocs/bcdocs2014_2/123586/1977.pdf.

Ostergaard, P. 2002. "Energy Resources in BC's Central Interior." *Western Geography* 12 (2002): 216–29.

Parks Canada. 2012. "History of the Establishment of Gwaii Haanas." pc.gc.ca/en/pn-np/bc/gwaiihaanas/info/histoire-history.

Pearse, P. 1982. *Turning the Tide: A New Policy for Canada's Pacific Fisheries*. Vancouver: Commission on Pacific Fisheries Policy.

Ross, W.M. 1987. "Fisheries." In *British Columbia: Its Resources and People*, ed. C.N. Forward, 179–96. Western Geographical Series, vol. 22. Victoria: University of Victoria.

Schwindt, R. 1979. "The Pearse Commission and the Industrial Organization of the British Columbia Forest Industry." *BC Studies* 41: 3–35.

Slaney, T.L., K.D. Hyatt, T.G. Northcote, and R.J. Fielden. 1996. "Status of Anadromous Salmon and Trout in British Columbia and Yukon." *Fisheries Magazine* 21, 10: 20–35.

Statistics Canada. 2011. "Population Rural and Urban by Province and Territory (British Columbia)." statcan.gc.ca/tables-tableaux/sum-som/l01/cst01/demo62k-eng.htm.

–. 2019. "British Columbia Labour Force Survey Estimates, by North American Industry Classification System, annual (× 1,000)." CANSIM Table 282-0008. statcan.gc.ca/t1/tbl1/en/tv.action?pid=1410002301.

Swenerton, D.M. 1993. *A History of Pacific Fisheries Policy*. Ottawa: Department of Fisheries and Oceans. dfo-mpo.gc.ca/Library/165966.pdf.

Taylor, S.W., and A.L. Carroll. 2004. "Disturbance, Forest Age, and Mountain Pine Beetle Outbreak Dynamics in BC: A Historical Perspective." In *Mountain Pine Beetle Symposium: Challenges and Solutions*, ed. T.L. Shore, J.E. Brooks, and J.E. Stone, 41–51. Natural Resources Canada, Canadian Forest Service, Pacific Forestry Centre, Information Report BC-X-399. cfs.nrcan.gc.ca/pubwarehouse/pdfs/25032.pdf.

Thompson, A.R., and Canada. 1978. *West Coast Oil Ports Inquiry: Statement of Proceedings*. Vancouver: The Inquiry. cmscontent.nrs.gov.bc.ca/geoscience/MapPlace1/Offshore/West-Coast-Oil-Port-Inquiry-Statement-of-Proceedings.pdf.

Vyse, A., D. Bendickson, K. Hannam, D.M. Cuzner, and K.D. Bladon. 2006. Chapter 5, "Forest Practices." In *Compendium of Forest Hydrology and Geomorphology in British Columbia*, Vol. 1, ed. R.G. Pike, T.E. Redding, R.D. Moore, R.D. Winkler, and K.D. Bladon, 111–32. for.gov.bc.ca/hfd/pubs/docs/Lmh/Lmh66/Lmh66_ch05.pdf.

West Coast Environmental Law. 2012. "Environmental Groups, First Nations Join in Opposition to Omnibus Bill C-45." ecojustice.ca/pressrelease/environmental-groups-first-nations-join-in-opposition-to-omnibus-bill-c-45/.

Young, N. 2008. "Radical Neoliberalism in British Columbia: Remaking Rural Geographies." *Canadian Journal of Sociology* 33, 1: 1–36.

The Twenty-First-Century Liberal Landscape

The NDP was voted out for a host of reasons and replaced by the Liberal Party, which was in power from 2001 to 2017. The new government came in with familiar promises: to grow the economy and bring prosperity to all, increased transportation infrastructure, and jobs. Largely a reconstituted Social Credit Party, the Liberals under Gordon Campbell (2001–11) and Christy Clark (2011–17) believed in restraint and austerity, privatization, and manipulation of the environmental assessment process, and that people wanted less government interference. In this chapter, you'll learn how this belief combined with unpredictable forces such as terrorism, global recession, resistance from First Nations, and mounting concerns about climate change to produce some of the most contradictory policies on resource development, Indigenous Title, and the environment in the province's history.

THE CAMPBELL YEARS

The Liberal Party under Gordon Campbell swept to victory in May 2001, winning all but two seats in the legislature. It inherited many issues from the NDP such as the collapsed softwood lumber agreement with the United States, a mountain pine beetle epidemic in the interior, and the need to come to some conclusion on how to provide electricity to Vancouver Island and the rest of the province. The party also had to come to terms with the concept of Indigenous Title, which it had promised to hold a referendum on as soon as it was elected.

By September, however, many of these issues took a back seat as fear of terrorism swept North America following the events of 9/11. The US reaction was immediate and led to greater scrutiny of the movement of people and goods across borders and more suspicion and delays. The bombing of the twin towers was a wake-up call when it came to global conflict, one that led to the realization that the world was considerably more vulnerable because of technologies that could shrink time and space and move knowledge in an instant.

In the aftermath of 9/11, Osama bin Laden and the Taliban became prime targets, and the War on Terror led to the invasion of Afghanistan in 2001. By 2003, the United States and other allied nations had also invaded Iraq to eliminate Saddam Husain and rid the country of

"weapons of mass destruction." The military action did not directly affect Canada or British Columbia, but Canadian lives (including some from BC) were lost in Afghanistan. Unrest in the Middle East, especially in the oil-producing regions, did, however, raise the price of oil and natural gas, which affected the economies of many countries and provinces such as British Columbia.

During the early Campbell years, the value of the Canadian dollar rose from below US$0.65 in 2002 to over US$0.93 in 2007, and the world market price of oil, natural gas, coal, and metals rose to historically high levels. While the rising dollar proved to be detrimental to the forest industry and tourism, mining interests in BC benefitted from rising prices in general. However, unease with respect to climate change grew apace with oil prices, particularly following the release of Al Gore's highly publicized film *An Inconvenient Truth* in 2006, which clearly demonstrated the link between fossil fuels, increased greenhouse gas emissions, and extreme weather events. Adding to the anxiety was the fact that a number of large carbon-emitting nations, including the United States and China, did not sign the Kyoto Protocol.

The Campbell government's final years in power were likewise shaped by an unpredictable economic catastrophe: the Great Recession of 2008. During the early to mid-2000s, the US dollar steadily declined, not only against the Canadian dollar but also against most currencies around the world. Many Americans shifted to investing in commodities such as oil that had rapidly rising prices, and American banks and investment dealers encouraged house buying as a form of investment. People heard ads on the radio promising "a mortgage for no money down." Lending standards were lowered so much that people with no jobs, income, assets, or credit ratings could get huge mortgages (Montana 2010). Speculative home buying resulted in huge, unsecured bank loans and an expectation that home prices would never decline, but they did, and by mid-2008 the recession mirrored the Depression. Bank and investment companies failed from subprime mortgages and from repossessing houses that were worth less than their mortgages. High unemployment was experienced throughout the United States, and the spinoff affected many economies around the world, including Canada's.

The price of oil, and thus gasoline, also spiked in May 2008 throughout North America. Although events such as Hurricane Katrina in 2005 and conflict in the Middle East interfered with world oil production, the depletion of oil fields in the United States, Indonesia, Mexico, and the North Sea restricted supply. Meanwhile, world demand, especially by developing nations such as China, was increasing, and these conditions led to exceedingly high oil prices. Finally, speculators pushed up the price of oil with the expectation of profit: "With hindsight, it is hard to deny that the price of oil rose too high in July 2008 and that this miscalculation was influenced in part by the flow of investment dollars into commodity futures contracts" (Hamilton 2009, 240). When the world market price of oil crashed at the end of 2008, so did the price of virtually all other commodities, with the exception of gold and silver.

POWER, PIPELINES, AND ACTION ON CLIMATE CHANGE

Electrical energy may not have reached the crisis stage when the Liberals took the reins in 2001, but important policy decisions had to be made on several fronts. The Liberals inherited from the NDP a policy to produce some 2,600 megawatts of electricity through cogeneration. The NDP believed this would leave less of an environmental footprint than large dams that impounded water, such as a proposed third dam on the Peace River, referred to as Site C. Moreover, natural gas cogeneration would solve the problem of having to replace obsolete undersea cables to Vancouver Island by 2007.

At first, the Liberal Party carried on with the previous government's plans, including the contract with Georgia Strait Crossing Pipeline, also known as GSX, to pipe natural gas under George Strait and erect a cogeneration plant in Port Alberni. But several developments caused the government to rethink its position on producing more electricity from natural gas. First, the federal government ratified the Kyoto Protocol in 2002, agreeing to reduce greenhouse gases below 1990s levels by 2012. But cogeneration plants increased greenhouse gases. By late 2002, the Liberals had formulated a new energy plan titled *Energy for Our Future: A Plan for BC* (Minister of Energy, Mines and Petroleum Products 2002). The plan stated that the following conditions would be met:

- Fifty percent of BC Hydro's incremental resource needs would be met through conservation by 2020.
- Clean or renewable electricity generation would account for at least 90 percent of total generation.
- New electricity-generation projects would have net zero greenhouse gas emissions.
- Existing thermal-generation power plants would have net zero greenhouse gas emissions by 2016.

Becoming self-sufficient on electrical energy by 2016 was the goal. The nuclear power option was out, but the Site C dam was back on the list. The government declared that "Site C is one of many resource options that can help meet BC Hydro's customers' electricity needs" (Minister of Energy, Mines and Petroleum Products 2002, 23). Another way to accomplish this goal was through the private sector or independent power producers (IPPs).

This plan launched British Columbia down an energy-responsible path with respect to curbing the production of greenhouse gases, and it also paid considerable attention to the negative effects of oil and gas, including petrochemical resources that may be offshore. One component of the plan was to reduce the flaring of natural gas and oil (a major contributor to greenhouse gases), which was to be cut in half by 2011 and fully by 2016. Another direction was an emphasis on reducing the demand side of energy use and stressing the importance of conservation. A further objective was the promotion of renewable sources of energy (wind, geothermal, tidal, solar, and biomass). The Liberal government was aware of the factors resulting in climate change and initiated policies to curb the province's carbon footprint.

Still, it had a contract with GSX to build an undersea gas pipeline to Vancouver Island. The price of natural gas had remained relatively low during the 1990s, averaging $2.50–$3.00/per million British thermal units (MMBtu). By 2003, however, the price was steadily increasing, which was another reason, in addition to adding carbon to the atmosphere, to rethink the GSX contract. To make matters worse, Port Alberni's citizens did not want a cogeneration plant in their backyard and expressed concern over increased greenhouse gases. The rezoning application necessary to locate the plant was rejected. Duke Point, close to Nanaimo, was selected as a new location. However, when natural gas prices hit the $7/MMBtu range in 2004, the government cancelled the GSX contract. The era of inexpensive natural gas appeared to be over.

In response, BC Hydro commissioned new undersea transmission lines to carry electricity to Vancouver Island, which sparked a new controversy. Even though BC Transmission Corporation used the existing transmission right-of-way, the transmission lines were upgraded from 138 to 230 kilovolts, which meant more corona discharge – that is, escaped electrical energy, which could have potentially adverse health effects. Residents of Tsawwassen and Salt Spring Island mounted protests over this decision but to no avail. BC Hydro made a one-time offer to buy their homes.

The ever-increasing price of natural gas certainly played a role in reversing the cogeneration option, not only for Vancouver Island but on the mainland as well. The provincial government lost a few hundred million dollars by breaking the GSX contract, but it gained considerably more from the revenues generated through the export of natural gas to the United States.

In 2006, when the Liberal government was in its second term and contemplating the 2009 election, it set out to accomplish a number of resource-related projects. Interest in developing Site C had increased, and it became part of the province's electrical energy self-sufficiency plan. Since the project had been rejected twice by the BC Utilities Commission (BCUC), the government prohibited the BCUC from doing another assessment. Moreover, the role of BC Hydro was restricted to upgrading existing hydroelectric generation stations and Site C. In other words, the Crown corporation lost its mandate to generate research and budgets for other electrical-energy options for the province. The private sector, through IPPs, was now in control of providing most new electricity.

Producing electrical power through IPPs was not without its critics. One of the fundamental questions raised by conservation organizations, environmental groups, and First Nations, among others, was whether more electrical energy was required. BC Hydro's forecasts were for a 1.7 percent increase in electrical demand each year, and between 2001 and 2006 the province did become a net importer of electricity for three of those years – a sure sign of demand outstripping supply. Many of the IPP proposals were for run-of-river installations, by which water is diverted from rivers or streams through a pen stock to a generator and then returned to the stream. One concern expressed was the "gold rush" mentality that had formed as IPPs rushed to harness rivers and streams and built transmission lines and roads. The overall environmental impacts could be severe.

There was certainly interest in watersheds throughout the province, but each IPP had to acquire an energy purchase agreement from BC Hydro after completing numerous environmental studies. They also had to enter into agreements with First Nations. Only thirty-four IPPs were accepted between 2001 and 2003, and by 2006 another thirty-eight had been approved (Clean Energy BC 2010). Most were small (10 MW or less) run-of-river projects. However, criticism mounted because the Liberals had modified the Environmental Assessment Act in 2002 to water down environmental requirements; the threshold for IPPs requiring an environmental assessment that studied "cumulative effects" (a detailed study of activities in a watershed and the IPP's expected impacts) was changed from 20 MW to 50 MW, meaning more IPPs could be built without this detailed environmental study. The Significant Projects Streamlining Act of 2003 likewise allowed the lieutenant-governor in council to assign special status to projects "deemed to be important to the economic, social or environmental well-being of British Columbia" (Government of British Columbia 2003). By 2006, regional districts no longer had a voice in rezoning applications for IPPs. The fear of many was that IPPs, and many other resource developments, would be approved with little attention to the environment or to local government concerns.

Another fear was that BC Hydro's IPP purchase agreements would lead to significant rate increases for the public. The purchase agreements were for rates higher than the rates for electricity produced by the heritage dams, but any means of producing electricity in the 2000s was going to be more expensive because new infrastructure was involved. What was often not acknowledged was that the IPP purchase agreements were contracts to buy electricity at a fixed cost for up to forty years (or more, in some instances). In other words, it was a higher price today but would likely mean considerably lower prices in the future.

The Liberal government responded to the Great Recession and rising concerns about climate change with a revised Energy Plan in 2007, which stressed policies to reduce greenhouse gases. Its 2008 BC Bioenergy Strategy emphasized the reduction of greenhouse gases through the conversion of wood wastes (from the mountain pine beetle epic especially) and the conversion of methane from landfills into useful electrical energy. Further, the Burrard Thermal plant on Burrard Inlet, a major emitter of greenhouse gases, would only be used for backup energy. The bioenergy strategy also set aside funds for research and included opportunities for First Nations.

The government's 2008 Climate Action Plan overlapped with the bioenergy strategy and established a number of guidelines to reduce greenhouse gases, including target dates for reductions. The plan also included "green" building codes, public transit use increases by 2020, and anti-idling policies. Perhaps the most influential policy to come out of the plan was the introduction in 2008 of a carbon tax to cut emissions. The tax was staged so that it would start low and increase until 2012, working out to about seven cents per litre of gas (Beaty, Lipsey, and Elgie 2014). It was also introduced as a "revenue neutral" tax as other taxes were cut to offset the carbon tax. The Campbell government was leading the nation in its contribution to reducing climate change, and it was a good platform leading up to the 2009 provincial election.

In 2010, as part of the government's policy to facilitate metal-mining ventures, BC Hydro began planning the Northwest Transmission Line, a multimillion-dollar transmission line that would follow Highway 37 to the remote Iskut-Stikine River system, where it would serve other users who assisted in paying for the investment. In 2011, the federal government made a sizable contribution ($130 million) to extend the line to the Tahltan First Nations at Iskut, whom BC Hydro had previously serviced using diesel generators (Terrace Standard 2011). Another contributor was AltaGas, a major IPP in the northwest region. Its main run-of-river project was Forrest Kerr on the Iskut River, but it had two other projects in the area – Volcano Creek and McLymont Creek – both of which were tributaries of the Iskut River. All three projects depended on the Northwest Transmission Line.

Overall, twenty-two IPP projects were approved by BC Hydro in its 2010 electrical purchase agreements. Seven were awarded to pulp mills under the federal government's subsidy program, two were for significant wind projects (Port Hardy and Tumbler Ridge), two were for small biogas installations, and the rest were run-of-river hydroelectricity facilities. This increase in electrical generation was still considered insufficient to meet the province's needs; in years with low snowpack, BC remained a net importer of electricity.

At this time, environmentalists became aware of another proposal that would put both land and water in jeopardy and had the potential to increase greenhouse gases. The Enbridge Northern Gateway Pipelines project had been proposed in the early 2000s to link Alberta's tar sands to Prince Rupert or Kitimat. The proposal led to two key concerns: the safety of the pipelines and oil tanker traffic in complex waterways that experience major storms. Memories of the catastrophic Pine River Pipeline rupture in 2000 produced some apprehension about a pipeline project that would cross so many salmon-bearing rivers and streams; moreover, First Nations were opposed to pipelines crossing their territories.

A major rupture of the Trans Mountain Pipeline in Burnaby in 2007 also raised public awareness of the seriousness of oil spills, while studies documented that there had been twelve thousand pipeline failures in Alberta between 1990 and 2005 (Levy 2009, 24). In BC, it was reported that thirty-eight landslides ruptured gas pipelines in the north between 1978 and 2008 (Levy 2009, 25). Undoubtedly, there was a high risk that the proposed oil pipeline could have catastrophic consequences to the environment and local populations.

Earlier, in 1972, when British Columbians expressed their fears regarding oil tanker traffic, the federal government responded with a moratorium. The Enbridge proposal would result in the moratorium being lifted. But an unfortunate tragedy in 2006 – the sinking of a BC ferry, the Queen of the North – led to considerable unease about the thousands of gallons of diesel and hydraulic fuel leaking into the aquatic environment. This relatively small spill only served as a reminder of the disastrous 1989 Exxon Valdez oil spill in Alaska, which destroyed a way of life for First Nations in that region. Coastal First Nations and many others who were opposed to oil tankers developed their own policies, such as moratoriums on tankers in Traditional Territorial Waters (McCreary 2010). The question remained: who benefits from the transportation of oil, and who pays the costs if there is an oil tanker disaster?

Because Stephen Harper's Conservative government supported the pipeline and lifting the moratorium, fear also rose that the government would allow offshore drilling for oil and natural gas. The provincial government was also keen to lift the moratorium. The catastrophic Gulf of Mexico oil spill in 2010 did not have an impact on the world market price for oil, but it did shine a light on the consequences of lax regulation and government policies that favoured exploration for offshore oil. The incident revealed that a culture of risk had emerged, one that shifted risk and responsibility away from industry and the federal government, and within that culture, little care had been taken to use inspection technologies such as blowout preventer units to eliminate risk. The costs were borne by the millions of people who lived in the Gulf Coast, the environment, and future generations (Davis 2012, 170). These disasters led many in British Columbia, and the world, to give greater consideration to the precautionary principle, which states that projects must take into account the short- and long-term costs to the environment and to the populations involved, including the increase of carbon dioxide in the atmosphere. The federal government's promotion of hydrocarbon development and the export of oil, coal, and natural gas also came under question, particularly whether it would jeopardize Canada's December 2009 endorsement of the Copenhagen Accord and its commitment to reducing greenhouse gases.

The BC government's 2010 Clean Energy Act updated and reinforced earlier policies to reduce the province's carbon footprint through the production of renewable energy sources. It also created a First Nations Clean Energy Business Fund that encouraged First Nations to engage in renewable energy projects. Although smart meters were to be installed by 2012, the role of the BCUC was weakened further. It had been excluded from any assessment of Site C, and now it was excluded from examining the impacts of the Northwest Transmission Line, IPPs, the use of Burrard Thermal, the smart metering program, or adding turbines to existing heritage dams.

The provincial government wanted no opposition to its energy polices.

LUMBER AND OTHER EXPORT COMMODITIES

When the Liberals came to power in 2001, they also had to address issues in the lumber industry. The Softwood Lumber Agreement with the United States, which included a quota, expired, and the Americans unilaterally set surcharges and duties on imported BC lumber. The fundamental issue for the Americans was that the BC tenure system resulted in an unfair tax system that put "subsidized BC lumber" on the US market and some of its sawmills out of business. In response, the BC government changed the tenure system in 2003 so that traditional tree farm licences and timber licences – which were held by large, often foreign corporations – were reduced by 20 percent and turned over to BC Timber Sales. The BC government also exempted raw logs from surcharges, duties, and taxes on exported wood. In response, the harvest and export of raw logs increased, and these exports represented a loss in value-added wood and much-needed employment in the province. The Liberals also abandoned the appurtenancy policy, which required timber harvested in a particular region to be processed by local sawmills. Many local sawmills, often in small communities, were no longer entitled to local logs and thus became vulnerable to shortages. Some mills shut down.

These closures took place at a time when the mountain pine beetle epidemic continued to destroy interior pine forests. The area affected grew by 480 percent between 1999 and 2001 (BC Ministry of Forests 2012, 2). In reaction to the destruction, the government's Mountain Pine Beetle Action Plan 2001 proposed to harvest increased amounts of dead pine, to curtail the spread of the infestations, to export pine to China, and to turn dead wood into fuel pellets for heating and creating electricity. A great deal of investment was put into salvaging dead pine and turning it into something of value. Left standing, it would simply become fuel for forest fires. By coincidence, 2003 was the worst firestorm year on record, not necessarily because of dead pine but because of drought conditions and interface fires that affected property: 334 homes and businesses were destroyed. It was widely recognized that the forest industry in the interior would be in jeopardy because it would have few trees to harvest in the future. What was not widely recognized was that climate change was ultimately responsible for mild winters, drought conditions, and wildfires.

During the NDP era, to address concerns about clear-cut logging and saving old-growth forests, the government had put the Forestry Practices Code in place to provide guidelines for the size of cutblocks and setbacks from streams. The Liberal government, however, believed in privatization and set out to streamline the bureaucracy. It replaced the code with the Forest and Range Practices Act, which "reduced regulatory requirements by 55 percent by eliminating unnecessary red tape and paperwork" but cut measures to protect the environment (West Coast Environmental Law 2004). Under the new deregulated and "results-based" scheme, professional company foresters became responsible for addressing environmental concerns.

While world market price increases had a negative impact on lumber, it had a positive impact on other industries. For example, the price of coal increased substantially, especially after 2003. Although Tumbler Ridge was close to becoming a ghost town with the closure of the Quintette mine in 1999 and Bullmoose in 2003, the Trend mine opened in 2005 and Wolverine in 2006, revitalizing the town. The world market price of metals (copper, gold, silver, molybdenum, lead, and zinc) also rebounded and led to the reopening of closed mines (e.g., Gibraltar in 2004 and Mount Polly in 2005). All of these commodity price increases translated into mineral-dependent communities being more secure. The rebound in commodity prices that "saved" communities such as Tumbler Ridge also reinforced the notion that resource development was a major engine of growth to BC's economy.

But the price of lumber did not rebound, and the failed Softwood Lumber Agreement dragged on as Canada made claims to the World Trade Organization and NAFTA. It won the claims, but the United States continued to appeal the decisions until the two countries finally reached an agreement in 2006, after the United States had collected approximately $5 billion in duties. Some, but not all, of the money was returned. However, lumber exports to the United States continued to decline, and the dismal price of lumber resulted in over thirty sawmill closures. Adding to the grief was the continued devastation caused by the mountain pine beetle epidemic: 4 million hectares of forest had been damaged by 2003, and by 2007 the area had increased to in excess of 13 million hectares (BC Ministry of Forests 2012, 4).

When the world market price of oil crashed in 2008, so did the price of virtually all BC export commodities, with the exception of gold and silver. Forest exports, which were already on the decline to the United States, fell off precipitously. The industry experienced the largest decline in seventy years: lumber outputs experienced a 44 percent reduction and pulp outputs a 27 percent reduction (Couture and Macdonald 2013). People feared that pulp-and-paper mills would close and increase unemployment for many small communities. However, international events often have an impact on local conditions. Faced with similar conditions south of the border, the US government initiated a subsidy scheme by which pulp-and-paper mills became energy producers: mills added some diesel to their black liquor, and the subsidy reduced their costs, making them the world's lowest-cost producers overnight (Hamilton 2009). The Canadian government responded with a comparable scheme, subsidizing the creation of electric power from wood and pulp wastes (hog fuel). Many pulp-and-paper mills became IPPs, which provided a significant infusion of capital to keep them solvent. The scheme also enhanced the province's bioenergy strategy.

However, community forest agreements became another area of discontent in the forest industry. Introduced in 1998, these agreements were increased substantially in 2003 with the 20 percent clawback of traditional forest tenures; however, they existed mainly on a probationary basis. In 2009, the government amended the Forest Act to make them permanent: it allowed existing five-year probationary agreements to transition to twenty-five year community forest agreements (BC Community Forest Association n.d.). The conflict was not so much about the length and security of tenure but rather about the fact that the blocks of forest land were some of the most contentious, caught up in local disputes over potable water supplies and the preservation of wildlife, recreation, tourism, and old-growth forests.

With the continued downward slide of the forest sector, the provincial government turned its attention to mining, but many of the proposed mines were on unceded

Crown lands. One proposal, for the Prosperity Mine, a gold and copper mine west of Williams Lake, pitted environmental concerns and First Nations rights against jobs and revenues. This large open-pit mine would trespass on unceded Tŝilhqot'in Territory, would increase heavy truck traffic on newly created roads, and would use Fish Lake as a tailings pond. Although the provincial and the federal governments usually came together to complete environmental assessments of projects, in this case they each did their own.

The province completed its assessment in January 2010. It recognized that there would be some impacts on fish and their habitats, especially if nearby Fish Lake were terminated and used as a tailings pond, but the executive director advised that any adverse effects would be justified by employment, other economic benefits, and a fish compensation program. The provincial government approved the mine. Again, political debates and media attention fed the notion that resource extraction was crucial to provincial revenues and employment. The federal government conducted its review in November 2010. Transparent and open, the public hearing process involved community members and technical experts and identified an array of negative impacts on fish, grizzly bears, navigation, tourism, grazing, trapping, First Nations, and future generations (Haddock 2011).

Policies to facilitate other metal mining ventures, which would depend on the Northwest Transmission Line, were also in the works. The Red Chris Mine would focus on metal deposits of copper and gold located at the headwaters of the Iskut River, but it required electricity and an environmental assessment to go into production. The BC government approved the project following an environmental assessment. The federal government "decided it didn't need to conduct a full assessment, opting instead for a screening process." Environmental groups took the issue to court, claiming that the federal government had neglected its responsibilities. In January 2010, the Supreme Court of Canada unanimously ruled that the mine could go ahead, despite the lack of federal environmental assessment (CBC News 2010).

FISHERIES

The government was also forced to deal with controversy in the fishing industry. The commercial salmon fishery was in decline, and fish farms, which the government licensed, were singled out as one of the causes. The NDP had placed a moratorium on fish farms in 1995, but the Liberals lifted the moratorium in 2002, walking into a major confrontation with the commercial salmon-fishing industry. The collapse of a major run of pink salmon in the Broughton Archipelago on northern Vancouver Island was blamed on sea lice from farmed salmon in the region. Even though the number of spawners decreased from 3.6 million to 147,000, no salmon farms were closed (Pacific Fisheries Resource Conservation Council 2002). The overall value of farmed salmon continued to rise while the value of wild salmon fell.

Another concern faced by many non-Indigenous fishers was that First Nations were going to be given a greater share as a result of the treaty process. The McRae-Pearse Report, commissioned by both the federal and provincial governments in 2004 and titled *Treaties and Transition: Towards a Sustainable Fishery on Canada's Pacific Coast*, projected that First Nations would eventually gain access to between 33 percent and 38 percent of the fishery through the negotiation of treaties (Nixon 2004). Towards this end, the McRae-Pearse Report proposed scrapping the gatekeeper model and establishing individual transferable quota (ITQ) licences, which would be renewable every twenty-five years. Again, the federal government was reluctant to abandon the gatekeeper model.

Another concern for the industry was the deterioration of salmon habitat. In 2006, a study found that more than 50 percent of the eighty-seven salmon stocks actively assessed and managed by government were at least 25 percent below target abundance or rapidly declining. In some cases, they were both. Habitat loss, particularly in estuaries, was to blame (David Suzuki Foundation 2006, vi). The protection of estuaries was turned over to professionals hired by the developers, another example of deregulation.

In fall 2009, however, when an important fourth-year run of sockeye salmon up the Fraser River failed to materialize, the federal government appointed the Cohen Commission to examine the cause of the collapse. One problem was that the federal government had authority over oceans and commercial fishing while the provincial government was the lead agency in licensing fish farms, which had previously been blamed for problems in the salmon fishery. A Supreme Court challenge and ruling in 2009 made it clear that the federal government was responsible for fish farm licences. This ruling led to another conflict of interest when the federal government began to encourage employment and resource development through fish farms while it was also responsible for maintaining and supporting increased runs of wild salmon.

The Cohen Commission recommended that the moratorium on fish farms, initiated by the provincial government in 2008, should be maintained. The commissioner determined that the gatekeeper model employed by the commercial fishery was questionable and suggested another direction: "Traditionally, the Pacific salmon commercial fishery has operated as a derby fishery. This model is not the only way of conducting a fishery and, in recent years, there has been interest in moving the commercial salmon fishery away from derby fisheries and toward share-based management" (Cohen 2012, 208). The commission recommended that a determination on the benefits of an individual quota or, alternatively, an individual transferable quota (ITQ) should be made soon. The report was shelved by the federal government, however, thus advancing the tragedy of the salmon fishery.

PRIVATIZATION AND TRANSPORTATION INFRASTRUCTURE

The Liberal government's conviction that the private sector was more efficient than the public sector was not confined to resource development. The service side of BC Hydro was privatized and contracted out to the US firm Accenture in 2003, and in that same year two other Crown corporations were affected. The operation of BC Rail (but not the track or land) was sold to Canadian National (CN), a highly controversial move. To gain the favour of northern voters, the Liberals had promised prior to the election that BC Rail would *not* be sold. More serious were charges of a corrupt bidding process as well as allegations that the Liberals had turned an asset into a liability. Privatization, or at least semiprivatization, also affected the BC Ferry system, which was handed over to the BC Ferry Authority, who were government appointees, to be run more efficiently. Increased fares, reduced sailings, and a contract with European builders for three new vessels angered most user groups and the local shipbuilding industry.

A further privatization scheme failed in 2003 when Premier Campbell announced the selling of the Coquihalla Highway, including its tolls. Resistance to such a move by most of the province caused the Liberals to back off from the plan. The Coquihalla was a major infrastructure project and a legacy of Expo 86. The Liberals required a new scheme for the upcoming 2005 election and launched Opening Up BC, a $1.1 billion transportation plan to expand local infrastructure over three years and position the province and its ports as gateways to Asian markets (Asia Pacific Foundation of Canada 2014).

It was also in 2003 that Vancouver-Whistler won the 2010 Winter Olympic bid. Significant infrastructure spending was required to construct Olympic venues, extend the SkyTrain to Richmond and the airport through the new Canada Line, and reconstruct the Sea-to-Sky Highway from North Vancouver to Whistler, a project that added $1 billion to the budget for roads. Winning the Olympic bid was cause for much jubilation and was one of the factors in the Liberal's triumph in the 2005 election.

The Liberals also promised, just in time for the election, to enhance their Opening Up BC plan with an investment of $12.1 billion over three years (Asia Pacific Foundation of Canada 2014). To the delight of the Liberals, this initiative caught the attention of the federal government, which came up with $590 million in federal funding (bumped up to $2.1 billion by 2007). The federal government referred to the project as the Asia Pacific Gateway and Corridor Initiative. The plan included the replacement of the Port Mann Bridge with a new toll bridge, increased lanes on the Trans-Canada Highway leading to the Fraser Valley, the South Fraser Perimeter Road Project, turning the port of Prince Rupert (Ridley Island) into a container facility, and improvements to the Trans-Canada and Okanagan Highways.

FIRST NATIONS AND INDIGENOUS RIGHTS

During this era, policies entitling First Nations to their Traditional Lands were anything but clear. The Liberal Party did not recognize Indigenous Title because it believed it threatened provincial control over Crown Lands and resources and that title had been surrendered at

Confederation. In 1998, as the Nisga'a Treaty was reaching its final stages, Campbell, then the leader of the Opposition, launched a law suit, claiming that the treaty was in opposition to the Constitution and that the self-government clauses in the treaty would make the Nisga'a another order of government on par with the provincial and federal governments. He insisted that a province-wide referendum was necessary to define future relationships with First Nations.

In July 1998, the acting chief commissioner of the BC Treaty Commission expressed the concern, shared by others, that, quite aside from the technicalities of the constitutional issue per se, a public referendum would be the wrong tool to use for the ratification of treaties for the following reasons:

1. The parties have agreed otherwise.
2. Treaties are about rights, not about voter preferences.
3. A referendum is too blunt an instrument to deliver meaningful input on an issue as complex as a treaty. (Hurley 2001, C.1.a.)

Campbell lost the court case in 2001 and wanted to appeal the decision, but he and the Liberals were now back in power, so the lawsuit was dropped.

Campbell, however, initiated a province-wide referendum in early 2002. There were eight statements, and Statement 6 regarding self-government proved to be the most contentious: "Aboriginal self-government should have the characteristics of local government, with powers delegated from Canada and British Columbia" (Elections BC 2002, 2). Indigenous leaders objected to the non-Indigenous majority voting on the substance of Indigenous Rights. Sixty-five percent of British Columbians refused to participate in the referendum; however, those who did vote gave Campbell what he believed to be a clear mandate to reverse the powers available to First Nations in the treaty process (Schouls 2005, 6). First Nations then turned to the courts when the provincial government proceeded to award resource-development permits on unceded Crown land.

However, previous Supreme Court decisions that had been on the side of First Nations – including *Delgamuukw* (1997), which confirmed Indigenous Title to the land

– could not be ignored. The government had to consult with or compensate First Nations whose rights were affected (BC Treaty Commission 1999). However, the provincial government awarded a tree farm licence on Haida Gwaii without consulting the Haida First Nation, and a mining company was permitted to build a road through lands claimed by the Taku River Tlingit First Nation. The court decisions in both cases once again stated that First Nations must be consulted and compensated.

The federal government, which was also at the negotiating table, had settled a number of treaties in the Northwest Territories and the Yukon and accepted the concept of Indigenous self-government. If the provincial government were to negotiate, it would need to change its stance on Indigenous Title. If it refused to negotiate, the result would be considerably more confrontation over resource harvesting and court cases in which the provincial government would suffer defeat. Over the course of four years, the Liberal government moved towards partial acceptance of the idea of self-government (Schouls 2005, 1).

In 2005, the provincial government did an about-face and announced a New Relationship with First Nations. Self-government would be on an equal basis, and First Nations must be consulted when developments took place on Indigenous Title Lands. The policy resulted in a number of interim agreements with First Nations throughout the province, mainly dealing with resource harvesting on Indigenous Title Lands. Through these agreements, many conflicts were resolved and assisted in the main objective – movement towards treaties.

The New Relationship produced some results. No treaties had been completed under the six-stage process set up by the NDP back in 1993, but the Lheidli T'enneh First Nation in the Prince George region completed all stages by 2006. Unfortunately, territorial conflict with other First Nations, including those in Treaty 8, resulted in the treaty not being ratified in 2007.

The Tsawwassen First Nation was more fortunate and signed the first urban treaty in the province. The treaty was ratified by the province in 2007 and the federal government in 2008. The agreement came into effect in April 2009, just before the provincial election. Similar to the Nisga'a Treaty, it included a cash settlement for surrendering Crown lands, exclusion from the Indian Act, the phasing out of tax exemptions, subsurface rights to their homeland, private home ownership, and recognition of the right to harvest fish and wildlife. One controversial portion of the agreement was the right to remove agricultural land for commercial and residential uses.

Overshadowing land-use issues during this time were a number of lawsuits launched by First Nations (mostly on an individual basis) over sexual and physical abuse experienced at residential schools across Canada. In 2005, the Assembly of First Nations launched a massive class-action lawsuit for victims of residential schools. It resulted in the Indian Residential Schools Settlement Agreement of 2006, which had a number of guidelines. In 2007, the federal government formalized a $1.9 billion compensation package for those who had been forced to attend residential schools: thus, all First Nations who had been at a residential school were entitled to a compensation package based on the number of years attended. There was further compensation for any physical and sexual abuse experienced while attending the schools. Conditions and experiences in the schools were documented under the agreement, which then transitioned into the Truth and Reconciliation Commission. In June 2008, Prime Minister Stephen Harper made an official apology in the House of Commons.

Two years later, when the provincial government turned its attention to mining in response to the decline of forestry, it reversed the New Relationship policy because many of the proposed mines were on unceded Crown lands. The Recognition and Reconciliation Act, which would have recognized Indigenous Title and implemented a comanagement model for land and resources, "collapsed before it was even fully drafted" (Wood and Rossiter 2011, 408). As geographers Patricia Burke Wood and David Rossiter have pointed out, the fundamental problem was that the federal government should have resolved Indigenous Title at the time of Confederation. But because it had not, the government of BC lacked territorial sovereignty over most of the province (Wood and Rossiter 2011, 423). Campbell and his Liberal government were back where they started; they did not want to recognize Indigenous Title.

THE CLARK YEARS
When Gordon Campbell and the Liberals went into the

2009 election, there was no discussion of the harmonized sales tax (HST), even though provincial and federal bureaucrats had been engaged in discussions about it long before the election (CBC News 2011). Campbell introduced the HST in 2009 after he won the election, and the public's demand for a referendum was immediate and province-wide. The controversy resulted in Campbell's resignation in 2010 and the rejection of the HST during the 2011 referendum.

When Christy Clark became the new leader of the Liberal Party in 2011, the world was still reeling from the European monetary crisis, which had started in Greece in 2009 but then spread to Ireland, Italy, Portugal, and Spain. Adding to the economic and political disruption was the "Arab Spring," which began in late 2010 with protests, violence, and civil war and involved Tunisia, Egypt, Libya, Bahrain, Algeria, and other countries. By 2011, this turmoil and other events, especially the curtailment of Libyan oil along with the prospect of increased conflict in other oil-producing Middle East nations, had affected the world market price of oil. Syria also erupted with protests that turned into a civil war.

Japan faced its own crisis in 2011 when an earthquake of magnitude 9.0 and massive tsunamis occurred several hundred kilometres north of Tokyo. This natural disaster killed over eighteen thousand people and destroyed whole communities. The economic ripples of the tragedy were confined to minor interruptions of trade and the flow of materials, but a number of lessons, perhaps warnings, emerged out of it, for British Columbia in particular. British Columbia has a geologic history of megathrust earthquakes off the southwest coast that result in gigantic tsunamis, and what happened in Japan is a warning to BC to be aware of this hazard and reduce the risk (see Chapter 3).

One of the events that made the Japanese tsunami even more catastrophic was the destruction of the Fukushima Daiichi nuclear power plant and the subsequent radioactive contamination of surrounding fields and water. The catastrophe brought the use of nuclear power into question once again. Nuclear power provided approximately 30 percent of Japan's electricity; however, as a result of the tsunami disaster, all fifty of the country's nuclear power plants were shut down and assessed in terms of safety. Korea, next door, also initiated an

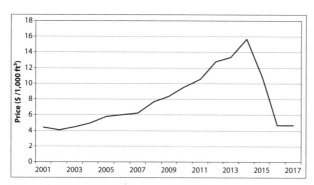

Figure 10.1 Price of US LNG exports, 2001–17
Source: US Energy Information Administration (2017).

examination of its nuclear power plants. One consequence of the reduction in electrical energy was increased imports of **liquefied natural gas (LNG)** to convert to electricity. The increased demand drove up the price of LNG and became a factor in setting a new resource-based direction for British Columbia that wasn't questioned until the price of LNG collapsed in 2014 (see Figure 10.1).

Christy Clark began her leadership by introducing policies to increase the minimum wage, proclaiming Family Day a new provincial holiday, and signing the Maa-nulth Final Agreement in April. The treaty involved five First Nations and two separate regions on the west side of Vancouver Island. It was similar to the Tsawwassen Treaty, but the Maa-nulth also negotiated an agreement on forestry revenue-sharing and water allocation, which made it an IPP. With a provincial election set for 2013, Clark also launched a new energy campaign centred on LNG and new infrastructure projects.

Although the federal government supported new energy proposals in this period, it also recognized that Canada was not keeping up with its commitments to reduce carbon emissions. At the end of 2011, the federal government withdrew from the Kyoto Protocol and stated: "Staying in Kyoto would force Canada to spend about $14-billion buying carbon credits abroad because the country is so far behind in meeting its target" (Curry and McCarthy 2011). That neither the United States or China had committed to the protocol was yet another rationale for Canada to pull out. However, the fact that Canada was considerably below its carbon-reducing targets was a

concern to many. In 2012, the federal government proposed new guidelines to reduce greenhouse gases for the future. Clark's victory in the 2013 election was a clear reversal of earlier policies to reduce greenhouse gases as well as confirmation of the public's perception that resource development was driving the provincial economy.

NEW LEADER, NEW PROMISES

The lead-up to the 2013 election saw the opening of major infrastructure projects, such as the Port Mann Bridge, which opened in 2012. Other new infrastructure projects were announced, including the Evergreen Line – a rapid transit Skytrain extension from Coquitlam and Port Moody to Burnaby and downtown Vancouver – which would be completed by 2016. Energy proposals introduced in the Campbell era, including the Northern Gateway pipeline, were also becoming more concrete, and a second proposal for Alberta's tar sands oil (bitumen) came to public attention. The Kinder Morgan proposal involved expanding the Trans Mountain Pipeline to Burrard Inlet, where bitumen and refined products would then be shipped to Asia by tanker. The transportation of bitumen is controversial. Because it is heavy, it sinks in water, making cleanup extremely difficult, if not impossible, if there is a spill.

First Nations, many of whom were caught up in the expensive and time-consuming treaty process – voiced strong objections to proposed resource developments such as oil pipelines, oil tanker traffic, metal mines, and the Site C dam. But the issue that took centre stage during the election campaign was pipelines. With respect to the Northern Gateway pipeline, the Clark Liberals stated that five conditions had to be met before the project could proceed:

1. Successful completion of the environmental review process;
2. World-leading marine oil spill response, prevention and recovery systems for B.C.'s coastline and ocean;
3. World-leading practices for land oil-spill prevention;
4. Legal requirements regarding Aboriginal and treaty rights are addressed; and

5. British Columbia receives a fair share of the fiscal and economic benefits. (Northern Gateway 2014)

The Kinder Morgan Pipeline was also hotly debated during the election, particularly because the opposition NDP rejected the pipeline. The Liberals' position was not to reject it; rather, they argued that it should go through the environmental review process as well as satisfy First Nations.

During the pipeline debates, the government said little about bitumen's or LNG's impact on the environment – from fracking for natural gas to the process of conversion and shipping LNG. Instead, Clarke emphasized that LNG held the key to a debt-free future. She promised that tens of billions of dollars would flow into a future prosperity fund, which would pay for improved health care and education. She also claimed that LNG exports would spark $1 trillion in economic development in just twenty to thirty years (Parfitt 2013). The link between LNG and climate change was disguised by presenting LNG as a "green" energy source that would replace higher carbon-producing coal in Asia. The provincial strategy, *Liquefied Natural Gas: A Strategy for BC's Newest Industry,* produced in 2012, extolled the virtues of LNG: "B.C. exports of liquefied natural gas (LNG) can significantly lower global greenhouse gas emissions by replacing coal-fired power plants and oil-based transportation fuels with a much cleaner alternative" (Ministry of Energy and Mines 2012).

The Liberals won the election by promising jobs, a strong economy based on LNG, and a five-year freeze on personal income tax and the carbon tax. Following the election, the premier, armed with a new four-year mandate, was already positioning the party for the next election with the announcement of another major infrastructure project – to replace the congested George Massey Tunnel with a ten-lane bridge. The project would begin in 2017, the year of the next provincial election.

RESOURCE DEVELOPMENT

If resource development was critical to the provincial economy, as Clark claimed, the news was not good. The forest industry was far from the number one industry in the province. Sawmills in the mountain-pine-beetle-

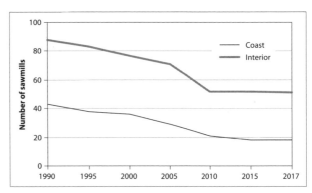

Figure 10.2 Number of coastal and interior sawmills, 1990–2017
Source: BC Ministry of Forests, Lands and Natural Resource Operations (2019).

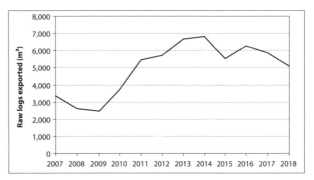

Figure 10.3 BC raw log exports, 2007–18
Source: BC Stats (2017); BC Ministry of Forests, Lands, Natural Resource Operations and Rural Development (2019); Penner (2019).

ravaged interior were closing, and companies that engaged in overharvesting for short-term gain were only delaying their reopening. In September 2014, the *Vancouver Sun* reported: "Companies operating around the sawmill town of Houston overshot the volume of green non-pine timber they were expected to log by the equivalent of almost 29,000 logging-truck loads between 2009 and 2013, after limits were set in 2008" (Penner 2014). That same year, hot weather combined with dead pine forests resulted in many forest fires (Hoekstra 2014). Between 2006 and 2016, nearly one hundred sawmills closed (see Figure 10.2) (Parfitt 2016).

With fewer mills and in response to these developments, Christy Clark's government removed barriers to exporting raw logs. Between 2013 and 2017, nearly

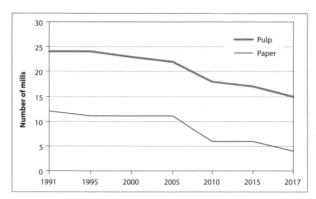

Figure 10.4 Pulp and paper mills in BC, 1991–2017
Source: BC Ministry of Forests, Lands and Natural Resource Operations (2019).

26 million cubic metres of raw logs (valued at $3.012 billion) were exported (see Figure 10.3). Journalist Ben Parfitt commented that "no government in BC history has sanctioned such a high level of valuable raw log exports on its watch or been so mute about the consequences" (Parfitt 2017).

The pulp-and-paper side of the forest industry also continued to shrink due to outdated mills, global competition, high labour rates, and the challenges of maintaining a wood supply. Although the 2008 recession gave the pulp mills a reprieve when they were allowed to become independent power producers (IPPs), thus diversifying their income, the number of paper mills continued to decline dramatically (see Figure 10.4).

The provincial government's promotion of mining did not fare well either. Given that the project was given the go-ahead provincially but rejected federally, the Prosperity Mine adjusted its plan and resubmitted its application to the federal environmental review panel under a revised name – the New Prosperity Mine. The provincial government strongly supported the project, and Bill Bennett, the energy minister, travelled to Ottawa twice to lobby for its approval (Robinson 2014). However, the federal government was not persuaded and concluded that the mine would cause significant adverse environmental effects that could not be mitigated and could not be justified under the circumstances (Canadian Environmental Assessment Agency 2014). In this instance, local environmental and social concerns won out; nevertheless, the provincial government's overt promotion of the

project raised a number of concerns. In particular, it was pointed out that since the Tŝilhqot'in and Secwepemc opposed the mine, the government's support of the project was "profoundly troubling and inconsistent with their commitments to First Nations" (Olsen 2014).

As resistance to the New Prosperity Mine indicates, there were a host of adverse environmental and social considerations associated with mine projects. The failure of the tailings pond at the Mount Polley Mine in August 2014 only heightened awareness and concern. When the pond failed, millions of cubic metres of heavy metals and other contaminants were flushed into Hazeltine Creek and Quesnel Lake, putting drinking water and salmon habitat at risk. When it came to light that inspectors had found a tension crack in the dam four years previously, attention focused on the inspection process and resulted in mandatory inspections of all tailing ponds (Huffington Post 2014).

In late 2013, the transportation of high-carbon American thermal coal (which is burned to produce electricity) through British Columbia to Asia also became an issue. When western states refused to export this coal because it would increase global greenhouse gases, it was proposed that BC could serve as an export port. Rail cars would carry the coal from Wyoming and Montana up through communities such as White Rock and Surrey to the Fraser Surrey Docks, where it could be loaded on to barges destined for Texada Island and then on to Asia. Delays south of the border created an opening for Fraser Surrey Docks to handle up to 8 million tonnes a year (Jang 2014). Concerned citizens, environmental organizations, regional districts, municipalities, and First Nations drew attention to the hazards related to escaping coal dust and the larger issue of increasing global greenhouse gases and were frustrated by federal government policies that left railways to regulate themselves. When Port Metro Vancouver assessed the project and concluded that it was "not likely to cause significant adverse environmental effects" (Port Metro Vancouver 2014), the lenient environmental assessment process resulted in more legal challenges and protests. The provincial government remained silent on the proposal.

On the positive side, a report completed in 2013 put an extremely positive spin on the carbon tax the government had initiated in 2008. CBC News reported that "B.C.'s consumption of fossil fuels has been reduced nearly 19 per cent per capita compared to the rest of Canada, while the province's gross domestic product has kept pace with the country's" (CBC News 2013). However, the good news did not translate into increasing the carbon tax, and the provincial government, federal government, and energy corporations continued to put considerable effort into promoting the export of LNG and bitumen – therefore adding to climate change.

In some instances, corporations were less than transparent in advancing the risks of exporting bitumen. For example, the Northern Gateway proposal included a somewhat deceptive map of the tanker route from Kitimat through the Douglas Channel that omitted islands that needed to be navigated, increasing the risk of spills. Kinder Morgan's application also stated it would transport considerably less bitumen than the carrying capacity of the pipeline. Although the application with the lesser amount would be subject to an environmental assessment as well as public review, this would not be the case if or when the corporation decided to ramp up production (Allen 2014).

Coming to terms with climate change and its principal contributor – fossil fuels – was not a strong suit of Stephen Harper's Conservative government, which dropped out of the Kyoto Protocol and then recalibrated timelines to give the appearance of reduced greenhouse gases. The National Energy Board restricted public input during reviews and then awarded conditional approval of the Northern Gateway pipeline proposal – subject to 209 recommendations

In this permissive environment, the provincial government set out to allow nonagricultural development of the Agricultural Land Reserve. The new policy divided the province into two zones. Zone 1 included regions with the greatest pressure to convert land – the Lower Mainland, Vancouver Island, and the Okanagan. No alterations could occur there. In Zone 2, the Agricultural Land Commission could now allow farmers more flexibility to support their farming operations (Gabriola Sounder 2014). In other words, agricultural land in Zone 2 could be converted for oil and gas development or other ventures more profitable than agriculture.

Major losses to relatively rare Class 1 and 2 agricultural land would also occur if BC Hydro went forward with the building of the Site C hydroelectric dam on the Peace River. An estimated 83 kilometres of productive valley bottom would be flooded. As with other environmental assessments, the provincial government restricted public input. For example, the Agricultural Land Commission was not allowed to make a presentation at the Joint Review Panel (JRP) hearings. Even more disturbing, even though the JRP recommended that the BCUC should review the economic and social cost of Site C, this was denied by the provincial government's 2010 Clean Energy Act. The JRP also noted that the provincial government policy had limited the mandate of BC Hydro to developing Site C and upgrading existing hydroelectric generation stations only; this restriction barred them from conducting a cost comparison between Site C and geothermal energy or any other renewable electrical energy options. (BC Hydro's mandate had historically included assessment of all means of producing electrical energy.) As the JRP stated, Site C is a megaproject with a megafootprint: "Site C would seem cheap, one day. But the Project would be accompanied by significant environmental and social costs, and the costs would not be borne by those who benefit" (Federal Minister of the Environment and the British Columbia Minister of Environment 2014). Many groups were in opposition to Site C, including First Nations.

FIRST NATIONS AND THE NEW REALITY OF CONSULTATION

In June 2014, the Supreme Court of Canada brought down a decision that would have profound implications for resource development and infrastructure projects. The Tŝilhqot'in First Nation was awarded Indigenous Title to some 200,000 hectares. The decision placed a greater burden on governments to justify economic development on Indigenous Lands, with one caveat. The court ruled that economic development could still go ahead without consent in cases "where development is pressing, substantial and meets the Crown's fiduciary duty" (Moore 2014). Claiming that development was in the "national interest" could be a loophole for either government, but the Tŝilhqot'in decision means much more consultation and negotiation for resource development to proceed on their territory. And the decision sets a precedent for other First Nations.

There was perhaps some reconciliation in the Tŝilhqot'in decision. The so-called Chilcotin War of 1864 had resulted in a number of Chiefs being hanged for defending their land and their people from smallpox and unwanted settlement and development. Chief Percy Guichon's statement with respect to the Supreme Court decision brought some closure: "I'm so thankful and grateful to say that 150 years later we see the Supreme Court of Canada's decision today as the final justice for six chiefs who died for their land, way of life and the future of the Tsilhqot'in people" (Moore 2014).

However, fighting for the recognition of Indigenous Title was not the priority for many First Nations – coming to terms with the residential school experience took precedence. Another class-action lawsuit – launched by those who attended residential schools as day scholars – was accepted by the Supreme Court. In 2015, another historical event also came to light that demanded a formal apology and compensation – the Sixties Scoop. CBC News reported: "The Sixties Scoop is the name given to the period of time between the 1960s and '80s when thousands of aboriginal children were placed with mostly non-native adoptive families" (CBC News 2015).

In this new context of social and political awareness, it became a challenge for the provincial government to launch resource development projects. For instance, construction on Site C started before First Nations had completed a legal action to stop the project on the grounds that it violated the terms and conditions of Treaty 8. The Liberal Party's victory in 2015 under the leadership of Justin Trudeau gave some hope to environmental organizations, First Nations, and others opposed to megaprojects, bitumen pipelines, and LNG terminals. In January 2016, there was some celebration by First Nations and those concerned with the Northern Gateway pipeline proposal when the Supreme Court ruled that First Nations had not been adequately consulted, which would delay construction. On the other hand, when the federal government approved the Woodfibre LNG proposal for Squamish, the Squamish First Nation did its own environmental assessment and also approved of the plant. The District Municipality of Squamish voted not to have this facility, but local governments were shut out of energy project decisions.

The treaty process was likewise complex and uncertain. The Yale First Nation, for instance, put a great deal of effort into the six-stage treaty process, which was completed in 2013, but it was suspended in 2016 due to fishing disagreements with other First Nations. Conversely, the Tla'amin First Nation of the Powell River area completed and ratified its treaty in 2016. The treaty was similar to other treaties with respect to doing away with the Indian Act, phasing out of tax exemptions, guaranteeing the First Nation a percentage of fish, and including compensation for ceded Indigenous Title Lands. But the treaty also included a unique fifty-year resource revenue–sharing agreement and the allocation of timber under BC Timber Sales, as well as the potential for the Tla'amin to become an IPP.

CLIMATE CHANGE

In late 2015, nations came together at the Paris Conference to make commitments to reduce greenhouse gases globally by "holding the increase in the global average temperature to well below 2°C above pre-industrial levels and to pursue efforts to limit the temperature increase to 1.5°C above pre-industrial levels" (New York Times 2015). Justin Trudeau signed the agreement in April 2016. By this time, however, many in British Columbia and in Canada were cynical about government commitments to climate change. In BC, the provincial government was promoting LNG, not increasing the carbon tax, and supporting bitumen pipelines from Alberta. At the federal level, the government had approved LNG and pipeline proposals and cancelled its commitments under the Kyoto Protocol. It appeared that governments were pledging carbon-reducing targets for the future and then, when they failed to live up to the commitments, simply adjusted the timelines.

Trudeau's signing of the agreement gave renewed hope that the government delay cycle had been broken. However, events that followed were contradictory. For example, a controversial proposal by Petronas to build an LNG export facility on Lelu Island off Prince Rupert was approved with 190 conditions. As had been the case with the Northern Gateway pipeline, these conditions gave the illusion that the approval could be reversed. And in November 2016 the federal government did in fact

Table 10.1

Provincial revenues (millions), 2000–01 and 2017–18

	2000–01		2018–19	
	$	%	$	%
Taxation revenue	13,933	56.7	32,714	57.3
Natural resource revenue	3,975	16.5	3,108	5.4
Other revenue	1,861	7.7	10,249	17.9
Contributions from government enterprises	1,500	6.2		
Contributions from the federal government	2,797	11.6	9,052	15.8
Commercial Crown corporation net income			2,005	3.5
Total combined revenue	24,066	100.0	57,128	100.0

Source: Ministry of Finance (2001, 2019).

terminate the Northern Gateway project, largely because it had promised a crude oil tanker ban for northern coastal waters during the election. The elation caused by this news was marred, however, when the federal government approved the Kinder Morgan Pipeline to Burnaby and the export of bitumen through Coal Harbour and the Gulf Islands.

Table 10.2

Populations of the ten largest municipalities, 2018

Municipality	Population
1. Vancouver, CMA	2,654,222
2. Victoria, CMA	396,509
3. Kelowna, CMA	208,864
4. Abbotsford/Mission, CMA	193,823
5. Nanaimo	112,949
6. Kamloops	111,646
7. Chilliwack	110,295
8. Prince George	91,969
9. Vernon	65,247
10. Courtney	58,234
Total	4,003,758
	(80.2%)

Note: CMA = Census Metropolitan Area.
Source: BC Stats (2018).

THE REALITY BEHIND THE RHETORIC

Although the provincial government supported resource development and energy projects, claiming that resources were the backbone of the province's economy, the reality was that resources were ceasing to be a major source of revenues (see Table 10.1) or jobs in the province, which became even more urban in the Liberal era. At the beginning of the Campbell years, natural resources accounted for 16.5 percent of revenues; by the end of the Clark era, that percentage had declined to 5.2 percent. The rural-to-urban transformation, along with the decline in resource-based employment, resulted, for the most part, in declining populations for rural regions (see Table 4.3, p. 80). By 2017, the population of the Vancouver Island and Coast and Lower Mainland–Southwest economic regions made up 78.5 percent of the total population of British Columbia; if the Thompson-Okanagan region is included, these three regions accounted for 90.2 percent of the population.

Data reveal that more and more people were moving not to the nearest small town but to larger urban centres. By 2017, Metro Vancouver dominated the province, accounting for 53 percent of the province's population, and over 80 percent of people lived in the largest ten communities (see Table 10.2). This pattern will likely continue as service-sector employment increases.

As Table 10.3 indicates, Vancouver Island and Coast along with the Lower Mainland–Southwest accounted for over 80 percent of employment by 2018. When the

Table 10.3

Employment, by economic regions, in thousands, 2001–18

	2001		2011		2018	
	N	%	N	%	N	%
Total	1,920.9		2,227.9		2,493.7	
Vancouver Island and Coast	307.7	16.0	357.8	16.1	396.6	15.9
Lower Mainland–Southwest	1,173.7	61.1	1,387.5	62.3	1,607.4	64.5
Thompson-Okanagan	210.4	11.0	250.0	11.2	255.2	10.2
Kootenay	70.5	3.7	70.6	3.2	71.2	2.9
Cariboo	79.2	4.1	81.1	3.6	80.6	3.2
North Coast and Nechako	46.8	2.4	43.5	2.0	43.1	1.7
Northeast	32.6	1.7	37.4	1.7	39.6	1.6

Source: Statistics Canada (2019b).

Table 10.4

Employment, census metropolitan areas, in thousands, 2001–18

	2001		2011		2018	
	N	*%*	*N*	*%*	*N*	*%*
BC	1,920.9		2,227.8		2,493.6	
Kelowna	71.5	3.7	92.9	4.2	101.9	4.1
Abbotsford-Mission	71.5	3.7	83.9	3.8	96.3	3.9
Vancouver	1,037.6	54.0	1,224.4	54.9	1,425.7	57.2
Victoria	151.0	7.9	178.0	8.0	195.8	7.9
Total		69.3		70.9	1,819.7	73.1

Source: Statistics Canada (2019a).

Thompson-Okanagan region is added in, the three regions accounted for 90.6 percent of jobs. Employment diminished (by percentage and in some instances by real numbers) in the four other economic regions between 2001 and 2018. Table 10.4 is also quite revealing: nearly three-quarters of jobs in the province reside in the four census metropolitan areas: Vancouver, Victoria, Abbotsford-Mission, and Kelowna.

As in previous eras, the percentage of employment related to agriculture declined, while jobs in forestry remained relatively even, if a minor portion of the goods-producing sector. Major changes occurred in construction and manufacturing, however (see Table 10.5). Infrastructure projects and urban construction accounted for the significant increase in construction jobs. Construction increased considerably in the Vancouver Island and Coast region, more than doubled in Lower Mainland–Southwest, and nearly doubled in Thompson-Okanagan. Conversely, manufacturing employment fell off in the other regions largely because of the closure of mines and mills.

These employment trends were likewise reflected in the census metropolitan areas (see Table 10.6). Overall, the service sector accounted for 80 percent of all jobs. Although construction and manufacturing represented a sizable share of employment in the goods-producing sector, only a small percentage of the "goods" produced were related to resource industries. Not surprisingly, the four metropolitan areas made up three-quarters of the province's service sector by 2018.

Clearly, by the time of the 2017 election, there was a major discrepancy between economic realities and the provincial government's presentation of them. The reality is that the distribution of the province's population is extremely uneven and balanced in favour of large urban areas, and resources are well down the list in terms of provincial employment. Even in the fairly rural economic regions – Kootenay, Cariboo, North Coast and Nechako, and Northeast – the goods-producing sector is higher than in urban regions but far from dominant. Yet the regions are where many of the forestry, mining, and oil and gas developments occur, and resource jobs remain important to these regions.

THE 2017 ELECTION

The lead-up to the 2017 election in BC was a fairly long, drawn-out affair. The Liberals continued to promote LNG and promised more tax cuts, a freeze on income taxes, and a freeze on the carbon tax. Some of the more contentious issues focused on the environment and potential conflicts of interest. When it came to light that the Liberals had accepted huge donations from private corporations and that the NDP had accepted sizable donations from unions, both parties contended that these donations were a conflict of interest. The Green Party rejected any corporate or union funding.

Citizens who lived south of the Fraser were concerned about a proposed ten-lane bridge to replace the George Massey Tunnel. They argued that the bridge would increase vehicles and carbon rather than reducing congestion. Also controversial was the Liberal's backing of the Kinder Morgan Pipeline, a project opposed by most in the Lower Mainland region and by the NDP and Green Party. Both parties, along with First Nations, concerned citizens, and farmers, also opposed the construction of Site C, especially after a UBC report stated the following: "Our analysis indicates that cancelling the Site C project as of June 30 would save between $500 million and $1.65 billion, depending on future conditions ... Suspending the project, allowing for future completion, would save $850 million to $2 billion" (Ducklow 2017).

The outcome of the election was unique. The Liberals gained the most seats – forty-three – but not a majority of the eighty-seven-seat Legislative Assembly. The NDP won forty-one, and the Greens won three. A power-sharing

Table 10.5

Employment by economic region, in thousands, 2001–18

	2001		2011		2018			2001		2011		2018	
	N	%	*N*	%	*N*	%		*N*	%	*N*	%	*N*	%
British Columbia							**Kootenay**						
Total	1,920.9		2,227.8		2,493.6		*Total*	70.5		70.6		71.2	
Goods	383.8	20.0	431.7	19.3	500.0	20.0	Goods	15.3	21.7	20.6	29.2	21.7	30.5
Agriculture	25.6	1.3	25.6	1.1	23.6		Agriculture	x		x		x	
Forestry, etc.	41.2	2.1	40.8	1.8	49.7		Forestry, etc.	2.8		5.5		8.3	
Utilities	10.6	0.6	11.3	0.5	13.9		Utilities	x		x		x	
Construction	112.3	5.8	197.1	8.8	238.4		Construction	5.1		8.2		5.8	
Manufacturing	194.1	10.1	156.8	7.0	174.3		Manufacturing	6.9		5.3		6.3	
Services	1,537.2	80.0	1,796.1	80.7	1,993.6	80.0	Services	55.1	78.3	50.0	70.8	49.5	69.5
Vancouver Island and Coast							**Cariboo**						
Total	307.7		357.8		396.6	396.6	*Total*	79.2		81.1		80.6	
Goods	56.1	18.2	60.7	17.0	66.2	66.2	Goods	23.9	30.2	20.9	25.8	23.9	29.7
Agriculture	4.5		3.7		3.6	3.6	Agriculture	x		x		x	
Forestry, etc.	10.5		7.6		10.0	10.0	Forestry, etc.	6.5		4.2		5.3	
Utilities	1.6		x		x	x	Utilities	x		x		x	
Construction	18.8		30.3		36.1	36.1	Construction	3.7		6.7		7.4	
Manufacturing	20.8		17.9		15.2	15.2	Manufacturing	12.1		8.4		10.3	
Services	251.6	81.8	297.1	83.0	330.4	330.4	Services	55.3	69.8	60.2	74.2	56.8	70.3
Lower Mainland–Southwest							**North Coast and Nechako**						
Total	1,173.7		1,387.5		1,607.4		*Total*	46.8		43.5		43.1	
Goods	212.7	18.1	250.5	18.1	306.0	19.0	Goods	14.0	29.9	11.5	26.4	13.1	30.4
Agriculture	13.7		12.3		15.4		Agriculture	x		x		x	
Forestry, etc.	7.7		11.2		8.2		Forestry, etc.	4.1		2.5		3.2	
Utilities	6.2		6.8		9.3		Utilities	x		x		x	
Construction	64.5		121.3		152.3		Construction	2.3		3.2		3.5	
Manufacturing	120.7		98.7		120.9		Manufacturing	7.3		5.1		5.6	
Services	961.0	81.9	1,137.0	81.9	1,301.3	81.0	Services	32.9	70.1	32.0	73.6	30.0	69.6
Thompson-Okanagan							**Northeast**						
Total	210.4		250.0		255.2		*Total*	32.6		37.4		39.6	
Goods	50.8	24.1	57.3	22.9	57.0	22.3	Goods	11.0	33.7	10.3	27.5	12.1	30.5
Agriculture	4.6		5.7		2.6		Agriculture	x		x		x	
Forestry, etc.	6.0		5.4		11.0		Forestry, etc.	3.8		4.2		3.7	
Utilities	1.5		2.5		1.9		Utilities	x		x		x	
Construction	14.7		24.0		27.3		Construction	3.2		3.3		6.0	
Manufacturing	24.1		19.7		14.2		Manufacturing	2.2		1.7		1.7	
Services	159.6	75.9	192.7	77.1	198.2	77.7	Services	21.6	67.3	27.1	62.5	27.4	69.5

Note: an "x" indicates that the population numbers were too small to record for privacy reasons.
Source: Statistics Canada (2019b).

Table 10.6

Employment in census metropolitan areas, in thousands, 2001–18

	2001		2011		2018			2001		2011		2018	
	N	%	*N*	%	*N*	%		*N*	%	*N*	%	*N*	%
British Columbia							**Vancouver**						
Total	1,920.9		2,227.8		2,493.6		*Total*	1,037.6	54.0	1,224.4	54.9	1,425.7	57.2
Goods	383.8	20.0	431.7	19.3	500.0	20.0	Goods	176.3		208.8		255.6	
Agriculture	25.6	6.7	25.6	5.9	23.6	4.7	Agriculture	6.6		7.2		8.2	
Forestry, etc.	41.2	10.7	40.8	9.5	49.7	9.9	Forestry, etc.	5.7		8.8		6.1	
Utilities	10.6	2.8	11.3	2.6	13.9	2.8	Utilities	5.6		6.1		7.8	
Construction	112.3	29.3	197.1	45.7	238.4	47.7	Construction	54.2		101.7		130.2	
Manufacturing	194.1	50.6	156.8	36.3	174.3	34.9	Manufacturing	104.2		85.1		103.3	
Services	1,537.2	80.0	1,796.1	80.7	1,993.6	80.0	Services	861.2		1,015.5		1,170.1	
Kelowna							**Victoria**						
Total	71.5	3.7	92.9	4.2	101.9	4.1	*Total*	151.0	7.9	178.0	8.0	195.8	7.9
Goods	15.0		19.7		20.5		Goods	17.7		20.5		25.2	
Agriculture	x		x		x		Agriculture	x		x		x	
Forestry, etc.	x		x		1.9		Forestry, etc.	1.7		1.6		x	
Utilities	x		x		x		Utilities	x		x		x	
Construction	6.4		10.7		12.0		Construction	8.0		11.8		17.2	
Manufacturing	6.5		6.3		5.3		Manufacturing	6.3		5.9		6.4	
Services	56.4		73.2		81.3		Services	133.3		157.5		170.6	
Abbotsford–Mission							**All census metropolitan areas**						
Total	71.5	3.7	83.9	3.8	96.3	3.9	*Total employment*	1,331.6	69.3	1,579.2	70.9	1,687.0	68.4
Goods	20.7		22.8		27.5		Goods	229.7	12.0	271.8	12.2	360.1	14.6
Agriculture	4.2		4.1		4.4		Services	1,101.7	57.4	1,307.3	58.7	1,376.4	55.8
Forestry, etc.	x		x		x								
Utilities	x		x		x								
Construction	5.1		10.3		11.3								
Manufacturing	10.6		7.5		10.3								
Services	50.8		61.1		68.9								

Note: an "x" indicates that the population numbers were too small to record for privacy reasons.
Source: Statistics Canada (2019a).

agreement between the NDP and the Greens defeated the Liberals, and a new coalition government was established by the end of June.

SUMMARY

Resource development dominated the Liberal era, largely because it was a controversial issue that loomed large in the media and because the provincial government remained resolute in insisting that resources were "the backbone of the economy."

Renewable resources – agriculture, fish, and forests – all made headlines but mainly for negative reasons. Although there is relatively little agricultural land in British Columbia, the government removed protections on a great deal of it to allow for resource development. The fishing industry, especially commercial salmon fishing, was a shadow of its former self and remained mired in conflicts over the harmful impacts of fish farms, mismanagement, and competition for fish with First Nations. Forestry, too, was mired in conflict as the mountain pine beetle epidemic continued to rage and the price of lumber and the Softwood Lumber Agreement with the United States collapsed, leading to unilateral duties and taxes on Canadian lumber. Although the government somewhat bailed out pulp-and-paper mills by allowing them to become independent power producers, mill closures and loss of jobs affected hinterland regions.

As forestry ceased to be the number one industry in the province, the provincial government turned its attention to mining, which has a significant environmental footprint, as the failure of the Mount Polley tailings pond made clear. As with lumber, coal mining was at the mercy of world market prices; mines opened when prices rose and then shut down as prices fell. But it was the export to Asia of low-grade thermal coal from the United States via the Fraser Surrey Docks that proved to be one of the most controversial developments, largely because of its contribution to climate change.

Energy resource development was without doubt the most controversial and divisive political issue. For example, the province manipulated the alleged need for electrical energy to push forward the Site C dam. The Liberal Party also promoted the benefits of LNG by labelling it "green" energy source, ignoring the fact that this fossil fuel contributes to climate change. It wasn't until the price of LNG crashed that the provincial government recalibrated its contribution to provincial revenues and employment. Pipelines carrying bitumen from Alberta's tar sands also provoked considerable disagreement and even legal action. Although the Northern Gateway Pipelines project was terminated, the federal government approved of the Kinder Morgan Pipeline to Burnaby.

Perhaps the most important development during this era was the recognition of climate change and the actions that were taken (and not taken) to curb its consequences. On the positive side, the Campbell government positioned the province as a leader in curbing greenhouse gases through policies to promote renewable energy, conservation measures, public transit expansion, and a carbon tax. On the negative side, both the federal and provincial governments continued to support initiatives and projects that were known to increase greenhouse gases. Climate change is somewhat analogous to cancer. Action can be taken or the problem can be ignored, at one's peril.

The struggles of First Nations to forge treaties and ensure that Indigenous Title was recognized and respected was also intimately bound up with resource development, and in many ways the province's approach to First Nations resembled its approach to climate change. The Liberals initially denied Indigenous Title, then accepted it. But they ultimately promoted resource activities on Crown Lands as if First Nations were not involved. The turmoil, demonstrations, and court cases that followed only increased the media's attention on resource development, once again fostering the idea that resources are the backbone of the economy, leading to a windfall of "good" jobs and bountiful provincial revenues.

REFERENCES

Allen, R. 2014. "Kinder Morgan Pipeline Expansion Designed to Carry Much More Oil." *The Tyee,* 28 May. the tyee.ca/Opinion/2014/05/28/Kinder-Morgan-Pipeline-Expansion/.

Asia Pacific Foundation of Canada. 2014. "A Brief History of Canada's Pacific Gateway." asiapacific.ca/asia-pacific-gateway-brief-history-canadas-pacific-gateway.

BC Community Forest Association. n.d. "A Brief History of Community Forestry in BC." bccfa.ca/a-brief-history-of-community-forestry-in-british-columbia/.

BC Ministry of Forests. 2012. *A History of the Battle against the Mountain Pine Beetle, 2000 to 2012.* Victoria: Government of British Columbia.

BC Ministry of Forests, Lands, Natural Resource Operations and Rural Development. 2019. *Major Primary Timber Processing Facilities in British Columbia, 2017.* gov.bc.ca/assets/gov/farming-natural-resources-and-industry/forestry/fibre-mills/2017_mill_list_report_final.pdf.

BC Stats. 2017. "Log Exports." gov.bc.ca/gov/content/data/statistics/business-industry-trade/industry/forestry.

–. 2018. "BC Census Metropolitan Areas and Census." catalogue.data.gov.bc.ca/dataset/bc-population-estimates/resource/087def5f-3500-4031-b8c0-a7143df5602e.

BC Treaty Commission. 1999. "A Lay Person's Guide to Delgamuukw." bctreaty.ca/sites/default/files/delgamuukw.pdf.

Beaty, R., R. Lipsey, and S. Elgie. 2014. "The Shocking Truth about B.C.'s Carbon Tax: It Works." *Globe and Mail*, 9 July.

Canadian Environmental Assessment Agency. 2014. "New Prosperity Gold-Copper Mine Project – Environmental Assessment Decision." ceaa-acee.gc.ca/050/document-eng.cfm?document=98459.

CBC News. 2010. "Ottawa Erred on B.C. Mine Review: Court." *CBC News*, 21 January. cbc.ca/news/canada/british-columbia/ottawa-erred-on-b-c-mine-review-court-1.898274.

–. 2011. "Timeline: The Battle over BC's HST." *CBC News*, 22 August. cbc.ca/news/canada/british-columbia/timeline-the-battle-over-b-c-s-hst-1.993242.

–. 2013. "B.C. Carbon Tax Cut Fuel Use, Didn't Hurt Economy." *CBC News*, 23 July. cbc.ca/news/canada/british-columbia/b-c-carbon-tax-cut-fuel-use-didn-t-hurt-economy-1.1309766.

–. 2015. "Sixties Scoop Victims Demand Apology, Compensation." *CBC News*, 18 June. cbc.ca/news/world/sixties-scoop-victims-demand-apology-compensation-1.3117552.

Clean Energy BC. 2010. "About." cleanenergybc.org/about_us/a_brief_history/.

Cohen, B. 2012. *The Uncertain Future of Fraser River Sockeye*. Ottawa: Minister of Public Works and Government Services Canada. publications.gc.ca/collections/collection_2012/bcp-pco/CP32-93-2012-1-eng.pdf.

Couture, L., and R. Macdonald. 2013. "The Great U.S. Recession and Canadian Forest Products." Analytical paper, Economic Insights, Statistics Canada. statcan.gc.ca/pub/11-626-x/11-626-x2013028-eng.pdf.

Curry, B., and S. McCarthy. 2011. "Canada Formally Abandons Kyoto Protocol on Climate Change." *Globe and Mail*, 12 December.

David Suzuki Foundation. 2006. *The Will to Protect: Preserving B.C.'s Wild Salmon Habitat*. davidsuzuki.org/wp-content/uploads/2019/02/will-to-protect-preserving-b.c.s-wild-salmon-habitat.pdf.

Davis, M. 2012. "Lessons Unlearned: The Legal and Policy Legacy of the BP Deepwater Horizon Spill." *Journal of Energy, Climate and the Environment* 155: 155–75.

Ducklow, Z. 2017. "Halting Site C Now Will Save Up to $2 Billion, Says UBC Report." *The Tyee*, 19 April. thetyee.ca/News/2017/04/19/Cost-of-Halting-Site-C/.

Elections BC. 2002. *Report of the Chief Electoral Officer on the Treaty Negotiations Referendum*. elections.bc.ca/docs/refreportfinal.pdf.

Federal Minister of the Environment and the British Columbia Minister of Environment. 2014. *Report of the Joint Review Panel, Site C Clean Energy Project, BC Hydro, May 1*. ceaa-acee.gc.ca/050/documents/p63919/99173E.pdf.

Gabriola Sounder. 2014. "BC Government Changes ALR Rules to Allow Development." *Gabriola Sounder*, 1 April.

Government of British Columbia. 2003. *Significant Projects Streamlining Act*. bclaws.ca/civix/document/id/complete/statreg/03100_01.

Haddock, M. 2011. *Comparison of the British Columbia and Federal Environmental Assessments for the Prosperity Mine*. Report for Northwest Institute for Bioregional Research. northwestinstitute.ca/images/uploads/NWI_EAreport_July2011.pdf.

Hamilton, J.D. 2009. "Causes and Consequences of the Oil Shock of 2007–08." *Brookings Papers on Economic Activity* Spring: 215–61. brookings.edu/~/media/files/programs/es/bpea/2009_spring_bpea_papers/2009a_bpea_hamilton.pdf.

Hoekstra, G. 2014. "Fires Rip through B.C.'s Tinder-Dry, Pine Beetle-Killed Forests." *Vancouver Sun*, 20 July.

Huffington Post. 2014. "Mount Polley Dam Tension Crack Detected 4 Years Ago, Says NDP Leader John Horgan." *Huffington Post*, 27 September.

Hurley, M.C. 2001. "The Nisga'a Final Agreement." publications.gc.ca/Collection-R/LoPBdP/EB/prb992-e.htm.

Jang, B. 2014. "B.C. Coal Export Plan Faces Resistance." *Globe and Mail*, 9 February.

Levy, D.A. 2009. *Pipelines and Salmon in Northern British Columbia: Potential Impacts*. Report for Pembina Institute. pembina.org/reports/pipelines-and-salmon-in-northern-bc-report.pdf.

McCreary, T. 2010. "Indigenous Protests Condemn B.C. Pipeline Project." *Climate and Capitalism*, 2 September. climateandcapitalism.com/2010/09/02/indigenous-protests-condemn-b-c-pipeline-project/.

Minister of Energy, Mines and Petroleum Products. 2002. *Energy for Our Future: A Plan for BC*. llbc.leg.bc.ca/public/PubDocs/bcdocs/357957/executive_summary.pdf.

Ministry of Energy and Mines. 2012. *Liquefied Natural Gas: A Strategy for BC's Newest Industry*. gov.bc.ca/assets/gov/farming-natural-resources-and-industry/natural-gas-oil/strategy_lng.pdf?bcgovtm=VC%20newsletter.

Ministry of Finance. 2001. "Revenue by Source." *2001 British Columbia Financial and Economic Review*. www2.gov.bc.ca/assets/gov/british-columbians-our-governments/government-finances/financial-economic-review/financial-economic-review-2001.pdf.

–. 2019. "Revenue by Source." *2019 British Columbia Financial and Economic Review*. gov.bc.ca/assets/gov/british

-columbians-our-governments/government-finances/
financial-economic-review/financial-economic-review
-2019.pdf.

Ministry of Forests, Lands, Natural Resource Operations and Rural Development. 2019. *Major Primary Timber Processing Facilities in British Columbia, 2017.* gov.bc.ca/assets/gov/farming-natural-resources-and-industry/forestry/fibre-mills/2017_mill_list_report_final.pdf.

Montana, S. 2010. "What Caused the Great Recession of 2008–2009?" economics-the-economy.knoji.com/what-caused-the-great-recession-of-20082009/.

Moore, D. 2014. "Supreme Court Ruling Grants Land Title to B.C. First Nation." *Huffington Post,* 26 June. huffingtonpost.ca/2014/06/26/supreme-court-decision-bc-first-nation_n_5533233.html.

New York Times. 2015. "Paris Climate Change Conference 2015." *New York Times.* nytimes.com/news-event/un-climate-change-conference.

Nixon, A. 2004. "Post-Treaty Fisheries Policy in British Columbia." Ottawa: Library of Parliament.

Northern Gateway. 2014. "Working Hard to Meet British Columbia's 5 Conditions." gatewayfacts.ca/five-conditions/.

Olsen, A. 2014. "B.C. Government Support for New Prosperity Mine Sends Confusing Message to First Nation." *Huffington Post,* 3 July.

Pacific Fisheries Resource Conservation Council. 2002. *2002 Advisory: The Protection of Broughton Archipelago Pink Salmon Stocks.* psf.ca/sites/default/files/Salmon Aquaculture-Broughton-Advisory_2002_0_CompleteR_20.pdf.

Parfitt, B. 2013. "Why BC Needs an LNG Plan B." *The Tyee,* 25 June.

–. 2016. "Political Leadership Needed to Revitalize BC's Forest Industry." *Policynote,* 4 October. policynote.ca/political-leadership-needed-to-revitalize-bcs-forestry-industry/.

–. 2017. "The Great Log Export Drain: BC Government Pursues Elusive LNG Dreams as More Than 3,600 Forest Industry Jobs Lost to Raw Log Exports." *Policynote,* 27 February. policynote.ca/log-export-drain/#.WLQ4jDulkh0.twitter.

Penner, D. 2014. "Future of Sawmill Industry Jobs Jeopardized by Overharvesting." *Vancouver Sun,* 29 April.

–. 2019. "Province Takes Starting Step to Crimp Log Exports from B.C. Coast." *Vancouver Sun,* 10 July.

Port Metro Vancouver. 2014. "Fraser Surrey Docks: Direct Transfer Coal Facility Project." portvancouver.com/development-and-permits/status-of-applications/fraser-surrey-docks-direct-transfer-coal-facility-project/.

Robinson, M. 2014. "Feds Reject Taseko's New Prosperity Mine over Environmental Concerns." *Vancouver Sun,* 26 February.

Schouls, T. 2005. "Between Colonialism and Independence: Analyzing British Columbia Treaty Politics from a Pluralist Perspective." Paper presented at the annual meeting of the Canadian Political Science Association, University of Western Ontario, 2005. cpsa-acsp.ca/papers-2005/Schouls.pdf.

Statistics Canada. 2019a. "Employment by Industry, Annual, Census Metropolitan Arres (x 1,000)." Table 14-10-0098-01. www150.statcan.gc.ca/t1/tbl1/en/tv.action?pid=1410009801.

–. 2019b. "Employment by Industry, Annual, Provinces and Economic Regions (x 1,000)." Table 14-10-0092-01. www150.statcan.gc.ca/t1/tbl1/en/tv.action?pid=1410009201.

Terrace Standard. 2011. "Power to Eventually Reach Iskut." *Terrace Standard,* 18 May.

US Energy Information Administration. EIA, Independent Statistics and Analysis. 2017. "Price of Liquefied U.S. Natural Gas Exports (Dollars per Thousand Cubic Feet)." eia.gov/dnav/ng/hist/n9103us3A.htm.

West Coast Environmental Law. 2004. "'Timber Rules': Forest Regulations Lower Standards, Tie Government Hands and Reduce Accountability." West Coast Environmental Law Deregulation Backgrounder. wcel.org/sites/default/files/publications/Deregulation%20Backgrounder%20-%20Forest%20&%20Range%20Practices%20Act_0.pdf.

Wood, P.B., and D.A. Rossiter. 2011. "Unstable Properties: British Columbia, Aboriginal Title, and the 'New Relationship.'" *Canadian Geographer* 55, 4: 407–25.

Conclusion

British Columbia's landscape was shaped by First Nations who were here long before the arrival of Europeans. But it was also altered drastically following the arrival of colonial powers. Many colonial policies were tied to resource exploitation, whether it be furs, coal, or gold. With Confederation, colonial policies evolved into federal, provincial, and municipal policies, and although federal policies were broad in scale and significant, the provincial government had the greatest impact on the region's peoples and landscapes.

Some policies were only indirectly related to resources. For example, racism was rampant in what was believed to be a white, British, privileged society. Because First Nations did not farm their lands, they were judged by the white "civil" society as not the owners of it. They were allocated small reserves, usually ten acres per family, which remained under the control of the federal government and governance of the Indian Act. Even then, some of these small reserves were whittled down further by other provincial and federal policies. Immigrants, such as the Chinese, Japanese, and South Asians, were denied the vote and worked under labour laws that allowed them to be paid half that paid to white workers. These immigrants were not wanted by the dominant white society, except as cheap labour. Policies of restricted immigration (e.g., head taxes, continuous-passage laws, and exclusion acts) were accompanied by restricted access to employment along with policies of residential segregation that restricted where minorities could live and work.

There were a number of driving forces that shaped who was attracted to the province, who was allowed to live in it, where they lived, and how they worked:

- the lure of resource wealth in combination with a belief that the so-called frontier contained an inexhaustible supply
- the need for infrastructure and transportation routes to shrink time and space and facilitate resource development
- land grants, licences, cash subsidies, and resource-harvesting rights awarded by colonial and provincial governments
- an ongoing belief that resource development represented the backbone of BC's economy and that the economic benefits of resource development outweighed environmental or social concerns.

As each chapter is this book shows, however, other transformative forces were at work, and many of them were unanticipated. Pandemics such as smallpox and Spanish Influenza and geophysical hazards such as earthquakes, volcanic eruptions, tsunamis, floods, avalanches, wildfires, violent windstorms, and landslides brought death and destruction to many. Other unforeseen events were policy-based and often global in scale. Wars, depressions, recessions, and acts of terrorism altered the political, economic, and social direction of the province and the world. More recently, climate-related hazards such as violent storms, flooding, droughts, wildfires, and mountain pine beetle epidemics have accelerated as climate change transforms the globe.

Technological innovations have also had an impact on the landscape. Over time, animal power was replaced by the steam engine, which was eclipsed by petroleum, all while the demand for electricity has increased. New innovations were not confined to energy. They permeated all aspects of the production process as industry after industry became more capital-intensive, increased productivity, harvested more resources, and produced more goods and services, but with fewer and fewer workers.

Where does this historical perspective take us? Over less than 250 years, since the beginning of European colonization, the province's population has increased greatly, but not evenly, and we know that resources were a significant factor in this pattern. However, since the 1970s, the digital and transportation revolutions have transformed the province into a "modern," urban society in which the majority of people work in the service sector. In fact, the natural resources of British Columbia have largely been squandered. To conclude, we'll return to the concept of the tragedy of the commons to summarize the developments that have overtaken the province and to drive home the need for a new political direction.

ANTI-ASIAN EXCLUSION AND IMMIGRATION LAWS

In the parable of the tragedy of the commons, peace and unchecked capitalism lead to the overgrazing of the commons and the ultimate downfall of a society dependent upon cattle. If British Columbia is the commons, then rules were created to restrict and even deny the use of the commons by Asians. The province exerted enormous pressure on the federal government, which has jurisdiction over immigration, to create barriers to Asian immigration, and provincial policies denied them the right to vote, dictated lower rates of pay, restricted their employment, and encouraged their segregation in ethnic enclaves such as Chinatowns. Racism was rampant and acceptable to the dominant white population.

The postwar economic boom, which lasted until the 1970s, led to the lifting of many of these restrictions and fundamentally reversed many racist values. However, it was not until 1967 that Canada adopted a nonracial point system for immigration. As a result, the racial diversity of British Columbia's (and Canada's) population has increased greatly. As well, Canada extended an official apology and financial compensation in 1989 to Japanese who were interned during the Second World War. It also made an official apology in 2006 to the Chinese who were forced to pay the $500 head tax; however, this compensation was paid only to those who had actually paid the tax (only twenty of whom were still alive) and not to their families. Sikhs who experienced the *Komagata Maru* incident in 1914 also received an apology in 2008, but it was announced in Surrey and not in the House of Commons. Many felt it was not an official apology.

Although overt racism against Asians has largely subsided, tensions remain over immigration. In recent decades, migration has been forced on many peoples around the world, in part because of economic and political upheavals such as civil war and in part because of climate change, which has increased the number of droughts, floods, and other intense weather events that cause countless deaths and the destruction of landscapes. In this context, there has been a backlash against refugees entering Canada.

FIRST NATIONS PURSUING THEIR SHARE OF THE COMMONS

First Nations have their own sets of values, organizations, and institutions, and they all derived from the land, or commons. But colonialism and Confederation inflicted a new set of values on the people and the land as they were relegated to a minority by disease, confined to reserves, and forced to assimilate through the Indian Act and residential schools. With only a few exceptions, there was no recognition of First Nations claims to the commons in British Columbia.

As was the case with Asians, attitudes towards First Nations changed following the Second World War. First Nations gained the right to vote and regained the right to perform traditional ceremonies. But the BC government remained adamant that they had no claim to the land. Slowly, this position changed, for the most part through the courts. Politically, the federal government recognized Indigenous Title and introduced comprehensive and specific claims in 1973. Provincially, a major breakthrough came in 1991 with the election of an NDP government that was willing to negotiate modern-day treaties and formed the BC Treaty Commission.

The treaty process is arduous, time-consuming, and costly, and few treaties have been completed (see Table C.1). But some have, and they have brought an end to land-use conflicts in those regions. First Nations that have completed or remain engaged in the treaty process represent only 104 of the 198 distinct Indian Act bands in the province. Nearly half the bands have decided, for a variety of reasons, not to participate; some simply do not agree with the process, while others believe it is too expensive. To engage in the treaty process, First Nations must be loaned money by the federal government. The BC Treaty Commission reported that as of 31 March 2018, "outstanding negotiation loans totalled approximately $550 million (excluding accrued interest)" (BC Treaty Commission 2018, 57).

Table C.1

First Nations engaged in the treaty process

Ratified treaties	4
Stage 6 but not ratified	2
Stage 5	7
Stage 4	42
Stage 3	2
Stage 2	6
Total	63*

* A number of First Nations are made up of more than one band (e.g., Maa-nulth First Nations consist of five former Indian Act bands).
Source: BC Treaty Commission (2017).

Land-use conflicts remain, although some resource-development issues have been resolved through interim agreements. Other conflicts may require legal measures to resolve. What can no longer be denied is the Supreme Court's ruling that provincial and federal governments *must* negotiate with and compensate First Nations for activities on unceded Crown lands. The Supreme Court's decision in 2014, in *Tsilhqot'in Nation v British Columbia*, in awarding Indigenous Title to a First Nation, set another precedent. If a First Nation deems that the province is not bargaining in good faith, it can launch a lawsuit to gain its land, and it just might win.

However, by 2005, Indigenous Title took a back seat for many First Nations when the Supreme Court accepted a class-action suit over the dreadful conditions experienced in Indian residential schools. Settlement occurred in 2006 and then transitioned into the Truth and Reconciliation Commission and a formal apology in the House of Commons. The commission's 2015 report documented many heart-wrenching stories of social, physical, sexual, and psychological abuses, including the deaths of thousands of children. Residential school abuses for day scholars has also been accepted as a class action by the Supreme Court, and it is likely that a similar process will occur to address wrongs associated with the "Sixties Scoop." Allegations of abuses suffered by First Nations in "Indian hospitals," where they were treated for tuberculous, are also surfacing. As well, a National Inquiry into Missing and Murdered Indigenous Women and Girls (MMIWG) was launched in 2016, with its final report – *Reclaiming Power and Place* – being released in June 2019. The final report reveals "that persistent and deliberate human and Indigenous rights violations and abuses are the root cause behind Canada's staggering rates of violence against Indigenous women, girls and 2SLGBTQQIA people" (MMIWG 2019). To emphasize this failure of Canadian society, the report labels these actions as genocide.

Several questions remain. Was the prime minister's 2008 apology accepted by First Nations, and do the Indian Residential Schools Settlement Agreement and the Truth and Reconciliation Commission's final report imply that First Nations have forgiven past abuses? Moreover, social, physical, and psychological abuse cannot be divorced from the abuse of the land. Reconciliation should unequivocally mean recognition of Indigenous Title,

but unilateral resource-development projects have been approved by the provincial government, allowing private corporations to develop unceded Crown lands. As conflicts continue, First Nations are forced to rely on the court system to pursue reconciliation. On the positive side, many more Canadians are now aware of past atrocities as well as ongoing inequities between Indigenous Peoples and others in society. And the introduction of a new provincial bill on Indigenous Rights – the BC Declaration on the Rights of Indigenous Peoples Act, in response to the United Nations' Declaration on the Rights of Indigenous Peoples – in late 2019 gives hope for reconciliation as well as greater consultation on any developments on unceded Crown lands in British Columbia. As more treaties are settled, more Crown lands and their resources will be transferred to First Nations.

LOSSES AND GAINS IN THE FISHING INDUSTRY

Commercial fishing – including for sea otters, fur seals, and whales – followed the trajectory outlined in the tragedy of the commons. All three fisheries resulted in near extinction of species, and only commercial extinction and treaties ended the exploitation. The recovery of sea otters, fur seals, and most species of whales occurred when they ceased to be labelled as resources to be harvested; now, they are perceived as more valuable as resources to view for the tourist industry. Another lesson to be learned is that these species, whales in particular, are a global management issue (similar to climate change).

The foundation of the commercial fishery on the West Coast has long been five species of salmon, and the gatekeeper model has long been the management tool for controlling the catch through openings and closings. The model has failed. Historically, salmon represented 80 percent or more of all commercial species caught. By 1951, that number was down to 72 percent; by 1991, it was 34.8 percent. Figure C.1 gives a sense of the declining importance of salmon.

The gatekeeper model applies only to trollers, gillnetters, and seiners, but the sustainability of the fishery also depends on other considerations, including the legal obligation to provide First Nations with salmon for food and ceremonial practices and meeting the needs of the sports fishery. Treaties have added another layer of complexity in that they break the gatekeeper model by

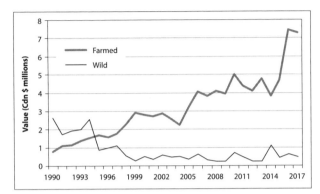

Figure C.1 Value of farmed salmon and wild salmon, 1990–2017
Source: Fisheries and Oceans Canada (2018, 2019).

treating salmon as private property resource a rather than a common property resource. Agreements render a percentage of salmon as a private property resource within specific runs within a Traditional Territory.

Aquaculture, or fish farms, first introduced in the mid-1980s, likewise poses a threat to wild salmon. In 2001, the value of farmed salmon was $271 million while the value of commercial salmon was only $37 million (see Figure C.1). Farmed salmon's value increased to $729 million by 2017, whereas wild salmon was only $46 million, and the total value of all commercial species was $398 million. Clearly, salmon raised in net pens in the ocean bring in greater revenues than all commercial species of fish combined, but farmed salmon are a foreign species (mostly Atlantic) that pose a threat to wild Pacific salmon. The federal government has the jurisdiction to manage both the commercial fishery and fish farms, and a simple solution would be for farmed salmon to be raised in land-based pens where effluent, diseases, sea lice, and escapees can be managed. The government could develop policies to avoid the conflicts but, to date, it has not done so.

Managing salmon is much more complicated than simply addressing the problems of the gatekeeper model. Salmon are a mobile resource that enter international and American waters in their migration cycles, and there are treaties to govern US catches of Canadian salmon. But salmon are also anadromous, meaning that most return to the streams where they were born to spawn and die. Freshwater habitat for salmon is critical, and many other land-use conflicts – related to logging, mining, pollution, urban development, and so on – interfere with salmon runs. Moreover, climate change is causing rivers and oceans to warm, which is detrimental to salmon. Tragically, salmon runs are nowhere near historical levels, and it will require many policies to increase them in the future.

THE EROSION OF FARMING

During the settlement of BC, many new arrivals were farmers and ranchers. Farming was mostly restricted to river valleys, where the soils are rich and farms productive. Ranching dominated the Interior Plateau regions, where natural grasses provided important grazing from spring until late fall. In agriculture, as with other commodities, transportation developments, competition, and even natural hazards such as floods created market price swings. And government policies likewise influenced farming. Marketing boards controlled the supply side of produce (for milk, butter, eggs, and so on), improving prices for farmers. By the 1970s, the creation of the Agricultural Land Reserve also made it difficult to develop good agricultural land for nonagricultural uses (e.g., residential, commercial, industrial, or recreational uses). Unfortunately, legislative amendments in 2014 watered down the concept of agricultural land being sacred by creating Zones 1 and 2 and allowing nonagricultural activities in the latter.

Farming in BC followed trends similar throughout North America. New technologies, especially after the 1950s and 1960s, resulted in a decline in the number of farms and farmers, but those farms that remained became larger, although with fewer employees. This development contributed to rural-to-urban migration. Few people today consider farming a significant occupation in the province (it accounts for about 1 percent of total employment), but there is general recognition of the importance of agricultural produce grown in British Columbia and the connection to the land, especially in the southern portions of the province, which have a unique climate. There, many small farms – two-thirds are less than sixty-nine acres (twenty-eight hectares) – produce specialty crops such as berries, fruit, organic vegetables, and herbs. Some of these farming special-izations also have strong links to other industries. For instance, grape production has an important tie-in with the wine and tourism industries. With a finite amount of agricultural land, along with technologies to increase productivity, land-based farms, similar to the commercial salmon fishery, are no longer a major employer in the province.

THE NEED FOR SUSTAINABLE FOREST MANAGEMENT

The forest industry has a long history in BC, a province that was richly endowed with coastal forests dominated by old-growth western redcedar, Douglas fir, and hem-lock, trees that were hundreds of years old and repre-sented an enormous volume of wood per hectare. Few old-growth stands are left, and logging relies on con-siderably smaller second-growth trees. The interior forest region is seven times larger than that of the coast and, while there are some Douglas fir and cedar, the dominant species are pine and spruce. The interior forests were largely untouched until truck logging came into its own. By the mid-1970s, however, the interior was outprodu-cing the coast by volume, and the gap has only increased since then. In the 1970s, the government produced a film titled *Fifty Cents of Every Dollar*. Essentially, the BC gov-ernment stated that fifty cents of every dollar generated in the province came from the forest industry. It also im-plied that forestry was the main employer.

Those days are over. Many factors have combined to send the forest industry well down the road to tragedy. A 2006 report warned that the coastal forest industry "is characterized by unacceptable rates of return on capital employed (ROCE); some of the highest production costs in the world; a product mix that is largely obsolete; mills that are undersized and outdated; and, not surprisingly, a chronic shortage of new capital investment" (Haley and Nelson 2006). The coastal mills were designed for large old-growth trees, but the industry had not invested in plant and equipment to harvest second-growth trees. In 2016, it was reported that not a single new sawmill of note had been built in more than a decade (Parfitt 2016). In-stead, Christy Clark's government encouraged the export of raw logs, most of which came from unceded Crown land. The solution is simple: put an end to their export and save the logs and jobs for future generations of Brit-ish Columbians.

In the interior, the forest industry has reinvested in state-of-the-art sawmills to harvest pine forests killed by the mountain pine beetle epidemic, and an immense volume of dead wood is available. It has been projected that 69 percent of mature lodgepole pines will be killed by 2024. This is 20 percent of the province's total timber harvest land base (Clinkard 2010). Somewhat akin to the commercial salmon-fishing industry, there has been plenty of talk about habitat restoration, in this instance reforestation, and how limits should be placed on over-harvesting. But a host of issues in this industry has left it reeling: years of service cuts to forestry management, an industry left to regulate itself, a shortage of fibre due to fires and mountain pine beetle kill, and the high cost of available fibre have meant that more than four thousand forestry workers have lost jobs in 2019 alone (BC Govern-ment and Service Employees' Union n.d.; Boynton and Palma 2019).

Tragically, there is a link between climate change and the mountain pine beetle epidemic. Cold temperatures used to keep the beetle in check, but winter temperatures are much warmer than they used to be. Moreover, increas-ingly hotter summers have led to repeated forest fires, with the summer of 2018 being the worst on record. These wildfires, in turn, have accelerated climate change by increasing greenhouse gases. In 2017, it was reported that the largest "wildfire season on record had emitted an estimated 190 million tonnes of greenhouse gases into the atmosphere – a total that nearly triples B.C.'s annual carbon footprint" (Hernandez and Lovgreen 2017).

Forestry was once the number one industry in the province, so there were (and continue to be) many com-munities dependent on the industry for employment. Although the provincial government continues to con-tend that "forestry is one of B.C.'s founding industries and a key driver of B.C.'s economy," it also recognizes that there will be further mill closures and hardships for communities (Ministry of Forests, Lands and Natural Resource Operations 2017). As a result, a number of pro-grams are available to assist communities to diversify their economies and offer skills training for individuals.

These forest-dependent communities are undergoing a major transition.

BC's forests are vast but not endless and need to be managed, not on a four-year political cycle but on the basis of forest-maturity cycles, which could be a fifty-to-sixty-year rotation on the coast or considerably longer in regions with harsher climates and terrain. Employment in logging, sawmills, and pulp-and-paper mills is (and will continue to be) an important part of the British Columbia landscape, particularly in the more remote regions of the province. However, as in agriculture, innovations in the forest industry will continue to reduce the need for workers. Sustainably managed forests can (and must) play a significant role in reversing the effects of greenhouse gases and reducing the threats of climate change.

UNCERTAINTY IN METAL MINING

Mining also has a long history in this province. In many ways, it was spurred by the promise of individuals becoming instantly rich, especially when gold was discovered during the era of placer mining. Over time, technologies evolved, first allowing miners to follow ore seams deep into mountains and then replacing individual miners with larger and larger corporations in capital-intensive megaprojects that vastly increased metal production.

Metal mining in BC is dominated by copper, although the province is a major producer of molybdenum and gold and some silver, lead, and zinc. However, putting a mine into production today is a high-risk venture for a variety of reasons. Fixed costs such as labour, equipment, energy, and transportation are high, especially if the ore deposit to be developed is in an isolated, northern location, where climates and terrain can be severe. Governments may help, as they did in building the Northwest Transmission Line to deliver much-needed electricity to the remote Iskut-Stikine River system. Still, there is the related cost of building roads, accommodations, and so forth. Another important consideration is that many of these deposits are on unceded Crown lands, and First Nations need to be consulted and compensated. Also, First Nations may be opposed to mining in their territory for a variety of reasons.

Adding to the uncertainty is the fluctuating value of the Canadian dollar, swings in the world market price for metals, and the increasing costs of instituting safeguards to avoid environmental risk. Moreover, investor confidence is often lacking because BC's copper deposits are good but not world-class. The idea that they are world-class has been propagated by politicians, the mining lobby, and companies. But when ranked against worldwide deposits, BC's are modest at best (Nelson 2015, 129). In addition, mining the low-grade copper results in massive acid rock tailings that need to be deposited securely and monitored.

The public's perception of mining is that it poses high risks for the local environment, a perception that was confirmed by the tailing pond failure at the Mount Polley mine in 2014 and by the federal government's failure to approve the Prosperity Mine and the subsequent New Prosperity Mine because of the environmental costs. The tailing pond failure and awareness of the potential impacts of a New Prosperity Mine have led to a concern that mining ventures mean private profit even as the burden of proof for environmental catastrophes falls on the local population. No fines were levied in the Mount Polley disaster.

More mines will open in the province because of the provincial government's tendency to portray BC as a resource-dependent province. When the forest industry waned, the government turned its attention to the mining industry and initiated a significant infrastructure project – the Northwest Transmission Line – without allowing the independent BC Utilities Commission (BCUC) to assess its merits. Again, the relatively remote northwest is not an empty landscape, and mining activity brings considerable disruption to the people and the land, even though some locals do gain good jobs.

ENERGY RESOURCES AND THE RESOURCE-DEPENDENCY MYTH

Energy is a foundational resource. It is essential to the production of everything. In recent decades, however, energy production has become a dominant theme or tool in provincial elections. British Columbia has considerable sources of energy, but many come at a cost to the environment, both locally and globally. What is disturbing is that the election-fuelled zeal for megaprojects has been accompanied by many contradictory messages.

Hydroelectricity was a major plank during the W.A.C. Bennett era (1950s–70s), when a number of heritage dams were built, and a few more were constructed during the energy crisis of the 1970s. However, as the province's population increased, so did the demand for electricity. Supply and demand were about even by 2001, but this was accomplished through the use of greater and greater amounts of gas-fired thermal generation, also known as cogeneration. However, even with the great reliance on hydro-generated power, the difference between net importing and exporting was (and still is) dependent on winter snow accumulation and runoff. Low-snow years resulted in net imports of electricity (Ministry of Finance 2017).

Several developments occurred in the production and consumption of electricity. Although the NDP and Liberal governments were committed in the 1990s and early 2000s to producing more electricity through cogeneration, a combination of high natural gas prices and concern over natural gas contributing to climate change led to a reversal of this position. Cogeneration plants were not pursued, and the old, inefficient Burrard Thermal plant was phased out. The Liberals under Gordon Campbell crafted a number of policies on conservation and to curb greenhouse gases, the most important being the introduction of a carbon tax. However, the link between developing, processing, and burning natural gas (a fossil fuel) and climate change was forgotten by the 2013 election.

Another direction initiated by the NDP was the production of electricity through independent power producers (IPPs). This option also fit well with the Liberal Party's ideology of privatization, and although the program originally encouraged small run-of-river hydroelectric projects approved by BC Hydro, it now allows for many types of electrical production, not all of them renewable (see Table C.2). Nevertheless, an interesting array of power options has developed, and IPPs now produce over 30 percent of energy consumed in the province.

With the increase in IPP production, British Columbia has become a net exporter of electricity since 2015. And the completion of the Site C Dam, slated for 2024, will give another major boost to electrical production. This project has a long history of being pushed forward by

Table C.2

Types of IPPs, 2017

Type	Energy (GWh/y)	%
Gas-fired thermal	3,205	15.4
Pulp mills = biomass	3,062	14.7
Storage hydro (e.g., Alcan at Kitimat)	4,905	23.6
Nonstorage hydro (e.g., run-of-river)	7,172	34.5
Wind	2,008	9.7
Municipal solid waste	166	0.8
Energy recovery generation = waste heat	141	0.7
Biogass	127	0.6
Solar	2	–
Total	20,788	100

Source: BC Hydro (2017).

provincial governments and BC Hydro and then rejected by independent bodies such as BCUC. It only went forward when the provincial government shut BCUC and the Agricultural Land Commission out of the decision-making process. It also restricted BC Hydro's ability to increase electricity production to existing dams and Site C. The Crown corporation no longer has the ability to research the feasibility of other sources of electricity such as geothermal or wind.

Frustration over the project was certainly felt by some local First Nations. The Prophet River and West Moberly First Nations argued that flooding the valley infringed on their constitutionally protected treaty rights. Although Site C is within Treaty 8 territory, a panel of three judges dismissed the appeal, contrarily arguing that the federal government could issue permits for projects such a Site C without first discovering if they violate treaty rights (Kurjata 2017). Now, with the NDP government's approval of Site C in 2018, five First Nations have signed benefit agreements.

Since it was discovered on Vancouver Island in the 1830s, coal has also been a source of energy in this province. The industry boomed when it fuelled the steam engine, but the rise of petroleum use in the early to mid-1900s led to a steady decline in the volume of coal used and the number of mines in operation. The industry underwent a metamorphosis in the 1970s, when the steel industry in Asia created a huge demand for metallurgical or coking coal, resulting in new coal mines and communities in the Kootenays and in the northeast, such as in Tumbler Ridge. However, a number of mines in the northeast have since closed, and no mines have been in operation in the Tumbler Ridge region since 2015. However, coal – in particular, low-grade thermal coal exported from the United States via Surrey, south of Vancouver – continues to be controversial, its use a contradiction to the goal of reducing greenhouse gas emissions.

British Columbia consumes more oil than it produces, and oil issues loom large in the province. Disagreements and controversies have centred on the transportation of bitumen from the Alberta tar sands. When the federal government cancelled the Northern Gateway Pipelines project and implemented a tanker ban for northern British Columbia waters, it sent a message that the transportation of bitumen is high-risk and that producing more oil contributes to global climate change. However, it then contradicted itself by approving the Kinder Morgan Pipeline. People are concerned not only about pipeline failure but also about a potentially catastrophic oil spill in Vancouver waters. Of course, more oil also means more greenhouse gases. Governments need to come to terms with the fact that emitting more carbon into the atmosphere is not in the local or global interest.

Natural gas has undergone the greatest transformation in political messaging in recent years. Although it was an important export to the United States from the 1950s to the early 2000s and slated to become part of the electricity stream as cogeneration, the new, efficient technology of fracking (also labelled "unconventional natural gas") transformed the United States into the largest producer of natural gas in the world. Fracking has also been used successfully to harvest a number of shale gas deposits in northeastern British Columbia. The Montney and Horn River shale basins, in the Fort St. John and Fort Nelson regions, respectively, have enormous reserves of unconventional natural gas deposits that "could account for fully 22 per cent of all of North American shale gas production by 2020" (Parfitt 2011, 12). However, as the supply of natural gas rises, prices fall (see Figure C.2).

Fracking involves drilling down vertically to a bed of coal (also known as coalbed methane) or shale gas deposits and then drilling horizontally. More than one well

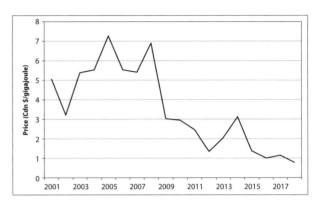

Figure C.2 Natural gas prices, 2001–18
Source: Ministry of Finance (2010, 2019b).

can be drilled from the same platform, thereby injecting the deposit with many drills. A high-pressure mixture of water, sand, and chemicals is then pumped into these deposits to fracture the seams so the natural gas can be extracted. One of the greatest concerns is that the process can contaminate both surface and underground sources of water. A lawsuit that gained some attention within Canada claimed that fracking had contaminated well water in Rosebud, Alberta, to the point where the tap water could be ignited (Nikiforuk 2011). Fracking has also been responsible for triggering earthquakes. Natural gas production – from fracking through to its transportation through pipelines – also produces greenhouse gases. And not all natural gas fields are equal in terms of carbon emissions. For example, the Horn River shale gas deposit contains approximately 12 percent carbon dioxide, whereas average gas pools in BC contain 2 percent carbon dioxide (National Energy Board 2009). British Columbia does not produce the cleanest gas in the world.

Contrary to government rhetoric, other problems are association with the conversion of natural gas to liquefied natural gas (LNG). When natural gas is cooled to minus 162 degrees Celsius, it liquefies, and its volume reduces six hundred times, making it economical to ship around the world in LNG tankers. But cooling the gas requires an enormous amount of energy that can be supplied by hydroelectricity or natural gas. If the conversion is done by hydroelectricity, then fewer greenhouse gases will be involved, and this is where the energy from Site C could be required. On the other hand, if the conversion from natural gas to LNG is done through burning natural gas, then greenhouse gases will increase considerably.

The provincial government failed to inform the public that the upstream production of LNG has a very large environmental footprint. And its downstream use also has many negative environmental consequences. Fugitive emissions can occur at all stages of the process, and it has been estimated that methane leaks in BC's natural gas industry are at least seven times greater than official numbers, "increasing the entire province's carbon footprint by nearly 25 percent" (Leahy 2013).

The federally approved Petronas LNG plant on Lelu Island was contested by First Nations who claim the island as their Traditional Territory. A major factor in the dispute is the proposed location of the facility in the Skeena Estuary, which will threaten the 100 million to 1 billion juvenile salmon that migrate through it every year (CBC News 2015). Although it has been glossed over by the federal and provincial governments, the project will also contribute to carbon emissions, especially since, according to Petronas, "plans call for the terminal to use natural gas turbines in the liquefaction process to super cool natural gas into liquid form" (Jang 2016).

The Liberal government also promised that LNG was a "green" energy source. There are, of course, no guarantees that LNG will replace coal in Asia. It is more likely to retard renewable sources of energy because it is so cheap. Moreover, if all potential emissions are taken into account, LNG may be dirtier than coal (Gillis 2015). Finally, the provincial government's estimates include only what the province produces – as if burning BC LNG in Asia is not its responsibility.

AVOIDING THE FINAL TRAGEDY

Understanding the issues that have shaped and reshaped the patterns of settlement and development of British Columbia is a way of unveiling the past to reveal the future. But the question remains: Why does the myth of resources being the backbone of the economy persist, contrary to the evidence? A significant portion of BC's population today grew up during the W.A.C. Bennett era (1950s–70s), when resources did "open up" the province (see Figure C.3). This demographic, as well as many others, believe that it is the export of goods such as fish, forests, minerals, and energy that generates

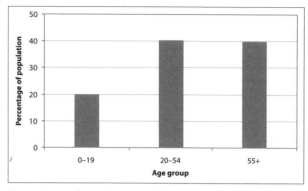

Figure C.3 Population by age group, 2018
Source: BC Stats (2019).

the most jobs and revenues for the province. Moreover, resources tend to dominate the news, largely because of their controversial nature, thus giving the illusion that they must be integral to the economy.

Politicians also play to this myth because there are relatively few economic projects that hold out the promise of providing hundreds of good jobs and more revenues. For example, infrastructure projects such as a new highway, bridge, or SkyTrain will certainly lead to jobs but few revenues. Events such as Expo 86 and the 2010 Winter Olympics provided both because they also brought in tourism dollars, but these types of events are relatively rare and will likely become even more so because of the costs of security. (For example, the province turned down hosting World Cup Soccer matches in 2019 for this reason.) But megaprojects such as a new LNG plant or a new copper-gold mine come with foreign investors and the promise of jobs and government revenues, which will pay for all the social costs.

Governments continue to promote the myth of a resource-dependent province even when 87 percent of the population lives in urban centres and resources account for only 5 percent of government revenues (Statistics Canada 2016, 2019a, 2019b; Ministry of Finance 2019b). All jobs, including those in the resource industries, are important, but British Columbia is no longer a resource-dependent province. It has an urban, service-based economy composed of many small businesses that employ a great number of people. The engines that propel BC's economy include educational institutions, conference services, the tourism industry, stock market transactions, retail sales, banking and financial institutions, a host of online services, legal services, health services, pharmaceutical technologies, the film industry, and software and high-tech companies.

Despite this reality, for both the public and politicians, the prospect of new resource development is appealing. These projects bring to mind the gold rush era and conjure up the notion of tremendous wealth for individuals and revenues for the government. The recent focus on LNG is simply the latest "gold rush." However, today, because of climate change, we don't have the luxury of engaging in major resource developments without understanding the costs, especially when projects involve fossil fuels.

In addition, come election time, the service sector, which represents over 80 percent of employment in the province, does not hold the same allure as megaprojects. The service sector is where many of the minimum-wage jobs reside, which is untenable in cities such as Metro Vancouver. There were few bragging rights associated with the opening of the Tsawwassen Mills shopping centre. It led to three thousand new jobs – but few were "good." Many service sector jobs are in fact under threat of being cut by the government. For example, educational services were constantly cut back until a Supreme Court ruling in 2016 forced the provincial government to adequately staff public schools. Public administration is another target for a provincial government that does not want "overregulation." And the media constantly run stories of problems in health care and social assistance: the shortage or absence of doctors and nurses, long wait times for surgeries, the inability to adequately handle the fentanyl crisis, huge caseloads for social workers, and aged-out youth in foster care.

Although more revenues accrue to the provincial government from the sale and transfer of property than from natural resources, real estate is also in a state of crisis and a fraught election issue. Foreign investment, and even money laundering, drove up the price of housing in the province. In response, the government implemented a foreign investment tax in 2018 and appointed an expert panel on money laundering, which in 2019 reported that "money laundering investment in BC real estate is sufficient to have raised housing prices and contributed to BC's housing affordability issue" (Ministry of Finance 2019a, 8). The price of housing is out of range and not affordable for the average person, which in turn results in qualified (and sometimes highly qualified) people rejecting opportunities for employment in the Vancouver area. The other side of the coin is that increasing numbers of homeless are demanding low-rent accommodations.

There is a pattern. Provincial governments have maintained that resource development is critical to the economy and have shown little tolerance for discussions of the costs of resource projects, especially those related to fossil fuels. As a consequence, government-orchestrated environmental assessments have rarely been objective. They have limited timelines and limited input by stakeholders; and they have limited the ability of arms-length organizations such as the BCUC to weigh in and limited the scope of assessments. Then, when the projects were approved with conditions, the conditions gave a sense that the approval could be overturned. The findings of the federal government's National Energy Board – which assessed the merits of oil pipelines through British Columbia and the development of LNG, Site C, and US coal exports – had the same failings.

The modern-day approach to resource development differs little from that of the colonial period, when it was believed that resources in the "frontier" needed to be liberated for the benefit of the exploiters, with little consideration for the environment or local populations. However, today, rural, out-of-the-way communities do not exist in isolation, and they are not dependent on the next wave of resource development. For example, the community of Kitimat turned down the proposed Northern Gateway Pipelines project, and the community of Squamish turned down the building of LNG facilities. However, municipalities and regional districts have little say in projects sanctioned by the provincial or federal governments.

Perhaps what is worse is that many of these resource development schemes are on unceded Crown Lands, putting First Nations in the position of having to use the courts to enforce recognition of their right to consultation and compensation. Moreover, the promotion of megaprojects is contrary to the spirit of reconciliation, even though governments insist they are in the national interest and required in the short run to meet greenhouse gases targets in the future. Is it in the national interest to increase our carbon output when emissions contribute to droughts, wildfires, infestations, food and water shortages, poor salmon runs, shrinking glaciers, coral bleaching, wilder storms, and coastal flooding from rising sea levels? Climate chaos is already here. It is not something that may happen in the future.

From a government policy perspective, there is talk about curbing greenhouse gases to meet targets in the future, but there is no recognition that action is required immediately. If the burning of fossil fuels continues to rise, the result will be the final tragedy of the commons, bringing ruin to all. The science behind climate change and its consequences is sound, and international environmental organizations such as 350.org have been lobbying governments for many years to reduce carbon emissions below 350 parts per million. The Intergovernmental Panel on Climate Change stated clearly: "Emissions, mainly from the burning of fossil fuels, may need to drop to zero by the end of this century for the world to have a decent chance of keeping the temperature rise below a level that many consider dangerous" (Intergovernmental Panel on Climate Change 2014). A 2018 report by the auditors general of nine provinces and the federal environment commissioner likewise concluded that "Canada is on pace to overshoot its emissions target for 2020 by nearly 20 per cent, and that more action is needed to meet the Paris Agreement target by 2030" (Ballingall 2018).

As the 2019 federal election loomed, the extremely controversial Trans Mountain Pipeline Expansion project – which would bring bitumen from Alberta to Burnaby, where it will be exported by tankers – was declared in the national interest by the federal government and purchased for $4.5 billion in 2018. Of course, the government will have to spend billions more to actually build the line. The election revealed some of the complications to political decision-making. The Liberals did not gain a majority, but continue to govern with the cooperation of other minority parties. The election results indicate that a majority of people across Canada are concerned about the environment and climate change in particular; on the other hand, polls have indicated that a majority of Canadians want to see the Trans Mountain Pipeline Expansion built (Joseph 2018).

What is needed, and posed by many, including the International Monetary Fund, is for countries to wean the fossil fuel industry off subsidies. In 2014, the International Monetary Fund estimated that governments in Canada pay more than $34 billion each year in subsidies to energy producers and uncollected taxes on externalized costs (Anderson 2014). Why are governments subsidizing a major source of climate change? Conversely, why are renewable energy sources not subsidized to facilitate their production and thus reduce our dependency on fossil fuels? Why has there been no class-action lawsuit by municipalities and provincial governments against fossil fuel companies, the leading contributor to climate change? Why are these corporations allowed to make huge profits while taxpayers pick up the tab for the cost of emissions? Keep in mind that class-action lawsuits have worked against the tobacco industry.

Reducing our greenhouse gas emissions is critical. However, another pillar of cleaning up the atmosphere is being aware of the carbon cycle and the importance of photosynthesis. Policies that encourage zero-tillage agriculture, the maintenance of range lands, and sustainable forestry practices also enhance the number of carbon cleansers in the environment because trees clean out air, wetlands filter water, and they both store enormous amounts of greenhouse gases. In other words, energy policies should also recognize the nonmarket value of the landscape, what scientists refer to as the ecosystem benefits. Reforestation would be compatible with this new policy direction (Moola 2013). Of course, the rising number of wildfires in British Columbia results in greenhouse gases moving in the opposite direction. There are also technologies to remove carbon from the atmosphere. One of them, pyrolysis, transforms biological carbon in plants and animals into a mineralized form that won't rot but will stay in the soil for hundreds or thousands of years, also improving it for food production (Flannery 2010).

Climate change is potentially a terminal condition like cancer. Recognizing that resources are no longer the backbone of BC's economy does not diminish the importance of jobs in the resource sector, particularly in relatively unpopulated portions of the province. Rather, it exposes half-truths that allow governments to place jobs and resource development before concern for climate change and the environment. Between 1952 and 1992, asbestos was mined in northwestern British Columbia. Today, most people today know of the carcinogenic qualities of asbestos and would protest future mining ventures. Successful class-action lawsuits have been launched against the industry. Why are we not taking similar actions against fossil fuels?

There is a political and economic model from the Second World War that may be appropriate in dealing with climate change. The Depression was the ruin of many people throughout Canada and the world, and during that era governments were reluctant to invest in projects that employed people. With the outbreak of the Second World War, however, the federal government declared the War Measures Act and immediately invested in building all the necessary implements of war, from planes, ships, tanks, and armaments to uniforms. All adults, including women, were employed and pursued the universal goal of winning the war.

It is time for the Canadian government and all governments of the world to declare a war on climate change and its cause – greenhouse gases. Governments need to invest in renewable energy sources along with technologies of conservation and carbon removal. Educational institutions should be employed to disseminate information on how to build and use these various technologies. Fossil fuel industries should be considered as heritage assets rather than stranded assets and left for future generations. It is all too easy for one province, such as British Columbia, or one nation, such as Canada, to put off policies that will reverse climate change. However, science and common sense tell us that "our influence on Earth is eroding our future, and that we cannot escape responsibility" (Flannery 2010, 272).

REFERENCES

Anderson, M. 2014. "IMF Pegs Canada's Fossil Fuel Subsidies at $34 Billion." *The Tyee*, 15 May. thetyee.ca/Opinion/2014/05/15/Canadas-34-Billion-Fossil-Fuel-Subsidies/.

Ballingall, A. 2018. "Liberal Climate Plans Not Enough to Meet 2030 Emissions Target, New Report Shows." *The Star*, 27 March.

BC Government and Service Employees' Union. n.d. "Forestry Cuts." former.bcgeu.ca/campaign/forestry-cuts.

BC Hydro. 2017. "Independent Power Producers (IPPs) Currently Supplying Power to BC Hydro." bchydro.com/content/dam/BCHydro/customer-portal/documents/corporate/independent-power-producers/independent-power-producers-calls-for-power/independent-power-producers/ipp-supply-list-in-operation.pdf.

BC Stats. 2019. "Total Population % Distribution by 5-Year Age Groups." gov.bc.ca/gov/content/data/statistics/people-population-community/population/population-projections.

BC Treaty Commission. 2017. "Negotiations Update." bctreaty.ca/negotiation-update.

–. 2018. *Annual Report 2018*. bctreaty.ca/sites/default/files/BCTC-AR2018.pdf.

Boynton, S., and J. Palma. 2019. "After a Devastating Series of Mill Closures, Can B.C.'s Forestry Industry Recover?" *Global News*, 13 September. globalnews.ca/news/5902266/bc-forestry-closures-future/.

CBC News. 2015. "Petronas LNG Terminal Set in Salmon's 'Grand Central Station.'" *CBC News*, 7 August. cbc.ca/news/canada/british-columbia/petronas-lng-terminal-set-in-salmon-s-grand-central-station-1.3181203.

Clinkard, J. 2010. "Lumber Supply, Employment in B.C. Will Feel Mountain Pine Beetle's Bite." *Journal of Commerce*, 5 August. forestindustries.eu/content/lumber-supply-employment-bc-will-feel-mountain-pine-beetle%E2%80%99s-bite.

Fisheries and Oceans Canada. 2018. "Aquaculture." dfo-mpo.gc.ca/stats/aqua/aqua-prod-eng.htm.

–. 2019. "Seafisheries Landings." dfo-mpo.gc.ca/stats/commercial/sea-maritimes-eng.htm.

Flannery, T. 2010. *Here on Earth: A Natural History of the Planet*. Toronto: HarperCollins.

Gillis, D. 2015. "Three Fibs Premier Clark Uses to Sell LNG Dream." *The Tyee*, 21 December. thetyee.ca/Opinion/2015/12/21/Premier-Clark-LNG-Fibs/.

Haley, D., and H. Nelson. 2006. "British Columbia's Crown Forest Tenure System in a Changing World: Challenges and Opportunities." Synthesis paper for BC Forum on Forest Economics and Policy. conservation-economics.com/pdf_pubs/synth_paper/SP0601_Forest_Tenure_System.pdf.

Hernandez, J., and T. Lovgreen. 2017. "'It's Alarming': Wildfire Emissions Grow to Triple B.C.'s Annual Carbon Footprint." *CBC News*, 24 August. cbc.ca/news/canada/british-columbia/it-s-alarming-wildfire-emissions-grow-to-triple-b-c-s-annual-carbon-footprint-1.4259306.

Intergovernmental Panel on Climate Change. 2014. "Concluding Instalment of the Fifth Assessment Report: Climate

Change Threatens Irreversible and Dangerous Impacts, but Options Exist to Limit Its Effects." News release. preventionweb.net/news/view/40214.

Jang, B. 2016. "More Than 90 Scientists Dispute LNG Project's Emissions Estimates." *Globe and Mail*, 30 May.

Joseph, R. 2018. "Majority of Canadians Support Trans Mountain Pipeline Expansion: Ipsos Poll." *Global News*, 2 May. globalnews.ca/news/4180482/majority-of-canadians-support-trans-mountain-pipeline-expansion-ipsos-poll/.

Kurjata, A. 2017. "Federal Court Dismisses First Nations' Challenge of Site C Dam." *CBC News*, 23 January. cbc.ca/news/canada/british-columbia/federal-court-dismisses-first-nations-challenge-of-site-c-dam-1.3948830.

Leahy, S. 2013. "BC LNG Exports Blow Climate Targets Way, Way Out of the Water." *The Narwhal*, 9 May. desmog.ca/2013/05/09/bc-lng-exports-blow-climate-targets-way-way-out-water.

Ministry of Finance. 2010. "Historical Commodity Prices." *2010 British Columbia Financial and Economic Review*. gov.bc.ca/assets/gov/british-columbians-our-governments/government-finances/financial-economic-review/financial-economic-review-2010.pdf.

–. 2017. "Revenue by Source." *BC Financial and Economic Review*. fin.gov.bc.ca/tbs/FinancialandEconomicReview_WEB.pdf.

–. 2019a. *Combatting Money Laundering in BC Real Estate*. gov.bc.ca/assets/gov/housing-and-tenancy/real-estate-in-bc/combatting-money-laundering-report.pdf.

–. 2019b. "Historical Commodity Prices." *2019 British Columbia Financial and Economic Review*. gov.bc.ca/assets/gov/british-columbians-our-governments/government-finances/financial-economic-review/financial-economic-review-2019.pdf.

Ministry of Forests, Lands and Natural Resource Operations. 2017. "Factsheet: Government Support for Forestry-Dependent Communities." gov.bc.ca/assets/gov/farming-natural-resources-and-industry/forestry/softwood-lumber/fact_sheet_government_support_for_forestry_dependent_communities.pdf.

MMIWG (National Inquiry into Missing and Murdered Indigenous Women and Girls). 2019. *Reclaiming Power and Place: The Final Report of the National Inquiry into Missing and Murdered Indigenous Women and Girls*, vol. 1a. mmiwg-ffada.ca/wp-content/uploads/2019/06/Final_Report_Vol_1a-1.pdf.

Moola, F. 2013. "Beyond Pipelines: Managing the Cumulative Impacts of Resource Development in BC." Paper presented at West Coast Environmental Law Forum, 26 February. wcel.org/sites/default/files/publications/Summary%20of%20dialogue%20proceedings.pdf.

National Energy Board. 2009. "A Primer for Understanding Canadian Shale Gas." cer-rec.gc.ca/nrg/sttstc/ntrlgs/rprt/archive/prmrndrstndngshlgs2009/prmrndrstndngshlgs2009-eng.pdf.

Nelson, J.L. 2015. "British Columbia Copper Mining Development: A Sixty-Year Economic and Political Retrospective." Phd diss., University of British Columbia.

Nikiforuk, A. 2011. "Albertan, Tired of Her Tap Water Catching Fire, Sues." *The Tyee*, 28 April. thetyee.ca/Opinion/2011/04/28/FrackingSuit/.

Parfitt, B. 2011. "Fracking Up Our Water, Hydro Power and Climate: BC's Reckless Pursuit of Shale Gas." Paper for Climate Justice Project, Canadian Centre for Policy Alternatives and UBC. policyalternatives.ca/fracking.

–. 2016. "Political Leadership Needed to Revitalize BC's Forest Industry." *Policynote*, 4 October. policynote.ca/political-leadership-needed-to-revitalize-bcs-forestry-industry/.

Statistics Canada. 2016. "Population and Dwelling Count Highlight Tables, 2016 Census." statcan.gc.ca/census-recensement/2016/dp-pd/hlt-fst/pd-pl/index-eng.cfm.

–. 2019a. "Employment by Industry, Annual, Census Metropolitan Arres (× 1,000)." Table 14-10-0098-01. www150.statcan.gc.ca/t1/tbl1/en/tv.action?pid=1410009801.

–. 2019b. "Employment by Industry, Annual, Provinces and Economic Regions (× 1,000)." Table 14-10-0092-01. www150.statcan.gc.ca/t1/tbl1/en/tv.action?pid=1410009201.

Acknowledgments

Many people are to be thanked in the production of this fourth edition. Charles Greenberg of Capilano University has been very helpful in providing ideas about liquefied natural gas as well as employment in the goods and services economy. Mike Smith at Langara College made a number of suggestions, especially with respect to evolving urban patterns. Cam Owen, along with his students at University of Victoria, was responsible for many updates in the third edition that also made their way into this edition. Thomas Hutton at the University of British Columbia has offered valuable ideas on a number of issues in this text. Thank you one and all.

Statistics Canada has been a valuable source of information, and people such as Marija Simjanovski have assisted me in acquiring necessary documents. I am grateful also to Lillian Hallan, who manages the accounts at BC Stats, for providing historic economic accounts on British Columbia.

Many of the students enrolled in the Sunshine Coast ElderCollege courses that I delivered offered suggestions and insight to British Columbia's settlement and development over time. Their personal experiences influenced a number of ideas in this text. A special thanks to Keith Maxwell, chair of the Programs Committee of Elder-College, who provided many helpful suggestions. I am grateful to Mike Smith and Katrina Erdros at Langara College for inviting me to "test run" this fourth edition in their Geography of BC courses.

UBC Press deserves a great deal of credit. James Mac-Nevin is to be thanked for accepting this somewhat "radical" departure from the other three editions. Katrina Petrik and her professional team of editors did an amazing job of organizing, reorganizing, and providing many helpful suggestions for the layout and each chapter of this fourth edition.

Most importantly, I want to thank my partner, CarolAnn, for being the first editor and enduring the many versions of the manuscript. Your suggestions and corrections were essential to this fourth edition.

Glossary

1969 White Paper. A white paper is proposed legislation (not an act, which majority governments can, and do, impose). The 1969 White Paper by the Pierre Trudeau government was viewed as a "solution" to the poor conditions suffered by most First Nations throughout Canada. The White Paper proposed to abolish the **Indian Act**, eliminate Indian Status, dissolve treaties, and divide reserve lands into private property that could be sold by the First Nation. This White Paper resulted in protest and opposition by First Nations and a wave of activism. It was withdrawn.

Agricultural Land Reserve (ALR). The British Columbia government established Agricultural Land Reserves in 1973, under the Land Commission Act, to stop the consumption of British Columbia's relatively rare agricultural land for uses other than farming. The loss of arable land to residential, commercial, industrial, and recreational uses was particularly high in the Lower Mainland, Southern Vancouver Island, and the Okanagan during the population expansion and urban sprawl of the 1960s and 1970s. The Canada Land Inventory uses a seven-point scale to grade soils in terms of agricultural capabilities, with class 1 being the best farmland and class 7 being land with no agricultural capability; all lands from class 1 to class 4 were designated as Agricultural Land Reserve. The Agricultural Land Commission is the body responsible for administering the Agricultural Land Reserve.

aquaculture. The raising of aquatic organisms (e.g., salmon and other fish species, shellfish, seaweed) in a controlled environment. In British Columbia, aquaculture has been confined mainly to raising salmon (salmon farming) and oysters in protected coastal bays and inlets. Most of the controversy surrounding aquaculture has centred on the farming of Atlantic salmon (a foreign species with inherent diseases) in net pens anchored to the foreshore. There are concerns about sea lice contamination, local pollution, the privatization of the foreshore, and the spread of diseases to migrating wild species of salmon.

burden of proof. This is mainly a legal concept that applies when rights have been violated (e.g., air, water, or land have been contaminated by individuals, a corporation, or government body), which leads to the question of responsibility for the violation (i.e., who has the obligation to prove the violation?). For many environmental violations, the burden of proof often falls on the public.

Calder. This was a landmark Supreme Court case. In 1973, the court was asked to rule on two questions: (1) did **Indigenous Title** exist, and (2) does Indigenous Title still exist? The seven judges unanimously agreed that Indigenous Title existed as a result of the **Royal Proclamation of 1763**. The second question resulted in a split decision: three judges ruled that Indigenous Title had been extinguished through Confederation, and three judges ruled that it still existed. The seventh judge threw the court case out on a technicality. What was precedent-setting is that the federal government sided with the three judges who stated Indigenous Title continues to exist. Out of that political decision evolved **comprehensive claims** (**modern-day treaties**) and **specific claims**.

carbon tax. A carbon tax is an effective way to reduce greenhouse gases from the burning of fossil fuels, a major contributor to **climate change**. In British Columbia, a $10 per tonne carbon tax was introduced in 2008, mainly on fossil fuels used for transportation and home heating. It was increased each year until it reached $30 per tonne in 2012. The tax scheme was unique in that was "revenue neutral," meaning that personal and corporate income taxes were reduced by the same amount as the carbon tax.

climate change. Climate change involves major shifts in both regional and global weather patterns. Although natural events such as volcanic eruptions can influence climate change, it is human activities, especially the burning of fossil fuels, that have the greatest influence. Extreme events – hurricanes, tornadoes, wildfires, drought, rising sea levels, flooding, receding glaciers, and the mountain pine beetle infestation – are more frequent and of a higher magnitude as a result of climate change.

cogeneration. The dual function of producing electrical power by burning natural gas to turn turbines and then capturing the heat to heat buildings, greenhouses, and other structures requiring heating.

Cold War. A period of tension between the Soviet Union and the United States and their respective allies, which began after the end of the Second World War (approximately 1947) and ended with the collapse of the Soviet Union in 1991. It was called a "cold" war because it didn't result in direct military combat between the two main participants; rather, the war was waged through political and economic means between Communist bloc nations and capitalist ones. The conflict led to an arms race (with atomic bombs and then ballistic missiles) between the two superpowers.

commercial extinction. This term is commonly used to describe the pursuit of renewable resources such as sea otters, fur seals, and whales, which were thought of as limitless in the past. These resources were harvested to near extinction, to the point where there were so few of the species left that it was not worth investing in harvesting them.

common property resource. These are resources that are not owned privately but are instead managed by various levels of government. The air we breath, water, most species of fish, the foreshore, and parks are examples of common property resources.

company town. A single-resource community created and controlled by a corporation. Company towns were common in British Columbia, particularly in the mining or smelting industries and forestry. The company was the only employer and controlled housing and the company store. Most of the communities were closed, meaning that they were not regulated under the province's Municipal Act.

comprehensive claims. Claims for compensation made by First Nations for **Indigenous Title** Lands that have not been covered by treaties or were never ceded by First Nations. The federal government recognized the need to negotiate comprehensive claims as a result of the *Calder* case (1973), although it was not until 1981 that it was willing to exchange the undefined Indigenous Land Right for concrete rights and benefits. The provincial government did not recognize these claims until 1992. Contrast with **specific claims**.

convection precipitation. The process by which incoming solar radiation heats the earth and air in proximity to the earth, causing the warm air, which contains moisture from evaporation and transpiration, to rise in the atmosphere, where it cools, condenses, forms clouds, and often results in thundershowers. In British Columbia, this is mainly a summer phenomenon, when solar radiation is at its most intense.

cultural region. An area of the populated world defined by having common cultural characteristics. These characteristics can be further subdivided into formal regions (e.g., common language areas), functional regions (e.g., timber supply areas), and vernacular/perceptual regions (e.g., sense of place).

deformation. General term to describe processes that produce folding, faulting, and changes to the surface of the earth.

Delgamuukw. The federal government recognized **Indigenous Title** as a result of the *Calder* case in 1973, but the province of British Columbia was adamant that First Nations had no claim to Crown lands. The Gitxsan and Wet'suwet'en Tribal Council filed a claim for their Traditional Lands at the Supreme Court of British Columbia. The ruling by Chief Justice McEachern, in 1991, was that Indigenous Title had been extinguished through Confederation. This was a setback. However, an appeal to the Supreme Court of Canada resulted in a precedent-setting decision in 1997. The ruling confirmed that Indigenous Title continues to exist in British Columbia and that it is a right to the Traditional Territory, not simply the right to hunt, fish, and gather. Moreover, the provincial government must consult with First Nations regarding activities on Crown lands that may affect the rights of First Nations, and the government may have to pay compensation.

digital revolution. This revolution is related to the development of the computer and digital electronics, which began following the Second World War and evolved, by the 1980s, to include technologies that allowed digital information to be transmitted globally and in an instant through coaxial cables and satellites. Often referred to as the Information Age, the digital revolution led to the reorganization of the production of good and services and the beginning of the global division of labour (i.e., the cheap labour components of production were contracted out to regions of the world with low labour rates). These technologies had profound implications on the application of **Fordism** in the production of resources in British Columbia

Douglas Treaties. The first treaties signed in British Columbia, although they occurred during the British colonial era in the new Crown colony of Vancouver Island. James Douglas, as governor of the new colony, signed fourteen small treaties between 1850 and 1854 in an attempt to avoid conflict between First Nations and the relatively few non-Indigenous settlers coming into the territory. The treaties guaranteed First Nations small plots of land and some British goods, along with an agreement that First Nations could continue to fish, hunt, and gather on unoccupied lands. The treaties were modelled after a treaty in New Zealand, not the **Royal Proclamation of 1763**.

earthquake activity. A tectonic vibration resulting from the clash of two crustal plates. These movements may be a result of volcanic eruption; a **transform fault**, in which two plates slide past one another; or **subduction**, in which one plate slides under another. The vibrations caused by this movement radiate from the epicentre of the earthquake, and the magnitude of the waves is calculated logarithmically on the Richter scale. A quake reading 3.0 produces little shaking, whereas one reading 8.0 produces major movement, especially at the epicentre or its vicinity.

erosion. The transportation of rock sediments and **weathering** of rock through the action of water, wind, glaciation, and mass wasting (gravity).

fiduciary trust. The legal obligation, under the **Indian Act**, for the federal government to act in the best interests of First Nations.

flash flooding. A result of intense and sustained rainfall; occurs in coastal watersheds, including Vancouver Island and Haida Gwaii, mainly in the winter months.

Fordism. The assembly line process of manufacturing standardized products, usually at centrally located plants. The term is derived from the process used by Henry Ford to assemble automobiles at the turn of the twentieth century.

gatekeeper model. This is a fisheries management model employed by the federal government, which has jurisdiction over the oceans and sea life of Canada. The government used the gatekeeper model to regulate the commercial salmon fishery, which is a **common property resource**. Allowing salmon to migrate up the streams and rivers of coastal British Columbia to their spawning grounds is essential to their survival. This model, similar to a gate that swings open and shut, allows commercial fishers to harvest as many salmon as possible during an opening but none during a closing, which is stipulated by the government.

geomorphology. The study of the processes that change the surface of the earth.

geophysical hazard. The assessment of risk from the earth's forces. These forces can be categorized in relation to tectonic, climatic, and gravitational forces. Geophysical hazards are also referred to as natural hazards. They are assessed in terms of the threat or risk to human property and/or life.

ghost town. A community, usually developed because of a single resource, in which the main employment base has terminated and most or all the residents have left.

goods-producing sector. As defined by Statistics Canada, this sector of the economy includes employment often related to resource development and includes five categories: (1) agriculture; (2) forestry, fishing, mining, quarrying, and oil and gas; (3) utilities, which can include energy distribution or municipal water and sewage systems; (4) construction, including building dams or pipelines, plants, or mills for resource extraction or employment in residential, commercial, and industrial developments related to urban construction and infrastructure such as highways and bridges; and (5) manufacturing, including resource-related employment in sawmills, pulp-and-paper mills, and primary metal manufacturing but also in the production of food and beverages, textiles and clothing, computer and electronic products, electrical equipment and appliances, transportation equipment, furniture, and products not directly related to resources.

gross domestic product (GDP). A measure of the total value of goods produced and services provided annually. For a more accurate comparison over time, GDP figures are calculated in chained dollars, or dollars fixed for one specific year, thus taking inflation into account.

head tax. A tax on individuals coming into a country. Within the context of British Columbia's history, the term applies specifically to the levy of a series of head taxes by the federal government in an attempt to reduce Chinese immigration. This became a barrier to the **spatial diffusion** of the Chinese to British Columbia.

ice jam flooding. This type of flooding can occur on any river that freezes over in winter. Ice frequently forms in layers, and where the stream slows, or where there are river obstructions such as bridge piers, the ice tends to be more stable or stuck firmly to the shore. When ice breaks up (usually in the spring, but unusual winter warming can also result in ice flows), the ice fragments jam where the ice remains stable and thus form a dam where the water backs up, flooding upstream areas. The pressure of the water behind the ice jam can result in a sudden release of the jam, thus causing rapid flooding of downstream areas.

igneous rock. Rock formed from the molten state, either rapidly through exposure to cooler surface environments

such as air or water or much more slowly if the material does not reach the surface of the earth. Extrusive igneous rock, such as basalt, has undergone the rapid cooling process and has high density and weight. The slow cooling process results in intrusive igneous rock, which has much larger grained rock structures with lower density and weight, such as granite.

independent power producer (IPP). In the 1990s, BC had a rapidly growing economy and was faced with the problem of having to increase the supply of electricity. The government opted for the production of hydroelectric power through the private sector, or through independent power producers. In the 1990s and early 2000s, IPPs were mainly small (5 to 10 MW) run-of-river systems that had to have a BC Hydro energy purchase agreement (i.e. BC Hydro owned the power generated), environmental assessments, and approval from First Nations. In 2008 and later, the private production of electricity was expanded to include other sources, including wind, pulp mills, and natural gas–fired thermal plants.

Indian Act. This act was introduced in 1876 by the federal government as a means of governing all First Nations in Canada. It has had a number of amendments over the years, and many were intended to assimilate First Nations (e.g., the Potlatch ban, rules regarding the election of Chiefs and band councils, and the creation of Status and Non-Status Indians). While a number of these discriminatory amendments have since been repealed, the Indian Act remains in place today for First Nations not governed by **modern-day treaties**.

Indigenous Rights. Also called Aboriginal Rights. Rights derived from the historical use of land that include the right to use the land and the right to self-government. These rights are constitutionally recognized by the Constitution Act, 1982, which protects them but fails to describe their nature.

Indigenous Title. Also called Aboriginal Title. Ownership or control of the historical territory of First Nations and Inuit. The geographic boundaries of Traditional Territories were not static; overlaps in territorial use for resource procurement were common. Identification of historical boundaries that define Indigenous title is important in establishing resource rights and compensation in the modern treaty process.

industrial economy. In an industrial economy, employment is based on the production of goods, or secondary industry. In **staples theory** and core-periphery (heartland-hinterland) analysis, a distinction is made between industry based on the manufacturing of consumer goods (e.g., automobiles), which is largely done by core regions, and industry based on the manufacturing of raw materials (e.g., logs to lumber). Industry, and its associated occupations, was a major impetus to urban growth until the 1970s.

instant town. A single-resource community created under the Instant Towns Act (1965). Major new investment in the resource frontier in the 1960s and '70s (especially forestry and mining) created a need for new communities. These single-resource communities were a response to the negative conditions of the **company town.** They were regulated under the Municipal Act and operated in a similar fashion to other villages and towns in the province.

isostasy. The balance between the weight of continental crusts pushing down into the mantle and the uplifting forces of the mantle itself. **Weathering** and **erosion** both wear down and reduce the weight of physiographic features such as the Rockies. This loss of weight, and height, results in an uplifting of the Rockies. The fairly rapid removal of the vast sheets of ice covering much of the northern portion of North America only 10,000 years ago resulted in isostatic rebound, or a similar uplifting of the earth with the removal of this enormous weight.

liquefied natural gas (LNG). Natural gas has been in production in British Columbia since the 1950s and distributed by pipeline throughout the province and to the United States. It can be liquefied by cooling it to minus 162 degrees Celsius; this process reduces its volume six hundred times, making it easier to ship via tankers. Recent controversies over LNG centre on the production of natural gas through fracking and growing awareness that greenhouse gases are emitted at all stages, from mining and transportation to processing and burning LNG to produce electricity. The amount of energy required to create LNG is also a growing concern.

lode mining. The process of crushing rock and extracting the minerals of value. Most metals occur in "bound" form: they are found along with other minerals within rock. Lode-mining technologies were employed in British Columbia in the late 1800s and early 1900s, mainly to extract silver, gold, and copper. Although some **placer mining** for gold continues today, nearly all metal-mining production in the province is through lode mining. The technologies of lode mining are sophisticated and costly, and they usually require large amounts of corporate financing.

magma. The liquid interior portion of the earth where extreme temperatures melt rock. Magma makes its way to the surface of the earth through **plate tectonics** and becomes **igneous rock.**

manifest destiny. A mid-nineteenth-century American belief that the United States was destined to expand through all of western North America and, eventually, according to some advocates, across the entire continent.

megaproject. Following the Second World War, Canada experienced a massive economic boom and major industrialization. Many manufacturing plants and assembly lines opened in southern Ontario and Quebec, creating tremendous demand for resources from provinces such as British Columbia. The province's megaprojects – from open-pit mines and clear-cut logging to pipelines and hydroelectric dams – were fuelled by new, ramped-up technologies and millions (even billions) of dollars in investment, often by foreign corporations.

metamorphic rock. Rock formed through intense pressure and heat from, or chemical infusion from, new molten material intruding into existing rock structures. Rocks undergoing this process can develop entirely new physical properties and chemical structures.

modern-day treaties. Contemporary treaties negotiated as compensation – in the form of land, resources, and resource management, or money – for the extinguishment of **Indigenous Title**. These treaties also include options for self-government and greater autonomy for First Nations. Treaty negotiations in British Columbia did not occur until 1992, at which point agreement between the provincial and federal governments resulted in a six-stage process.

New Caledonia. The name given to the central Interior Plateau region of British Columbia by Simon Fraser in 1806. The Americans referred to this territory, and the area south of it, as the Oregon Territory. Fraser's name remained after the **Oregon Treaty** (1846) extended the US-Canada border west along the forty-ninth parallel from the Rockies to the coast, but in 1858 Queen Victoria named the area British Columbia to avoid confusion with a French colony in the South Pacific also named New Caledonia.

Oregon Treaty. An 1846 boundary agreement between the United States and Britain to continue the border along the forty-ninth parallel from the Rockies to the Pacific coast. Vancouver Island, which extends south of the forty-ninth parallel, was allowed to remain part of British North America. British sovereignty over the area, then known as the Oregon Territory, began with this agreement.

orogeny. The process of mountain building caused by tectonic forces that fold and fault land masses through compression.

orographic effect. Precipitation resulting from relatively warm, moist air being forced up mountain barriers, where it cools and condenses. It is analogous to a saturated sponge being squeezed.

overcapacity. Refers, in the commercial salmon fishery, to fishing technologies, such as seine boats, that catch a great number of salmon. As a result of overcapacity, Royal Commissions have recommended the reduction in the number of commercial fishing vessels in an attempt to keep salmon sustainable.

place-name geography. The placing of names on settlements, regions (such as British Columbia), and physical landscapes (such as mountains, oceanic and fresh water bodies, peninsulas). In British Columbia, places have traditional Indigenous names, but during colonization settlers gave these places mainly British names. However, in recent years, the Indigenous names of these places have begun to be recognized and used by settler-colonial society (for example, Haida Gwaii is now used instead of the colonial name Queen Charlotte Islands).

placer mining. The process of mining stream beds for gold. Since gold can be found in a pure state as dust or nuggets and is one of the heaviest elements, it tends to settle in stream beds. Once there, it can be recovered by some of the simplest and cheapest technologies: a shovel and a gold pan. The discovery of gold in British Columbia, and the gold rush that followed in the mid-1800s, centred on placer mining. Many other technologies were also employed – including dams and sluices and hydraulic systems and dredges – but they cost more and were less common.

plate tectonics. A combination of two older hypotheses, continental drift and seafloor spreading, plate tectonics theory is essential to the understanding of **geomorphology** and geology. It asserts that the earth's crust is made up of large and small plates, which move in a manner somewhat analogous to a conveyor belt as **magma** is forced to the surface of the earth in some geographic locations and destroyed in others. Regions where molten material comes to the surface and pushes plates apart are known as **rift zones**, and regions where plates are pushed under or over one another are known as **subduction** zones. Regions where plates simply push past on another in a parallel manner are referred to as **transform faults**. Plate movement results in earthquakes, **volcanic activity**, and mountain building.

post-Fordism. The new technologies and economic conditions that dominated the period after Fordism. Since the mid-1960s, multinational corporations have employed techniques of flexible specialization (short-run production through contracting out) to manufacturing goods at a global scale. The result has been the bankruptcy, merger, or restructuring of many corporations and uncertainty for all.

precautionary principle. This concept mainly refers to the introduction of something new, whether a product or resource development, stating that it may have negative consequences that should be assessed prior to its introduction. The principle stipulates that the **burden of proof** lies with the developer, that adequate funding must be set aside by the developer in case of disaster, that adequate environmental- and health-hazards assessments must occur, and that compensation must take into account the long-term effects of negative impacts on the land or people.

private property resource. A resource that can be held, or controlled, by private interests.

rain shadow effect. Relatively low precipitation in a given region because of mountain barriers. In southwestern British Columbia, westerly winds are forced to rise over the Insular Mountains of Vancouver Island. This **orographic effect** wrings out much of the moisture from the air mass before it passes over the mountains. As it descends on the lee (east) side of the mountains, the air mass expands and absorbs moisture, leaving the region from Victoria to Vancouver considerably drier than the west side of Vancouver Island.

region. An area of the surface of the earth that can be distinguished through physical or human characteristics.

regional geography. A subfield of the discipline of geography in which spatial phenomena are described and studied by dividing the world into areas that have common physical or human characteristics.

residential segregation. This type of segregation occurred throughout the settlement process when visible minorities located in separate locations or ethnic enclaves such as Chinatowns. Residential segregation can also be based on income – for example, wealthy neighbourhoods are distinct from middle- and lower-class neighbourhoods.

resource. Any naturally occurring substance of value to a society. This definition implies that resources are culturally defined.

rift zone. A region of earth where **magma** reaches the crust's surface and splits crustal plates apart. The ocean floors are where the earth's crust is thinnest and where most rift zones occur, resulting in seafloor spreading.

rock cycle. The process of rocks constantly being recycled as part of the earth's physical processes. **Igneous rock** created from the molten state are subjected to **weathering** and **erosion** and may become **sedimentary rock**; heat, pressure, and chemical action produce **metamorphic rock**; tectonic processes may cause these rocks to go back to the molten state.

Royal Proclamation of 1763. This proclamation by the British government formally recognized that Indigenous Peoples had been in North America first, that they had forms of self-government that should continue, that they should not be abused, and, most importantly, that they had **Indigenous Title**. The British also recognized that, over time, this right would have to be extinguished (ceded or surrendered) and the means to do so was through a treaty with the Crown.

sedimentary rock. A relatively soft rock formed through the bonding, or cementing together, of sedimentary materials (e.g., limestone, shale).

service economy. Employment from the provision of services, including knowledge and information services, as opposed to manufacturing. Services were categorized as tertiary industry until 1961, when service occupations were divided into tertiary and quaternary industries. Within this system, service workers who deal directly with the public, such as through counter service in retail and fast food outlets, are categorized as tertiary workers. Quaternary workers, by contrast, carry out transactional services and are sometimes referred to as transactional workers. Lawyers, accountants, government employees, and those who work in tourism, real estate, and the expanding information and education sectors are examples. Census Canada also divides quaternary employment into producer services (or government and corporate services) and consumer services (or services to individuals). The shift to quaternary industry employment lies at the heart of the service economy, which is based in large urban centres that house global communications and many economic, political, and cultural functions.

silviculture. The management of forests, which includes the scientific reforestation of a range of trees once the forest has been harvested, management of forest diseases and pests, and implementation of forest enhancement procedures such as limbing lower branches and residual spacing. More recently, it also includes recognition of values other than the fibre forests can provide – such as water quality, wildlife, preservation of old growth stands, and aesthetics.

snow-melt flooding (spring runoff flooding). This type of flooding is associated with the interrelated factors of drainage basin size, snowpack over the winter season, and the spring weather conditions responsible for the rate of snowpack melt. Spring runoff flooding affects communities and built environments adjacent to the many rivers draining the interior of the province.

spatial diffusion. A concept of movement through time and space employed to trace the spread (or adoption) of ideas or innovations, people, and goods from one geographic location to another. Also known as the spread effect, spatial diffusion identifies the barriers (forces that prevent movement) and carriers (factors assisting movement) that produce the spatial distribution of phenomena.

specific claims. Claims for compensation by individual bands and tribal councils based on an alleged breach of fiduciary duty or responsibility on the part of Canada. These claims are often for reserve lands that have been taken without compensation. Seizure of such land occurred in a number of ways in British Columbia, from the outright annexation of reserve land to the construction of roads, railways, hydroelectric lines, and pipelines through reserve lands. Contrast with **comprehensive claims**.

staples theory. A theory of Canada's economic development that focuses on the exploitation of five resources: fish, furs, timber, wheat, and minerals. Regional economic growth, according to the theory, occurs through the discovery, development, and export of these resources, and some regions also undertake resource manufacturing. Economic historian Harold Innis suggested this theory in the 1930s to account for the regional development of Canada.

subduction. Plate tectonic activity that occurs when plates collide and one plate overrides the other. The overridden plate – usually the heavier oceanic plate – bends downward and descends, or subducts, into the mantle. Mountain building occurs as a result of compression along the boundary where two plates collide. Subduction zones also result in deep oceanic troughs and continental **volcanic activity** because of the friction of subduction.

sustainability. Also referred to as sustainable development. "Development" implies the use of resources, but the term "sustainable development" recognizes that renewable resources can be exploited beyond their ability to reproduce and that nonrenewable resource use often has major negative effects on the ecosystem. This concept, which emerged in the early 1980s, has important implications for appropriate technologies in resource harvesting and use.

tenure. A system of allowing private corporations access to publicly held land and resources. Tenure involves various types of arrangements: licences, leases, and grants. The main issue of tenure in British Columbia is allowing private corporations timber rights to provincially controlled forest land.

terrane. A fragment of oceanic or continental plate. When plates collide, these fragments attach themselves to the adjacent continental plate. Much of British

Columbia is made up of attached, or accreted, terranes, making the geology of this province very complex.

territory. The boundaries of a geographic area within which political control is exerted.

time-space convergence. Change in transportation technologies that reduces the time required to move or communicate between geographic locations, also referred to as time-space compression or collapse. Expressions such as "the world is shrinking" recognize that modern satellite communications, airline flights, and expressway systems allow communication and movement on many geographic scales. It is important to recognize that changes to movement are not equally distributed, however, and some geographic locations are therefore more isolated and remote than others.

transform fault. An area of the earth's crust where one crustal plate pushes past another crustal plate in a parallel manner. Both the Pacific Plate and North American Plate are moving north, for example, but the Pacific Plate is moving more rapidly. The two plates are often in a "stuck" position along the transform fault from Haida Gwaii north to Alaska, and a major earthquake occurs when they move.

volcanic activity. The result of the eruption of molten material, or **magma**, that has come to the surface of the earth's crust. This activity is frequent in **rift zones**, where magma comes to the surface under the ocean. Volcanoes are also associated with **subduction** zones, such as the one off the west coast of British Columbia. Many parts of the interior and west to the coast have been active volcanic areas for the past 150 million years.

weathering. Breaking down. Two broad divisions, chemical weathering and mechanical weathering, categorize the agents involved in this process. Chemical weathering is the decomposition of minerals through agents such as water, carbon dioxide, and oxygen, which form acids. These agents sometimes combine with organic materials during this process. Mechanical weathering is the breaking down of rocks, mainly through water running into rock cracks or fissures and then freezing, expanding, and breaking the rock into fragments.

Further Readings

INTRODUCTION

Garrett Hardin's article "The Tragedy of the Commons," *Science* 162 (3859): 1243–48, science.sciencemag.org/content/sci/162/3859/1243.full.pdf, is well worth reading and discussing. References to the many tragedies in British Columbia history are made throughout Part 2 of the textbook.

PART 1: GEOGRAPHICAL FOUNDATIONS

Chapter 1: British Columbia, a Region of Regions

To gain an in-depth understanding of the various regions of British Columbia, explore the BC government's Statistics website, gov.bc.ca/gov/content/data/statistics. For example, the profiles of the eight development regions can be found under "People, Population and Community."

Chapter 2: Physical Processes and Human Implications

Natural Resources Canada's website "Geology and Geosciences," nrcan.gc.ca/earth-sciences/geography/atlas-canada/selected-thematic-maps/16876#rocks covers all of Canada. Considerable insight on physical processes in British Columbia can be accessed by exploring the following sections: "Rocks," "Surficial Materials and Glaciation," and "Land."

Sydney Cannings, JoAnne Nelson, and Richard Cannings's *Geology of British Columbia: A Journey through Time* (Vancouver: Greystone Books, 2011) is an excellent, readable account of the complexity of British Columbia's geology.

For more on the provincial government's commitment to addressing climate change, consult the climate change page on its website, gov.bc.ca/gov/content/environment/climate-change.

A short visual introduction to British Columbia's weather and climate can be viewed in the short video "Weather in British Columbia: Insights into Vancouver Weather and Vancouver Winter," youtube.com/watch?v=ZyPCSVZdAps.

Chapter 3: Geological Hazards and Their Risks

The PreparedBC website, gov.bc.ca/gov/content/safety/emergency-preparedness-response-recovery/preparedbc, offers information and advice on emergency preparedness for geophysical hazards. Explore "Know Your Hazards," gov.bc.ca/gov/content/safety/emergency-preparedness-response-recovery/preparedbc/know-your-hazards. The provincial government, through the Wildfire Service, also provides guidelines for corrective and preventative measures in the case of wild fires, gov.bc.ca/gov/content/safety/wildfire-status

For an in-depth assessment of flooding in British Columbia, see Engineers and Geoscientists BC, *Professional Practice Guidelines – Legislated Flood Assessments in a Changing Climate in BC*, 2018, egbc.ca/getmedia/f5c2d7e9-26ad-4cb3-b528-940b3aaa9069/Legislated-Flood-Assessments-in-BC.pdf.aspx.

Up-to-date and historical earthquake data and information can be accessed through the Earthquakes Canada website, earthquakescanada.nrcan.gc.ca/index-en.php. For a more personal assessment of earthquake risk, submit your postal code to the Institute for Catastrophic Loss Reduction's Earthquake Risk Mapping Tool, iclr.org/earthquake-risk/.

Chapter 4: Resource Development and Management

Although it's not an easy read, Melville H. Watkins's "A Staple Theory of Economic Growth," *Canadian Journal of Economics and Political Science/Revue canadienne d'Economique et de Science politique* 29, 2 (1963): 141–58, explains the basis of Harold Innis's staples theory, which has been so influential in explaining the economic development in Canada.

Anup Shah's "Sustainable Development Introduction," *Global Issues,* 18 November 2019, globalissues.org/article/408/sustainable-development-introduction#globalissues-org, provides a relatively brief history of the concept of sustainability.

For an in-depth view of restructuring and its impact on coastal British Columbia communities, see Sulan Dai and S. Martin Taylor's "Socio-economic Restructuring and Health: A Qualitative Study of British Columbia Coastal Communities," *Western Geography* 17–19 (2007–9): 5–38, geog.uvic.ca/dept/wcag/dai_taylor.pdf.

PART 2: THE ECONOMIC GEOGRAPHY OF BRITISH COLUMBIA

Chapter 5: "Discovering" Indigenous Lands and Shaping a Colonial Landscape

For a good, easy-to-read overview, see Robert J. Muckle, *The First Nations of British Columbia, An Anthropological Overview,* 3rd ed. (Vancouver: UBC Press, 2014).

For an excellent article on the historical spatial diffusion of smallpox in North America, see Robert Boyd's "Smallpox in the Pacific Northwest: The First Epidemics," *BC Studies* 101 (1994): 5–40, ojs.library.ubc.ca/index.php/bcstudies/article/viewFile/864/905.

For a down-to-earth description of the Fraser River and Cariboo Gold Rushes, see M. Poncelet, *BC Gold Rush Press: Stories of the Fraser River and Cariboo Gold Rushes* (blog), bcgoldrushpress.com/.

Chapter 6: Boom and Bust from Confederation to the Early 1900s

For an examination of anti-Chinese sentiment in California and British Columbia, see Jay Martin Perry, "The Chinese Question: California, British Columbia, and the Making of Transnational Immigration Policy, 1847–1885" (PhD diss., Bowling Green State University, 2014), etd.ohiolink.edu/!etd.send_file?accession=bgsu1394761542&disposition=inline.

Douglas M. Swenterton examines the history of Pacific fisheries policy in *A History of Pacific Fisheries Policy* (Ottawa: Department of Fisheries and Oceans, 1994), dfo-mpo.gc.ca/Library/165966.pdf.

More information on residential schools can be accessed online at Erin Hanson, "The Residential School System," *Indigenous Foundations,* indigenousfoundations.arts.ubc.ca/home/government-policy/the-residential-school-system.html. Arthur Ray's 1999 article "Treaty 8: A British Columbia Anomaly" describes how BC's First Nations in the northeast sector of the province became the only First Nations in the province to be included in the Numbered Treaties: *BC Studies* 123 (Autumn), ojs.library.ubc.ca/index.php/bcstudies/article/view/1508/1551.

Chapter 7: Resource Dependency and Racism in an Era of Global Chaos

On attitudes and policies towards First Nations during the McBride era, see Patricia E. Roy, "McBride of McKenna-McBride: Premier Richard McBride and the Indian Question in British Columbia," *BC Studies* 172 (Winter 2011–12): ojs.library.ubc.ca/index.php/bcstudies/article/view/2059/2309.

Steven Hume's short article from the *Vancouver Sun* gives a sense of what the First World War meant to British Columbians: Steven Hume, "B.C. Streets Filled with Patriotism, Then Flooded Like Rivers of Grief as First World War Began," *Vancouver Sun,* 2 August 2014, vancouversun.com/streets+filled+with+patriotism+then+flooded+like+rivers+grief+First+World+began+with+photos/10083538/story.html.

One of the factors that enhanced Vancouver's economic position was the building of grain elevators and the transportation of Prairie grain west instead of east. However, this infrastructure was not a sure thing. See John Everitt and Warren Gill, "The Early Development of Terminal Grain Elevators on Canada's Pacific Coast," *Western Geography* 15–16 (2005–06): geog.uvic.ca/dept/wcag/everitt06.pdf.

Enemy Alien (Ottawa: National Film Board, 1975), directed by Jeanette Leman, remains one of the better accounts of Japanese internment in British Columbia.

Chapter 8: Changing Values during the Postwar Boom

Stephen Tomblin gives some insight into the W.A.C. Bennett era in "W.A.C. Bennett and Province-Building in British Columbia," *BC Studies* 85 (Spring 1990): ojs.library.ubc.ca/index.php/bcstudies/article/view/1346/1389.

Susan Toller and Peter Nemetz provide more detail about the benefits and costs of the Columbia River Treaty to the Kootenays in "Assessing The Impact of Hydro Development: A Case Study of the Columbia River Basin in British Columbia," *BC Studies* 114 (Summer 1997): ojs.library.ubc.ca/index.php/bcstudies/article/view/1710/1756.

John Bradbury's PhD dissertation on instant towns reveals many of the concerns and vulnerabilities in creating new communities. See "Instant Towns in British Columbia, 1964–1972" (Simon Fraser University, 1977), mackenziemuseum.ca/wordpress/wp-content/uploads/Bradbury-1977-Thesis-on-Instant-Towns-in-BC.pdf.

Stuart Nelson and Bruce Turris provide an overview of commercial salmon fishing, including conditions prior to the Bennett era and after. See *The Evolution of Commercial Salmon Fisheries in British Columbia*, report to the Pacific Fisheries Resource Conservation Council, 2004, psf.ca/sites/default/files/EvolutionCommercial Fisheries-BC_2004_0_Complete.pdf.

Ben Bradley offers an excellent account of the influence of road building in British Columbia during the W.A.C. Bennett era and how it fits into Fordism in *British Columbia by the Road: Car Culture and the Making of a Modern Landscape* (Vancouver: UBC Press, 2017). Another recent book published by UBC Press, Jonathan Peyton's *Unbuilt Environments: Tracing Postwar Development in Northwest British Columbia* (2017), focuses on developments, both proposed and developed, for the northwest region of the province. Although a number of the projects described occurred during the 1950s and '60s, Peyton also brings a number of the developments up to the present.

Chapter 9: Resource Uncertainty in the Late Twentieth Century

For a brief history of the formation of the Agricultural Land Commission and Agricultural Land Reserves during the 1970s NDP era, see "How the ALR Was Established," alc.gov.bc.ca/alc/content/alr-maps/alr-history. The scientific basis of classifying agricultural land was derived from the Canada Land Inventory, which can be accessed at omafra.gov.on.ca/english/landuse/classify.htm#defclass.

For an assessment of the impact of the 1980s recession, see Trevor Barnes and Roger Hayter, "British Columbia's Private Sector in Recession, 1981-86: Employment Flexibility without Trade Diversification?," *BC Studies* 98 (Summer 1993), ojs.library.ubc.ca/index.php/bc studies/article/view/1457/1501.

For those interested in commercial fishing during this era, see Douglas M. Swenerton's *A History of Pacific Fisheries Policy* (Ottawa: Department of Fisheries and Oceans, 1994), dfo-mpo.gc.ca/Library/165966.pdf. The final chapter assesses the fishery from 1969 to the 1990s.

Protests at Clayoquot Sound to save old-growth forests on Vancouver Island resulted in the greatest number of protesters being sent to jail to stop the logging. See Michaela Killoran Mann, "'Clearcut' Conflict: Clayoquot Sound Campaign and the Moral Imagination" (master's thesis, Saint Paul University, 2013), ruor.uottawa.ca/bitstream/10393/24042/1/Mann_Michaela%2Killoran _2013_thesis.pdf.

For a brief overview of the *Delgamuukw* ruling, see the BC Treaty Commission's "A Lay Person's Guide to DELGAMUUKW," bctreaty.ca/sites/default/files/delgamuukw.pdf.

Chapter 10: The Twenty-First-Century Liberal Landscape

The Mountain Pine Beetle epidemic became widely known during this era, and the BC government's forestry website, with its many links, gives an overview of the history and impact of the beetle. See "Mountain Pine Beetle in B.C.," gov.bc.ca/gov/content/industry/forestry/managing-our-forest-resources/forest-health/forest -pests/bark-beetles/mountain-pine-beetle.

Tim Schouls's paper "Between Colonialism and Independence: Analyzing British Columbia Treaty Politics from a Pluralist Perspective," cpsa-acsp.ca/papers-2005/Schouls.pdf, provides context for the Gordon Campbell Liberals' rather perplexing attitudes towards the Nisga'a and First Nations more generally.

On the impact of the subprime mortgage collapse and recession on the forest industry in Canada, see Lydia Couture and Ryan Macdonald, "The Great U.S. Recession and Canadian Forest Products," Economic Insights, Analytical Paper, no. 028, July 2013, publications.gc.ca/collections/collection_2013/statcan/11-626-x/11-626 -x2013028-eng.pdf.

For a comparison of the process of completing environmental assessments by the provincial and federal governments, see Mark Haddock, "Comparison of the British Columbia and Federal Environmental Assessments for the Prosperity Mine," report for Northwest Institute, Vancouver, July 2011, stopajaxmine.ca/files/documents/NWI_EAreport_July2011.pdf.

Chapter 5 in Johnathan Peyton's *Unbuilt Environments: Tracing Postwar Development in Northwest British Columbia* (Vancouver: UBC Press, 2017) gives an excellent account of what happened to the landscape when the government built the Northwest Transmission Line in a relatively wilderness landscape.

CONCLUSION

One need only scan through Jacqueline Nelson's PhD dissertation to get a sense of the history of metal mining in British Columbia as well as some of the politics surrounding the opening of new mines. See "British Columbia Copper Mining Development: A Sixty-Year Economic and Political Retrospective" (UBC Faculty of Engineering, 2015), open.library.ubc.ca/cIRcle/collections/ubctheses/24/items/1.0167712.

Ken Davidson outlines government interference in the acquisition of electrical energy in "Zapped: A Review of BC Hydro's Purchase of Power from Independent Power Producers Conducted for the Minister of Energy, Mines and Petroleum Resources," BC government report, February 2019, gov.bc.ca/assets/gov/farming-natural-resources-and-industry/electricity-alternative-energy/electricity/bc-hydro-review/bch19-158-ipp_report_february_11_2019.pdf.

For more on the environmental and political side of fracking, see Ben Parfitt, *Fracking Up Our Water, Hydro Power and Climate: BC's Reckless Pursuit of Shale Gas* (Vancouver: Canadian Centre for Policy Alternatives, 2011), policyalternatives.ca/fracking.

For a science-based look at the impact of climate change (and what politicians are not doing to curb greenhouse gases), see reports by the Intergovernmental Panel on Climate Change, ipcc.ch.

Photo Credits

Index

Note: page references where glossary terms appear in **bold** are also marked in **bold**; "(f)" after a page number indicates a figure; "(t)" after a page number indicates a table.